Proceedings of the 3rd Brazilian Technology Symposium

Yuzo Iano · Rangel Arthur
Osamu Saotome · Vania Vieira Estrela
Hermes José Loschi
Editors

Proceedings of the 3rd Brazilian Technology Symposium

Emerging Trends and Challenges in Technology

Springer

Editors
Yuzo Iano
Faculty of Electrical
and Computer Engineering
University of Campinas
Campinas, São Paulo
Brazil

Rangel Arthur
Faculty of Technology
University of Campinas
Campinas, São Paulo
Brazil

Osamu Saotome
Divisão de Engenharia Eletrônica
Instituto Tecnológico de Aeronáutica (ITA)
São José dos Campos, São Paulo
Brazil

Vania Vieira Estrela
Universidade Federal Fluminense (UFF)
Duque de Caxias, Rio de Janeiro
Brazil

Hermes José Loschi
Faculty of Electrical and Computer
Engineering
University of Campinas
Campinas, São Paulo
Brazil

ISBN 978-3-030-06592-8 ISBN 978-3-319-93112-8 (eBook)
https://doi.org/10.1007/978-3-319-93112-8

Softcover re-print of the Hardcover 1st edition 2019

This Springer imprint is published by the registered company Springer Nature Switzerland AG
The registered company address is: Gewerbestrasse 11, 6330 Cham, Switzerland

Foreword

It is with deep satisfaction that I write this Foreword to the Proceedings of the 3rd Brazilian Technology Symposium: Emerging Trends and Challenges in Technology (BTSym'17), held at the University of Campinas-SP, Brazil, in December 2017. This event is in its third edition and has consolidated to become an excellent opportunity for researchers, professors and students to present and discuss the results of their research works.

The 2017 edition of BTSym is characterised by the broad scope of the areas exposed, with papers dealing with current and essential topics for Brazilian and world technological development, including subjects related to the various branches of engineering, architecture and computer science.

Events such as BTSym are an essential part of the research and innovation process. First, these events contribute to the promotion of research activities, which are key to a country's technological development. The dissemination of research results, as promoted by BTSym, contributes to the transformation of research findings into technological innovation. In addition, these events facilitate the sharing of findings, leading eventually to the formation of research networks, which accelerate the achievement of new results. Therefore, I would like to congratulate the BTSym General Chair, Prof. Yuzo Iano and his group of collaborators for the important initiative of organising the BTSym 2017 and for providing the opportunity for authors to present their work to a wide audience through this publication. Last but not least, I congratulate the authors for the quality of the work presented in this Proceedings.

Campinas, Brazil

Prof. Paulo Cardieri
Former President of the Brazilian
Telecommunications Society (2014–2017)
University of Campinas

Preface

This book contains the refereed *Proceedings of the 3rd Brazilian Technology Symposium: Emerging Trends and Challenges in Technology*, held in Campinas-SP, Brazil, in December 2017.

The Brazilian Technology Symposium is an excellent forum for presentations and discussions of the latest results of projects and development research, in several areas of knowledge, in scientific and technological scope, including Smart Designs, Sustainability, Inclusion, Future Technologies, Architecture and Urbanism, Computer Science, Information Science, Industrial Design, Aerospace Engineering, Agricultural Engineering, Biomedical Engineering, Civil Engineering, Control and Automation Engineering, Production Engineering, Electrical Engineering, Mechanical Engineering, Naval and Oceanic Engineering, Nuclear Engineering, Chemical Engineering, Probability and Statistics.

This event seeks to bring together researchers, students and professionals from the industrial and academic sectors, seeking to create and/or strengthen the linkages between issues of joint interest. Participants were invited to submit research papers with methodologies and results achieved in scientific level research projects, completion of coursework for graduation, dissertations and theses.

The 32 full papers accepted for this book were selected from 102 submissions, and in each case, the authors were shepherded by an experienced researcher, with a rigorous peer-review process. Among the main topics covered in this book, we can highlight Artificial Neural Networks, Computational Vision, Security Applications, Web Tool, Cloud Environment, Network Functions Virtualization, Software-Defined Networks, IoT, Residential Automation, Data Acquisition, Industry 4.0, Cyber-Physical Systems, Digital Image Processing, Infrared Images, Patters Recognition, Digital Video Processing, Precoding, Embedded Systems, Machine Learning, Remote Sensing, Wireless Sensor Network, Heterogeneous Networks, Unmanned Ground Vehicle, Unmanned Aerial System, Security, Surveillance, Traffic Analysis, Digital Television, 5G, Image Filter, Partial Differential Equation, Smoothing Filters, Voltage Controlled Ring Oscillator, Difference Amplifier, Photocatalysis, Photodegradation, Cosmic Radiation Effects, Radiation Hardening Techniques, Surface Electromyography, Sickle Cell Disease Methodology, MicroRNAs, Image

Processing Venipuncture, Cognitive Ergonomics, Ecosystem Services, Environmental, Power Generation, Ecosystem Services Valuation, Solid Waste and University Extension.

We hope you enjoy and take advantage of this book and feel motivated to submit us your papers in the future to Brazilian Technology Symposium.

Best wishes,

Campinas, Brazil

Prof. Hermes José Loschi
Technical and Finance Chair of Brazilian Technology Symposium

Organizing Committee

Organizing and Executive Committee

Yuzo Iano, LCV/DECOM/FEEC/UNICAMP, General Chair BTSym & WSGE
Osamu Saotome, ITA, Associate-General Chair BTSym
Rangel Arthur, FT/UNICAMP, Vice-General Chair BTSym
Evaldo Gonçalves Pelaes, UFPA, Vice-Associate-General Chair BTSym
Hermes José Loschi, LCV/DECOM/FEEC/UNICAMP, Technical Program and Finance Chair
Telmo Cardoso Lustosa, LCV/DECOM/FEEC/UNICAMP, Local Arrangements Chair
Çamila Santana Domingues, LCV/DECOM/FEEC/UNICAMP, Registration Chair
Luiz Antonio de Sousa Ferreira, LCV/DECOM/FEEC/UNICAMP, Proceedings Chair
Reinaldo Padilha, LCV/DECOM/FEEC/UNICAMP, Communication Chair
Ana Carolina Borges Monteiro, LCV/DECOM/FEEC/UNICAMP, Marketing Chair
Douglas Aguiar do Nascimento, LCV/DECOM/FEEC/UNICAMP, Institucional Relationship Chair

Scientific and Academic Committee

Osamu Saotome, ITA
Vania Vieira Estrela, UFF, Rio de Janeiro
Luiz Cézar Martini, DECOM/FEEC/UNICAMP
David Bianchini, PUC/Campinas
Luis Geraldo Pedroso Meloni, DECOM/FEEC/UNICAMP
Ana Cláudia Seixas, PUC/Campinas

Cristiano Akamine, Universidade Presbiteriana Mackenzie
Luiz Vicente Figueira de Mello Filho, Universidade Presbiteriana Mackenzie
Guillermo Leopoldo Kemper Vásquez, USMP & UNI-INICTEL
Lucas Heitzmann Gabrielli, DECOM/FEEC/UNICAMP
Edgard Luciano Oliveira da Silva, EST/UEA
Talía Simões dos Santos, FT/UNICAMP
Janito Vaqueiro Ferreira, DMC/FEM/UNICAMP
Vlademir de Jesus Silva Oliveira, UNEMAT/Sinop
Hugo Enrique Hernandez Figueroa, DECOM/FEEC/UNICAMP
Marcos Antonio do Nascimento Guimarães, UNIP/CAMPINAS, JUNDIAÍ
Maria Thereza de Moraes Gomes Rosa, Universidade Presbiteriana Mackenzie
Angela del Pilar Flores Granados, FEA/UNICAMP
Paolo Bacega, Faculdade Anhanguera
Marcos Fernando Espindola, IFSP São Paulo
Polyane Alves Santos, Instituto Federal Da Bahia
Jude Hemanth, Department of Electrical and Computer Engineering, Karunya University, Coimbatore, India

Technical Reviewers Committee

Adão Boava, Universidade Federal de Santa Catarina, UFSC
Ana Carolina Borges Monteiro, LCV/DECOM/FEEC/UNICAMP
Amilton Lamas, PUC-Campinas
Agord de Matos Pinto Júnior, LCV/DECOM/FEEC/UNICAMP
Angela del Pilar Flores Granados, FEA/UNICAMP
Silvio Renato Messias de Carvalho, LCV/DECOM/FEEC/UNICAMP
Joaquim Marcos Santa Rita da Silva, Instituto Nacional de Telecomunicações
José Alexandre Nalon, Centro Universitário Salesiano São Paulo, UNISAL
Murilo Cesar Perin Briganti, LCV/DECOM/FEEC/UNICAMP
Luigi Ciambarella Filho, Universidade Veiga de Almeida/Develop Biotechnology
Ingrid Araujo Sampaio, Universidade Estadual de Campinas
Hermes José Loschi, LCV/DECOM/FEEC/UNICAMP
Daniel Rodrigues Ferraz Izario, LCV/DECOM/FEEC/UNICAMP
Mariana Carvalho, DCA/FEEC/UNICAMP
Diego Pajuelo, LCV/DECOM/FEEC/UNICAMP
Douglas Aguiar do Nascimento, FACTI/LCV/DECOM/FEEC/UNICAMP
Edson José Gonçalves, LCV/DECOM/FEEC/UNICAMP
Marcos Fernando Espindola, IFSP São Paulo
Polyane Alves Santos, Instituto Federal Da Bahia
Rangel Arthur, INOVA/FT/UNICAMP
Reinaldo Padilha, LCV/DECOM/FEEC/UNICAMP

Kelem Christine Pereira Jordão, LCV/DECOM/FEEC/UNICAMP
Euclides Lourenço Chuma, LCV/DECOM/FEEC/UNICAMP
José Yauri, DCA/FEEC/UNICAMP
Julío Humberto León Ruiz, LCV/DECOM/FEEC/UNICAMP

Sponsor

Beta Telecommunications

Acknowledgements

Our appreciation goes to a lot of colleagues and friends who assisted in the development of this book: *Proceedings of the 3rd Brazilian Technology Symposium: Emerging Trends and Challenges in Technology.*

First of all, I would like to thank all the members of the Organizing and Executive Committee, for the commitment throughout the year, several meetings were held and many challenges were overcome for the accomplishment of the BTSym 2017. Also, and with great merit, I would like to thank all the members of the Scientific and Academic Committee, and Technical Reviewers Committee for their excellent work, which was essential to ensure the quality of our peer-review process, and collaborating with the visibility and technical quality of the BTSym 2017.

The Brazilian Technology Symposium is an event created by Laboratory of Visual Communications of the Faculty of Electrical and Computer Engineering of the University of Campinas (UNICAMP). In this way, I would like to thank the UNICAMP, especially the UNICAMP Cultural Development Center for the support and hosting of the BTSym 2017, which was fundamental for the successful their accomplishment.

Beta Telecommunications played a key role in holding the BTSym 2017; due to the financial support from it, it was possible to consolidate with quality many BTSym 2017 organization aspects, which ensured the quality to support the authors and Speakers.

Finally, I thank all the authors for their participation in the BTSym 2017, I sincerely hope to have provided an experience that was very useful and enriching in the personal and professional life of everyone, and my special thanks go to Profa. Vania Vieira Estrela. In my almost 50 years of academic career in UNICAMP, few were the opportunity to work with a professional such her. She is definitely one of a kind, extremely efficient, hardworking and the BTSym 2017 certainly has much to thank for Prof. Vania Vieira Estrela.

Best wishes,

Prof. Yuzo Iano
General Chair of Brazilian Technology Symposium

Contents

Cloud Detection for PERUSAT-1 Imagery Using Spectral and Texture Descriptors, ANN, and Panchromatic Fusion

Giorgio Morales, Samuel G. Huamán and Joel Telles

Abstract The cloud detection process is a prerequisite for many remote sensing applications in order to use only those cloud-free parts of satellite images and reduce errors of further automatic detection algorithms. In this paper, we present a method to detect clouds in high-resolution images of 2.8 m per pixel approximately. The process is performed over those pixels that exceed a defined threshold of blue normalized difference vegetation index to reduce the execution time. From each pixel, a set of texture descriptors and reflectance descriptors are processed in an Artificial Neural Network. The texture descriptors are extracted using the Gray-Level Co-occurrence Matrix. Each detection result passes through a false-positive discard procedure on the blue component of the panchromatic fusion based on image processing techniques such as Region growing, Hough transform, among others. The results show a minimum Kappa coefficient of 0.80 and an average of 0.94 over a set of 25 images from the Peruvian satellite PERUSAT-1, operational since December 2016.

Keywords Cloud detection · High-resolution · Artificial neural networks
Texture analysis

1 Introduction

In the context of the recent launch of the Peruvian satellite PERUSAT-1, several applications in the areas of agriculture, environmental monitoring, cartography and security have been proposed, taking advantage of the high-resolution that these images

G. Morales (✉) · S. G. Huamán · J. Telles
National Institute of Research and Training at Telecommunications (INICTEL-UNI), National University of Engineering, San Luis 1771, Lima 15021, Peru
e-mail: gmorales@inictel-uni.edu.pe

S. G. Huamán
e-mail: shuaman@inictel-uni.edu.pe

J. Telles
e-mail: jtelles@inictel-uni.edu.pe

Y. Iano et al. (eds.), *Proceedings of the 3rd Brazilian Technology Symposium*,
https://doi.org/10.1007/978-3-319-93112-8_1

can provide. However, the presence of clouds in acquired satellite images can make these analyses difficult. This is why, after atmospheric correction, cloud detection is an important preliminary step for any subsequent study. Previous works on cloud detection involve multi-temporal analysis of images with close acquisition dates [1], threshold-based cloud detection [2, 3], extraction of spectral and texture information using intelligent classifiers [4, 5] and even more recent methods that propose the use of convolutional neural networks [6].

In this paper, we propose a methodology for the detection of clouds in high-resolution multispectral images of PERUSAT-1 using spectral and texture descriptors processed by an Artificial Neural Network and discarding false positives by morphological analysis on a panchromatic fusion, as shown in Fig. 1. Spectral information is extracted from the red, green, blue, and near-infrared (NIR) bands; besides, we highlight the use of the blue normalized difference vegetation index (BNDVI) [7]. Spectral information is often not sufficient for classification tasks, so we will take advantage of the spatial information provided by high-resolution images [8]. The texture descriptors are extracted using the Gray-Level Co-occurrence Matrix. The rest of the paper is organized as follows. Section 2 explains the proposed cloud detection methodology; Sect. 3, the results obtained and Sect. 4 summarizes the conclusions of the work.

2 Proposed Method

2.1 Experimental Data

A PERUSAT-1 image has four spectral bands: red (0.63–0.7 μm), green (0.53–0.59 μm), blue (0.45–0.50 μm) and NIR (0.752–0.885 μm). The spatial resolution of the multispectral bands is 2.8 m per pixel and the panchromatic band, 0.7 m per pixel. A total number of 32 composite images of variable area and from different geographies (e.g., rainforest, desert, agriculture and urban areas) have been used: seven for the training and validation of the ANN and 25 for the validation of the method.

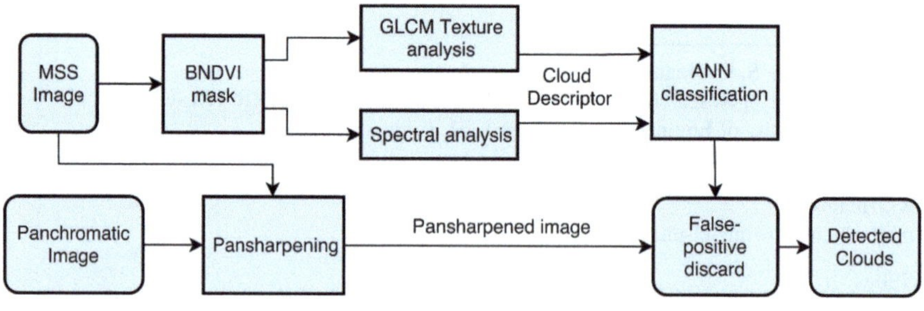

Fig. 1 Cloud detection procedure

2.2 Feature Extraction

The descriptor vector X is composed of the following spectral and texture features:

Spectral and color descriptors. Since there are four spectral bands, the four DN values of each pixel are part of the cloud descriptor. In addition, three more features are added to the descriptor: H (Hue), S (Saturation) and V (Value). Furthermore, we consider BNDVI [7] as an additional descriptor taking into account that clouds have relatively high and constant reflectance values between the blue and NIR bands, which means that their BNDVI index would be lower than those of other objects such as vegetation or silt loam.
Texture descriptors. We calculate the texture descriptors from the Gray-Level Co-occurrence Matrix (GLCM), which is calculated from the blue band, where there is more contrast between clouds and soil. Six statistical indicators are extracted from the GLCM matrix: Mean, variance, contrast, correlation, energy, and homogeneity [6] using a kernel size of 10 pixels for four directions: 0°, 45°, 90° and 135°, which sums a total of 24 texture descriptors.

2.3 Artificial Neural Network Training

First, we create a data-set which consists of 22,386,038 descriptor vectors as patterns (‘*clouds*’ and ‘*non-clouds*’) extracted from seven manually annotated images of the 32 available images. Then, we split 90% of the data to create the training set, 5% to the validation set and 5% to the test set. To select an optimal model, we perform a variable selection step and a configuration selection step:

Feature selection. Joining the spectral and texture features, the X descriptor has a dimension of 32 elements. We chose a neural network with two hidden layers of seven and two neurons, respectively, to initialize a training model with all the 32 variables available. Then, we remove one variable at a time and perform 10 trainings to evaluate the causality. Finally, it is verified that when removing the ‘correlation’ variable the changes in the identification precision of the model vary from −0.08 to 0.01%, so it is decided to remove this variable from the model, which is left with 28 variables in total (four variables of ‘correlation’ were removed, one for each direction).
Model tuning. We tested heuristically six different network configurations and executed each one 10 times to check its stability and trend. Table 1 shows the performance of five models of ANNs, with different configurations, using the validation set. The calculus of the average F1-score reaches its maximum value when the ANN has seven neurons in the first hidden layer and two neurons in the second one, so that we choose this ANN for classification. The F1-score over the test set for the best model was 0.975.

Table 1 Comparison between different ANNs

Type	Architecture		Performance (minimum)	F1-score over validation set (mean)
	Hidden layer 1	Hidden layer 2		
1	4	2	0.0122	0.9328
2	5	2	0.0103	0.9328
3	6	2	0.0108	0.9743
4	7	2	0.0100	0.9754
5	8	2	0.0094	0.9660

2.4 *Classification*

We first calculate BNDVI values of all pixels from each image after radiometric and atmospheric correction and set a threshold of 0.4 to distinguish between 'non-clouds' and candidates to clouds (lower than 0.4). Then, for each pixel is calculated an X descriptor of 28 elements, which is processed in the neural network to decide if it corresponds to cloud or not. Figure 2 shows the result of applying the described classification over four different scenes.

2.5 *False-Positive Discard Method*

Figure 2 shows an effective classification over those zones covered by clouds, but presents some errors over some high-reflective buildings. Clouds shapes are more natural and irregular, while building shapes have more regular shape outlines; thus,

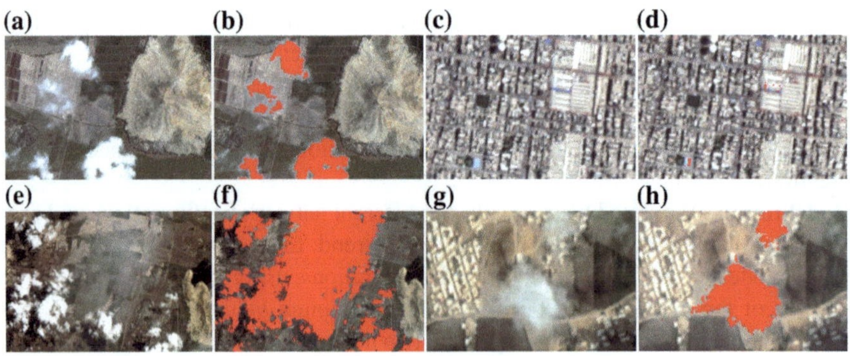

Fig. 2 Cloud classification using the neural network. **a**, **c**, **e**, **g** Original images. **b**, **d**, **f**, **h** Detected clouds in red

we propose a method to exploit this information using a panchromatic fusion, which increases the spatial resolution and, therefore, the detail of the buildings.

We apply the ISH method, one of the fastest pansharpening algorithms [9], and take only the blue component of the pansharpened image, as it is the component where there is more contrast between high-reflective buildings and soil, as it was observed. Then, the first step is to analyze each small object classified by the ANN as a cloud, take as reference the pixel with the greatest intensity of the object (so that we make sure not to take any pixel that is very near the edge of a building) and create a 50×50 window around it (Fig. 3a), which will be used to perform the panchromatic fusion (Fig. 3c). Then, with this high-resolution image, we are able to determine whether or not the central pixel is located on the rooftop of a building; that is, within a regular shaped object. To do this, the central pixel is taken as seed point to perform a simple region growing algorithm. The region is iteratively grown by comparing all unallocated neighboring pixels to the region (Fig. 3d).

To determine if the region surrounding the central pixel is regular, the Hough transform is used over the edge mask of the region (Fig. 3e) to detect lines of at least 13 pixels (9 m) as shown in Fig. 3f. If there is a straight line, the analyzed object will be removed from the detected cloud mask. Finally, the resulting mask is dilated.

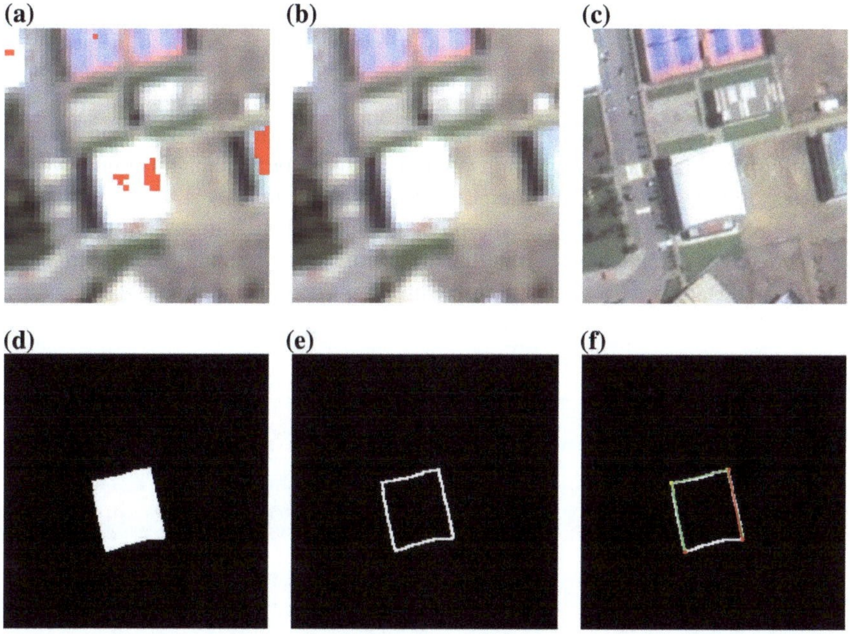

Fig. 3 Cloud classification using the neural network. **a, c, e, g** Original images. **b, d, f, h** Detected clouds in red

3 Results and Discussion

Figure 4 shows how the application of the explained method removes effectively regular shaped objects. The results for the 25 test images are shown in Table 2, where we obtained a minimum Kappa coefficient of 0.8065, a maximum of 0.9889, a mean of 0.9430 and a standard deviation of 0.0513. The percentage of false positives is significantly lower than false negatives. The false positives are mainly due to the presence of high-reflective buildings whose regular edges could not be detected by the proposed method either because of lack of contrast with their surroundings or because of a very small size. False positives are also due to the presence of high-reflective soil, which even for the human eye can be confused by clouds, and to the sea foam.

The Kappa coefficient is reduced mainly by the false negatives, which are mostly due to the fog and cloud areas with low density, which in some cases border the large clouds. Nevertheless, the proposed method proves to be able to recognize much of the areas affected by fog (Fig. 2f) and even clouds of very low density and small size. If the satellite images had a very different context than the images used for this work, the artificial neural network must be trained again to get similar good results.

4 Conclusion

Despite the different cases of false negatives, the results of cloud detection in high-resolution satellite images show high values of Kappa coefficients. In simple words, it means a high grade of concordance between cloud images selected by human observer and images detected by the proposed method. It is important to note the cloud coverage percentages of satellite images used to test vary from 0 to 90% approximately and not have relationship with Kappa coefficient or with the accuracy.

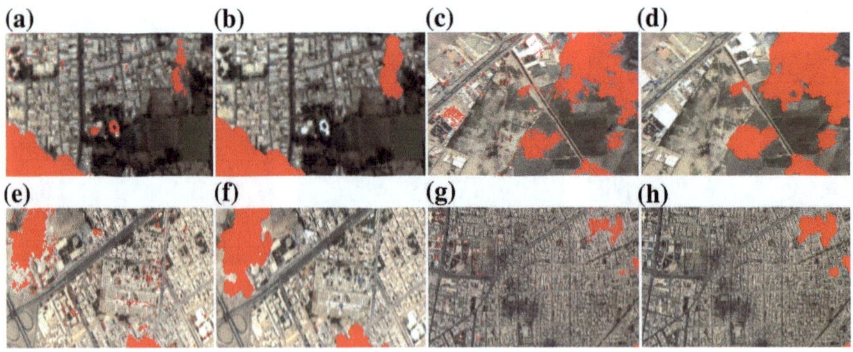

Fig. 4 Cloud classification using the neural network. **a**, **c**, **e**, **g** Original images. **b**, **d**, **f**, **h** Detected clouds in red

Table 2 Validation results

Image	Kappa	Accuracy	Cloud coverage (%)	Image	Kappa	Accuracy	Cloud coverage (%)
1	93.0324	99.87	1.04	14	96.4787	98.5	31.64
2	94.7298	99.96	0.44	15	94.276	97.61	30.8
3	97.6722	99.51	12.11	16	97.4476	98.79	39.17
4	97.4228	99.18	20.24	17	98.891	99.46	41.5
5	80.6516	96.08	13.36	18	97.3919	98.72	43.19
6	80.0462	91.32	35.82	19	97.8001	98.91	45.47
7	92.9886	97.9	19.34	20	96.2772	98.28	37.03
8	85.365	96.26	86.87	21	98.1937	99.21	32.61
9	93.2113	97.21	30.32	22	97.1358	98.57	49.79
10	94.6574	99.58	4.27	23	97.4612	98.73	53.28
11	90.0928	99.97	0.16	24	95.8069	97.9	52.32
12	97.9754	99.42	17.66	25	95.4834	98.97	13.56
13	97.1734	98.61	43.81				

References

1. Tseng, D.C., Tseng, H.T., Chien, C.L.: Automatic cloud removal from multi-temporal spot images. Appl. Math. Comput. **205**(2), 584–600 (2008)
2. Hang, Y., Kim, B., Kim, Y., Lee, W.H.: Automatic cloud detection for high spatial resolution multi-temporal. Remote Sens. Lett. **5**(7), 601–608 (2014)
3. Marais, I.V.Z., Du Preez, J.A., Steyn, W.H.: An optimal image transform for threshold-based cloud detection. Int. J. Remote Sens. **32**(6), 1713–1729 (2011)
4. Li, P., Dong, L., Xiao, H., Xu, M.: A cloud image detection method based on SVM vector machine. Neurocomputing **169**, 34–42 (2015)
5. Bai, T., et al.: Cloud detection for high-resolution satellite imagery using machine learning and multi-feature fusion. Remote Sens. **8**(9), 715 (2016)
6. Shi, M., et al.: Cloud detection of remote sensing images by deep learning. In: 2016 IEEE International Geoscience and Remote Sensing Symposium (IGARSS), pp. 701–704. IEEE, Beijing (2016)
7. Wang, F., et al.: New vegetation index and its application in estimating leaf area index of rice. Rice Sci. **14**(3), 195–203 (2007)
8. Tsai, F., Chou, M.J.: Texture augmented analysis of high resolution satellite imagery in detecting invasive plant species. J Chin. Inst. Eng. **29**(4), 581–592 (2006)
9. Vivone, G., et al.: Critical comparison among pansharpening algorithms. IEEE Trans. Geosci. Remote Sens. **53**(5), 2565–2586 (2015)

Implementation of a Digital Image Processing Algorithm on a Colibri IMX6 Embedded Industrial System for Optical Mark Recognition

Carlos Herrera, Stephany Del Campo, Abel Dueñas, Julio León, Guillermo Kemper and Christian del Carpio

Abstract This paper presents the implementation of digital image processing algorithms on Toradex's Colibri IMX6 module and Iris carrier board. These algorithms will be used for the extraction, detection, and recognition of optical marks on digital images. Images will be obtained from an academic evaluation booklet through a webcam. Likewise, the computer will control an electrical/mechanical enclosure that stores and transports the evaluation booklets to be registered. The system proposed responds to the need to increase the efficiency of the grade recording process as previously designed for the Academic Coordination Department of the San Martín de Porres University's School of Engineering and Architecture.

Keywords Linux · Embedded · Toradex · ARM · Images · Backpropagation Python · OpenCV · GTK+3 · Database

C. Herrera · S. Del Campo · G. Kemper · C. del Carpio (✉)
Faculty of Engineering and Architecture, School of Electronic Engineering, San Martín de Porres University, Av. La Fontana 1250, La Molina, Lima, Peru
e-mail: cdelcarpiod@gmail.com

C. Herrera
e-mail: carlos.herrera.trujillo@gmail.com

S. Del Campo
e-mail: sandreadelcampo@gmail.com

G. Kemper
e-mail: guillermo.kemper@gmail.com

A. Dueñas · J. León
School of Electrical and Computer Engineering (FEEC), University of Campinas, Av. Albert Einstein – 400, Barão Geraldo, Campinas-SP, Brazil
e-mail: aduenas.rodriguez@gmail.com

J. León
e-mail: j.leon@decom.fee.unicamp.br

Y. Iano et al. (eds.), *Proceedings of the 3rd Brazilian Technology Symposium*,
https://doi.org/10.1007/978-3-319-93112-8_2

1 Introduction

At present, there are several options when it comes to embedded systems. One of the most used is the Raspberry Pi minicomputer, which has a Linux-based operating system that is compatible with many software programs and a wide variety of peripherals. The disadvantage of this embedded system is, however, that optimal performance cannot be ensured if used for too long or with applications requiring a high level of processing. What is more, it has no protection measures for the system and peripherals.

Another option in terms of embedded systems is offered by Toradex and its wide range of devices for industrial use, capable of giving comprehensive solutions thanks to their high performance, compactness, scalability, open-source embedded Linux system development and compatibility with several peripherals. They are already used in several industries such as the automotive, defense, laboratory, and measurement, e.g., the iGuide mapping system used to generate 3D virtual environments [1].

Due to all the features offered by the brand, it was decided to use Toradex's Colibri IMX6 DL computer-on-module [3] that, along with its Iris carrier board, performs well in the implementation of digital image processing algorithms and hardware control.

2 Description of the System Proposed

2.1 Preparation of the Linux Embedded System

For the implementation of the enclosure control and processing system, it is required to choose the necessary elements. First, the operating system to be embedded must be run in the Colibri IMX6 module through the following steps:

- Create the Linux OS development environment (Ubuntu 14.04) on the computer.
- Synchronize files from the "layers.openembedded.org" repository containing the layers developed by Toradex for this module.
- Set up the device tree to add hardware.
- Add the necessary compatible software to the system image configuration for the IMX6 module (main applications: Python 2.7 and OpenCV 3).
- Build the embedded Linux system for Toradex by cross-compilation with a GNU Compiler Collection (GCC) tool.
- Update the system on the computer-on-module.
- Create an application development environment on the minicomputer [4].

2.2 Detailed Description of the System

Figure 1 shows the pictorial block diagram of the integrated system, which consists of a main component—the Colibri IMX6 module and the Iris carrier board—and four secondary components—the graphical interface, the image capture device, the mechanical enclosure and the MySQL database.

The main component, i.e., the Toradex single-plate computer (Fig. 2), was placed inside a transporting acrylic box. In addition, the processor was provided with a maxiFLOW heatsink [5] whose performance is better compared to that of normal heatsinks. Find below a list of peripherals used for the Iris carrier board [6] to connect to secondary devices.

(A) USB port to connect to the webcam.
(B) RS232 connector (B) to connect to the mechanical enclosure.
(C) Expansion port to touch control the LCD screen.
(D) RJ45 Internet Port to connect to the database.
(E) RS232 Port (A) to access the system console.
(F) maxiFLOW heatsink.

For the implementation of image processing and recognition algorithms, the Python programming language [7] and the OpenCV image processing library were used.

The optical marks were obtained from the evaluation booklet through a web camera. Figure 3 shows the booklet's yellow regions containing the optical marks that students will use to enter their code and grade.

Figure 4 shows the enclosure where booklets are stored and transported through a mechanical system. Inside the enclosure, images of the booklet are also obtained through a webcam. The handling of both the enclosure and the camera are carried out with the Colibri IMX6 computer-on-module.

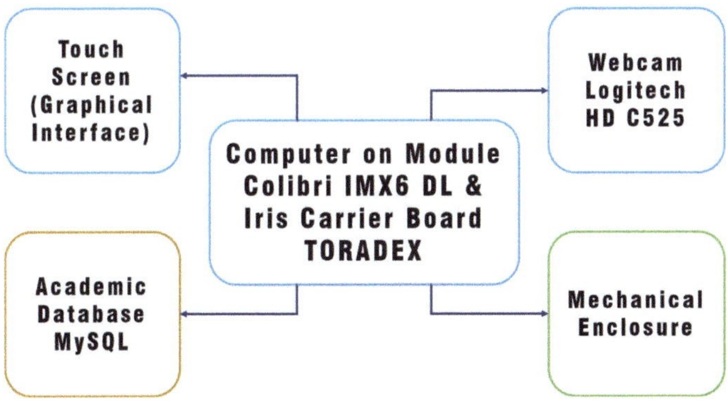

Fig. 1 Pictorial block diagram of the system developed

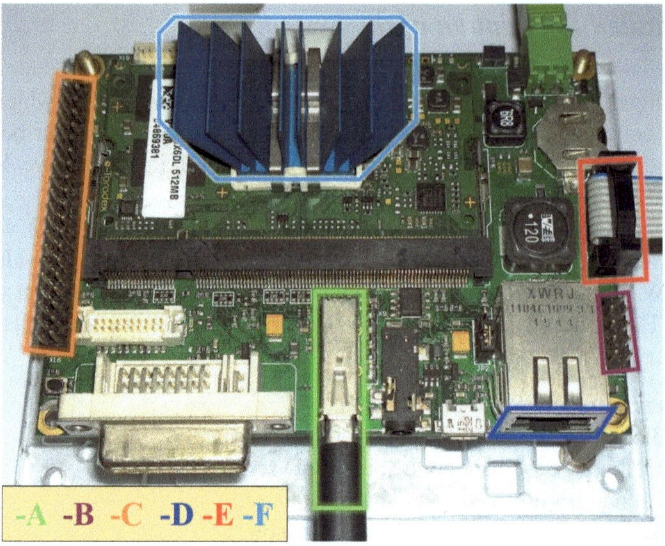

Fig. 2 Personalized single-plate Toradex computer

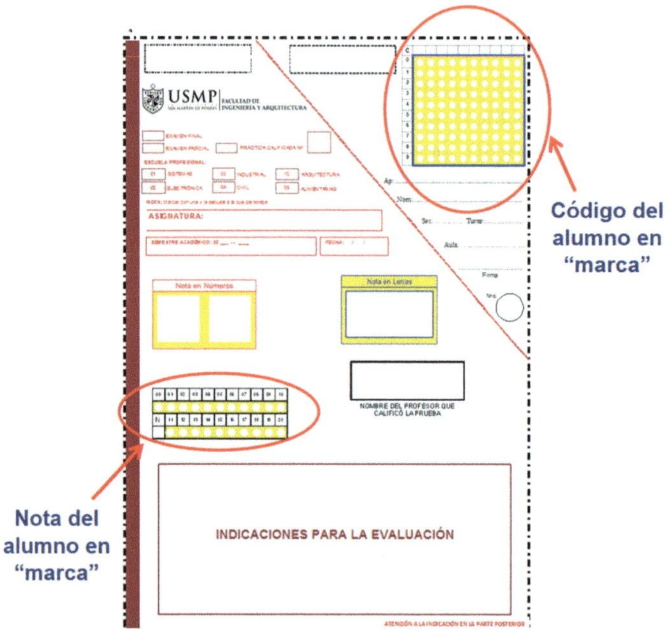

Fig. 3 Evaluation booklet to be processed

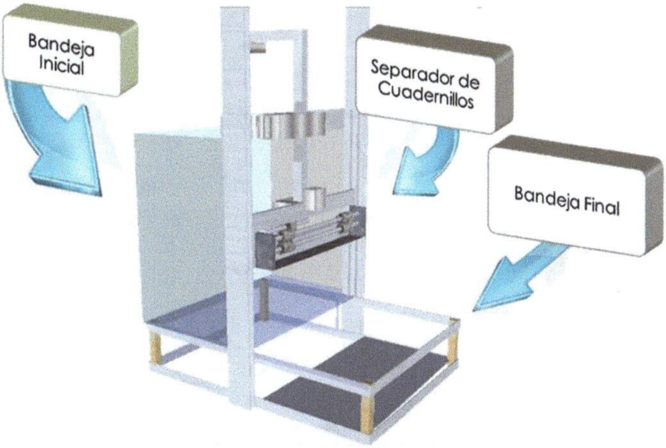

Fig. 4 Mechanical enclosure for booklet storage and transport

2.3 Process for the Recognition of the Code and Grade on the Booklet

This project is based on a method that recognizes the code and grade on the booklets [2] using image processing techniques in Python, with the OpenCV computer vision library [8] and the NumPy scientific library.

Figure 5 shows the steps taken in image processing to obtain the patterns required for the trained neural network. First, a video capture object (C.1) is created using the camera ID number—in this case, 0. Thus, images are obtained from the camera, while setting up the parameters required, such as resolution (1024 × 576 pixels) and focus.

```
captura = cv2.VideoCapture(ID)
```
(C.1)

Then, the image of the object (***captura***) is imported by the ***read*** method, returning an 8-bit integer numeric matrix in the BGR color model, which is saved under ***frame*** (C.2).

```
_,frame = captura.read()
```
(C.2)

At this point, the captured color image is ready, and the booklet, in this case, can be verified for the first time. To do so, the quantity of yellow pixels is to be determined.

First, the BGR color model (Blue, Green, Red) is converted to HSV (Hue, Saturation, Value) (Eqs. (1), (2) and (3) and C.3).

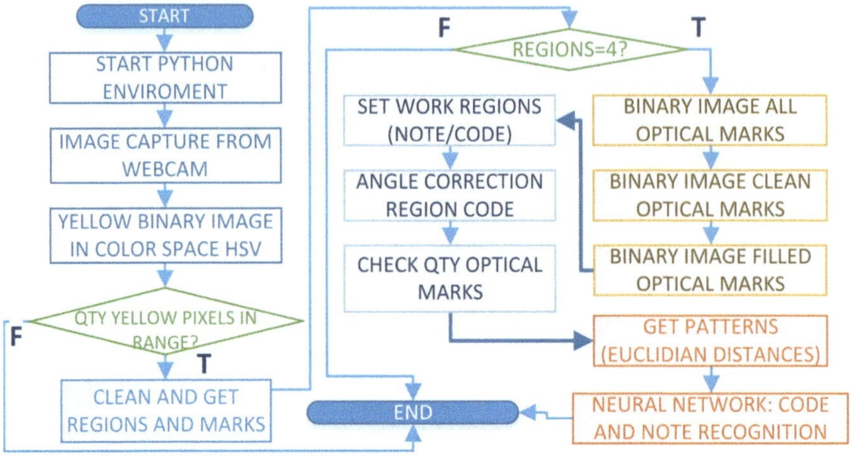

Fig. 5 Flow diagram of the recognition algorithm

$$V \leftarrow max(R, G, B) \tag{1}$$

$$S \leftarrow \begin{cases} \frac{V - min(R,G,B)}{V} & If V \neq 0 \\ 0 & otherwise \end{cases} \tag{2}$$

$$H \leftarrow \begin{cases} 60(G - B)/V - min(R, G, B) & Si V = R \\ 120 + 60(B - R)/V - min(R, G, B) & Si V = G \\ 240 + 60(R - G)/V - min(R, G, B) & Si V = B \end{cases} \tag{3}$$

If: $H < 0$, then $H = H + 360$

where R, G, and B are red, green, and blue pixel values, and H, S and V contain values for hue, saturation and value.

```
img_hsv = cv2.cvtColor(imagen_bgr,cv2.COLOR_BGR2HSV)
```
(C.3)

where ***imagen_bgr*** is a copy of the original color image, and ***img_hsv*** is the numeric matrix corresponding to the HSV image.

Then, a binarized image is created by thresholding (Eq. (4)), where ***lowerb*** and ***upperb*** are the lower bound and the upper bound corresponding to each pixel layer, and ***dst (I)*** is the resulting binary image. If the HSV pixel value is within the ranges, this will return the value 255; otherwise, it will be zero.

$$\begin{aligned} dst(I) = lowerb(I)_0 \leq src(I)_0 \leq upperb(I)_0 \wedge lowerb(I)_1 \leq src(I)_1 \\ \leq upperb(I)_1 \wedge lowerb(I)_2 \leq src(I)_2 \leq upperb(I)_2 \end{aligned} \tag{4}$$

Command C.4 performs Eq. (4) with the lower bound (***nivel_bajo***) and the upper bound (***nivel_alto***) and the resulting binary image (***bin_amarillo***).

```
bin_amarillo = cv2.inRange(img_hsv,nivel_bajo,
nivel_alto)
```
(C.4)

The next step is to verify that the quantity of yellow pixels is within the range [25000, 50000]; for this, the value of all the matrix elements in the binarized image (**bin_amarillo**) is added and then divided by 255, as shown in Eq. (5).

$$cant\_pixeles = \frac{\sum_{j=0}^{1023} \sum_{i=0}^{575} pixel_{i,j}}{255} \tag{5}$$

After the first booklet verification, a cleanup of the binarized image (**bin_amarillo**) is done using command C.5 in Python, which will perform a morphological close operation with a 3 × 3 (8-bit integer) kernel.

```
bin_amarillo = cv2.morpholgyEx(bin_amarillo,
cv2.MORPH_CLOSE.kernel)
```
(C.5)

In addition, objects whose regions are greater than the threshold value are removed by segmentation. For such purpose, a find contours function is used, offering different types of data, two of which will be used in this algorithm: the region of the object and the centroid. In command C.6, the external retrieval mode (***retr_external***) and the simple approximation type (***chain_approx_simple***) are established.

```
_,contornos,_= cv2.findContours(bin_amarillo.copy(),
″modo″,″aproximación″)
```
(C.6)

Once the binary image of the matrix (***bin_amarillo***) is totally clean, all the objects will be filled to obtain a clearly defined binary image with work regions (***bin_regiones***) so that an image of optical marks can be generated when executing the "exclusive or" function (Eq. (6), C.7) between the binary image (***bin_amarillo***) and the image with work regions (***bin_regiones***).

$$dst(I) = src1(I) \oplus src2(I) \tag{6}$$

where ***src1*** (***I***) corresponds to the first binary image, ***src2*** (***I***) to the second binary image, and ***dst*** (***I***) is the resulting image.

```
bin_marcas = cv2.bitwise_xor(bin_amarillo,
bin_regiones,mask = None)
```
(C.7)

Then, a second booklet verification is made by counting the number of external contours (C.6), considering that the regions obtained must be equal to four (4): the code region, the written grade, and the two grade regions (optical marks). It should be noted that the find contours function returns the objects in order, from the bottom to the top.

The next step is to obtain all unfilled optical marks (***bin_marcasblancas*****).** For this, a binary image in white is created by thresholding (Eq. (7)) from the original color image. Then, function Y (Eq. (8), C.8) is executed between the binary image obtained and the binary image with the yellow regions.

$$M(x, y) = \begin{cases} 255,\ B(x, y) \le \mu_0 \wedge G(x, y) \le \mu_1 \wedge R(x, y) \le \mu_2 \\ 0,\ otherwise \end{cases}, \quad (7)$$

where M(x, y) is the resulting binary image, μ_0, μ_1, μ_2 are the thresholds for each pixel layer corresponding to colors B (blue), G (green), R (red).

Finally, logic operation XOR (Eq. (6), C.7) is executed between the binary image of all the optical marks (***bin_marcas***) and the optical marks in white (***bin_marcasblancas***), which returns the optical marks necessary for the recognition of the code and the grade (***bin_marcasnegras***).

$$dst(I) = src1(I)\ and\ src2(I) \quad (8)$$

```
bin_marcasblancas = cv2.bitwise_and(bin_regiones,
bin_marcasblancas,mask = None)                        (C.8)
```

At this point, the main binary image for the recognition of code and grade has been obtained. The process to obtain mark patterns (relative distance between centroids) is detailed below.

To process the code and grade regions, a new work region is established for each region adding approximately 25 pixels for each extreme bound. For this, the extreme coordinates x_m, x_n, y_m, y_n of each object are obtained using the ***cv2.boundingRect*** command, in which the input argument is each contour to be processed and the return values are the coordinates of the upper left corner, the width and height of the object. In addition, it is necessary to correct the angle of inclination of the code region using command C.9 to find the minimum rectangular region of the segmented object.

```
datos_rect = cv2.minRegionRect(contorno)       (C.9)
```

where ***contorno*** is the input argument of the command representing the contour object of the code region, and ***datos_rect*** represents a vector with the resulting data: coordinates of the upper left corner, the width and height of the rotated object and the angle of inclination. Once this angle is obtained, rotation of the code region is carried out (Eq. (9), C.10).

Rotation matrix:

$$\begin{bmatrix} \alpha \;\; \beta \;(1-\alpha)\times centro\_x - \beta \times centro\_y \\ -\beta \;\; \alpha \;\; \beta \times centro\_x + (1-\alpha)\times centro\_y \end{bmatrix}, \tag{9}$$

where

$$\alpha = escala \times Coseno(\theta) \;\; ; \;\; \beta = escala \times Seno(\theta), \tag{10}$$

where ***escala*** is the size of the object, θ is the angle of rotation, **centro_x** and **centro_y** are the center of the object.

The corresponding command in Python is the following:

```
bin_codigo = cv2.getRotationMatrix2D((centro_x,
centro_y),angulo,escala)
```
(C.10)

The next step is to count the number of objects (optical marks) corresponding to the code region (value of 100) and the grade region (value of 21). If a number different than expected is obtained for each case, the recognition process of such region is canceled. To count, the find contours function (***cv2.findContours***) will be executed; it will return the objects in order, and this will allow to know the number of optical marks in the corners, which will, in turn, reveal the four reference centroids (Eq. (11)).

$$\bar{x} = \frac{\sum_{i=1}^{N} x_i}{N} \;\; ; \;\; \bar{y} = \frac{\sum_{i=1}^{N} y_i}{N}, \tag{11}$$

where $\bar{x}$ and $\bar{y}$ are the centers of each mark, determined by the set of pixels $\{(x_i, y_i), i = 1, 2, \ldots, N\}$

To obtain the necessary patterns to input in the neural network, we determine the Euclidean distances (Eq. (12)) between the centroids of each filled optical mark to each of the references previously obtained (Fig. 6). This process applies to the area of the grade region.

$$d = \sqrt{(\bar{x}_m - \bar{x}_n)^2 + (\bar{y}_m - \bar{y}_n)^2}, \tag{12}$$

where $\bar{x}_m$ and $\bar{y}_m$ correspond to the centroids of the filled optical mark, and $\bar{x}_n$ and $\bar{y}_n$ correspond to the centroids of one of the optical marks located in the "corners".

2.4 Backpropagation Nerual Network

In this stage, the code and the grade are recognized with "Artificial Backpropagation Neural Networks" [9]. Serrano [10] says: "The computing power of a neural network derives mainly from its structure of distributed parallel computing. This

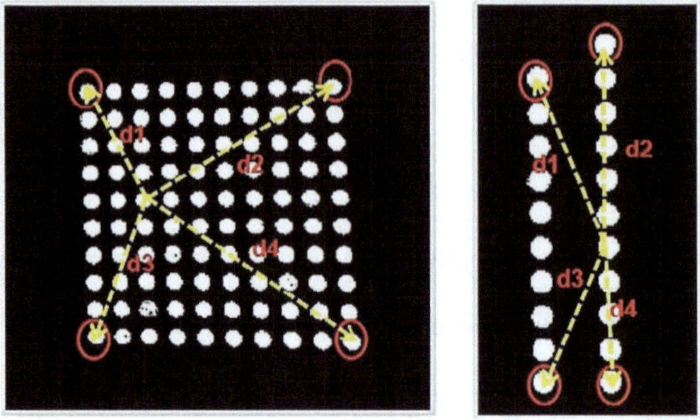

Fig. 6 Relative distances to white points in the "corners" of the total points (code region in the left, grade region in the right)

structure allows to solve problems that otherwise would take a lot of time in "classic" computers". Thanks to this, its implementation, together with the Colibri IMX6 embedded system, showed great efficiency. Two networks were previously trained with an error margin under 10^{-23} in each case, which granted the respective "weights" and "bias" to be applied in the transfer functions.

First, the weights and bias were imported and stored in different program variables; then, the input networks for neural network layer (Eq. (13)) were obtained.

$$z\_in_j = v_{0j} + \sum_i i_i v_{ij}, \tag{13}$$

where i_i are the input units, in this case, the patterns obtained (relative distances) in the image processing; v_{ij} is the weight of each input layer; v_{0j} is the bias of the hidden layer, and z_in_j is the result of the activation function.

Subsequently, the ***log-sigmoid*** transfer functions were executed (Eq. (14)) to obtain the codes and the coded grades.

$$f_1(w) = \frac{1}{1 + exp(-w)}, \tag{14}$$

where w is the vector resulting from the activation function.

Finally, the results obtained at the output are rounded to the nearest integer value, which will result in a 20-bit code for the optical marks of the code, and another 21-bit code for those of the grade. Once all the booklets are processed, the data obtained will be stored in a MySQL database [11] through the local region network.

2.5 Graphical Control Interface

For the adequate development of the project, it was necessary to create a graphical interface allowing to control the functions of the enclosure and the interconnection to the database in the system developed. However, Python does not have its own tools for the development of Graphical User Interfaces. As a result, this project was developed with the GTK+3 tool [12], which, unlike others found, has a free use license.

The graphical interface is subdivided into three main sections. The first main section (Fig. 7) displays the booklet configuration, where the "NUEVO" button allows to choose a new selection of parameters such as subject, section and type of test (T.P.); the "APLICAR" button performs a previous validation before establishing the configured data.

Figure 8 shows the main section of the interface, which is responsible for the control of the functions of the enclosure. The "INICIAR" button allows the start of the mechanical transport of booklets, while the "PAUSAR" and "CANCELAR" buttons pause and cancel the entire process, respectively. On the other hand, the "ESTADO DE LA MÁQUINA" section displays the following features: "Estado de la cámara" shows if the camera is connected, "Estado de la bandeja" indicates the presence of booklets, and "Estado del proceso" indicates whether the mechanical processing of the booklets is activated.

The third main section has two groups. The first one is the state of connection to the database, while the second is the configuration of the synchronized booklets. Finally, there is a secondary section that shows the results of the booklets in process, and two hidden sections, one for configuration and another for image processing observation.

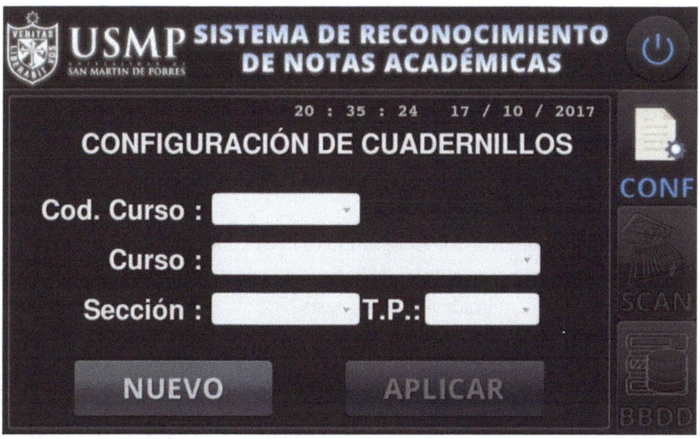

Fig. 7 Configuration screen to set data before the processing of booklets

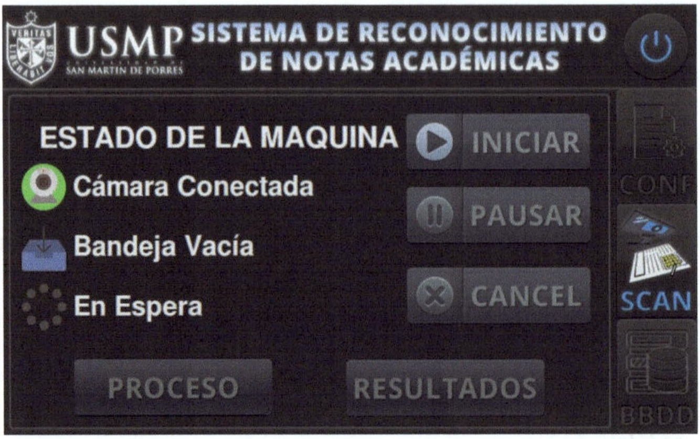

Fig. 8 Basic System Control Screen

3 Results

Statistical tests were conducted using the mechanical system, which is fully implemented with the personalized individual computer (Colibri IMX6 & Iris carrier board) to quantify the probability of error when recognizing the code and the grade using the algorithm developed.

The conditions adopted to conduct the different tests were:

- The user correctly placed the booklets into the enclosure's box.
- A maximum of 40 booklets were placed into the enclosure.
- Optical marks must have been filled out correctly.
- The camera must be correctly centered to capture the entire region of the booklet.

Table 1 shows the number of errors when recognizing the code or grade. In such cases where neither were recognized, it was because of the incorrect transport of the booklet inside the enclosure, possibly due to the lack of cleaning of the rollers responsible for the extraction of booklets.

Table 1 Test of probability of error for recognition of codes and grades on booklets

Experimental tests	#° of Errors in code recognition	#° of Errors in grade recognition	Experimental tests	#° of Errors in code recognition	#° of Errors in grade recognition
1° TEST	0 of 40	0 of 40	6° TEST	2 of 40	2 of 40
2° TEST	0 of 40	0 of 40	7° TEST	0 of 40	0 of 40
3° TEST	1 of 40	0 of 40	8° TEST	0 of 40	0 of 40
4° TEST	0 of 40	0 of 40	9° TEST	0 of 40	1 of 40
5° TEST	0 of 40	0 of 40	10° TEST	0 of 40	0 of 40

In total, 10 tests were carried out with 40 booklets per test. Equations (15) and (16) show the probability of error in each region of recognition.

$$\begin{aligned} \begin{matrix} Probability\,of\,error \\ in\,Code\,Recognition \end{matrix} &= \sum \frac{Total\,Errors\,in\,Code\,Recognition}{Total\,booklet\,Test \times Total\,Experimental\,Test} \\ &= \frac{3}{40 \times 10} \rightarrow 0.75\% \end{aligned} \tag{15}$$

$$\begin{aligned} \begin{matrix} Probability\,of\,error \\ in\,Grade\,Recognition \end{matrix} &= \sum \frac{Total\,Errors\,in\,Grade\,Recognition}{Total\,booklet\,Test \times Total\,Experimental\,Test} \\ &= \frac{3}{40 \times 10} \rightarrow 0.75\% \end{aligned} \tag{16}$$

Finally, the average processing time for the recognition of code and grade was 0.35 s per booklet.

4 Conclusions

Given the difference between programming languages MATLAB and Python, as well as the lack of functions not included by default, it was necessary to look for alternative solutions that meet the requirements, i.e., the use of digital image processing libraries—OpenCV.

Increasing the yellow physical region between the marks and the sides of the code region allows for a better processing; it was estimated and verified that 1 additional millimeter per side is the minimum necessary to achieve this.

At the beginning, an Oracle database was developed. After performing different tests, a better compatibility was obtained between the MySQL database and the Colibri IMX6 embedded system, so it was decided to opt for the latter.

Finally, special care was taken in the use and/or addition to the system of certain programs, as it is necessary to know the type of license of each software. The use of a LGPL V2 license is recommended, as it offers development without restrictions unlike other types of license.

References

1. Toradex, iGuide Camera System by Planitar. https://www.toradex.com/industry-segments/laboratory-test-measurement/planitar. Last accessed 27 Feb 2018
2. Del Carpio, C., Kemper, G.: Un Método Automático de Extraccion de Información de Imágenes Digitales Basado en la Detección de Posicionamiento de Marcas Opticas a partir de Distancias Euclidianas y Redes Neuronales. In: Proceedings of the 2015 Brazilian Technology Symposium, UNICAMP, Campinas (2015)

3. Toradex, NXP®/Freescale i.MX 6 Computer on Module - Colibri iMX6. https://www.toradex.com/computer-on-modules/colibri-arm-family/nxp-freescale-imx6
4. Toradex, How to setup environment for embedded Linux application development. https://developer.toradex.com/knowledge-base/how-to-setup-environment-for-embedded-linux-application-development. Last accessed 27 Feb 2018
5. ATS, maxiFLOW™ Heat Sinks w/ superGRIP™ Attachment. https://www.qats.com/eShop.aspx?q=maxiFLOW/superGRIP. Last accessed 27 Feb 2018
6. Toradex, Iris V1.1 Datasheet. http://docs.toradex.com/100069-iris-datasheet.pdf. Last accessed 27 Feb 2018
7. Python Software Foundation, Python 2.x Docs, https://www.python.org/downloads/. Last accessed 27 Feb 2018
8. Mordvintsev A., Abid K. OpenCV-Python Tutorials. https://opencv-python-tutroals.readthedocs.io/en/latest/py_tutorials/py_tutorials.html
9. Fausett, L.: Fundamentals of Neural Networks: Architectures, Algorithms, and Applications. Prentice Hall, Upper Saddle River (1994)
10. Serrano A., Sorja E., D. Martín. J.: Redes Neuronales Artificiales, Universidad de Valencia
11. Oracle, MySQL Connector/Python Developer Guide. https://dev.mysql.com/doc/connector-python/en/. Last accessed 27 Feb 2018
12. The GNOME Project, GTK+3 Reference Manual. https://developer.gnome.org/gtk3/stable/. Last accessed 27 Feb 2018

IEEE 802.11 De-authentication Attack Detection Using Machine Learning on Unmanned Aerial System

Gustavo de Carvalho Bertoli and Osamu Saotome

Abstract This paper presents a system to detect and alert events of de-authentication attacks on Unmanned Aerial System (UAS). The platform is a Parrot AR.Drone that uses the IEEE 802.11 protocol for command and control. De-authentication is a functionality available on IEEE 802.11 Wireless protocol which is widely deployed on consumer products and can be misused by malicious agents for Denial of Service Attack. The UAS is a real-time system and is normally used in critical applications. The detection approach for this type of event is based on Machine Learning algorithm presenting the best metrics (high accuracy, low latency, and low false positive) to detect and alert between typical algorithms comparison.

Keywords Unmanned aerial system · Security · Machine learning

1 Introduction

Unmanned Aerial Systems (UAS) are an enabling technology being using for typical tasks in areas like delivery, agriculture, inspection, rescue, catastrophe response among others. Initially used as military vehicles, nowadays they are increasing its presence as commercial and consumer machines.

Despite its benefits and being an evolving technology, it embraces aspects like complex architecture and mission-critical applications requiring a rigorous development process to guarantee operational success during its whole life cycle.

Recent news reports that security-related events are becoming commonplace, compromising systems aspects like confidentiality, integrity, and availability [1, 2].

G. de Carvalho Bertoli (✉) · O. Saotome
Electronic Engineering Division, ITA—Instituto Tecnológico de Aeronáutica,
São José dos Campos, São Paulo, Brazil
e-mail: gubertoli@gmail.com

O. Saotome
e-mail: osaotome@gmail.com

Y. Iano et al. (eds.), *Proceedings of the 3rd Brazilian Technology Symposium*,
https://doi.org/10.1007/978-3-319-93112-8_3

This paper presents an algorithm responsible for detection and alert of de-authentication attacks on a UAS platform using IEEE 802.11 Wireless Protocol for Communication. The platform considered for this study is the Parrot AR.Drone a consumer quadcopter that has security assessments as published providing the security vulnerabilities that shall be fixed to guarantee secure operation [3–6].

Prior publications [7, 8] presented machine learning detection systems but they were focused on infrastructure application of IEEE 802.11 Wireless Protocol and not UAS. The novelty of this proposed approach resides on machine learning algorithm development and deployment on the UAS platform, this contribution aims to improve operational performance when IEEE 802.11 controlled UAS be exposed to Denial of Service (DoS) attacks, specifically, de-authentication attacks.

This work is organized into the following sections. Section 2 presents the background and related work into UAS security and machine learning application. Section 3 introduces the methodology adopted for this study, Sect. 4 presents the conclusion and highlights for future work.

2 Background and Related Work

2.1 Security Principals

Security relates to events with intentionally misuse. Instead of well-known concepts of failure using probability of components failures and its impacts leading to redundancy and architectural changes.

Security analysis must "think as a malicious attacker" raising possibilities of misuse of available functions (scenarios) and its impacts on the whole system. After its findings, a trade-off shall be conducted to analyze impact of these scenarios (e.g. catastrophic) and its likelihood aiming to support design decisions.

Security Aspects or Goals can be classified in three major categories ("three pillars")

- Confidentiality: Only those with rights/privileges shall be capable of access information/data (e.g., Authentication)
- Integrity: Confidence information/data—content and sender/receiver (e.g., Hashes)
- Availability: Information, Data or System shall be available when required (e.g., redundancy).

All these aspects have many corresponding security threats or attacks, some examples [9]:

- Confidentiality: Traffic Analysis
- Integrity: Data Injection
- Availability: Denial of Service Attack, Flood Attacks and Malware.

2.2 *Unmanned Aerial Vehicle (UAS)—Parrot AR.Drone*

The UAS being used is the Parrot AR.Drone (Fig. 1) which is an accessible quadcopter that has been successfully used by many researchers. Commercially available as a toy, it has many data available to support modifications like the manufacturer (Parrot) Software Development Kit (SDK).

It is a UAS that has a platform based on ARM Cortex A8 1 GHz microprocessor with DSP video 800 MHz, 1 GB RAM at 200 MHz, USB Port, Wi-Fi IEEE 802.11 b/g/n, BusyBox Embedded Linux Operational System—Kernel 2.6.32, 3-axis Accelerometer (accuracy $\pm$ 50 mg), 3-axles gyroscope (2000°/second accuracy), magnetometer (3 axles, accuracy of 6°), pressure sensor (accuracy of $\pm$ 10 Pa), highly efficient propellers, carbon-fiber structure, lithium-polymer battery (1500 mAh), 4 "inrunner" brushless motors (28,500 r/min, 14.5 W), front and vertical camera [10].

Deligne [3] conducted a full security assessment on Parrot AR.Drone Platform, in this publication the Platform was a prior version but all architectural findings are present in the current platform. These findings relate to DoS, Wireless Injection and Hijacking Video Capturing.

Vasconcelos et al. [4] presents the impacts of DoS attacks on the Parrot AR.Drone platform but focusing on its effects on the video streaming functionality only.

Hooper et al. [5] based on a newer Parrot UAS platform, Bebop, alerts that vulnerabilities expand to all Wi-Fi-based commercial UAS compromising flight due to attacks resulting in catastrophic impact. Also proposes a security framework for hardening Wi-Fi-based commercial UAS.

Pleban et al. [6] also presents the Parrot AR.Drone vulnerabilities and highlights its large application nowadays due to low acquisition costs of this platform. Despite prior references, it already proposes a hardening task for the platform to overcome vulnerabilities, it proposes an encryption and authentication scheme through a Wi-Fi communication protocol change.

None of related works on Parrot AR.Drone platform proposes Machine Learning technique to overcome the vulnerabilities presented on this platform. This paper contributes on this field by a Machine Learning approach to secure the Parrot AR.Drone platform and strength the communication protocol without major changes on platform.

Fig. 1 Parrot AR.Drone 2

2.3 Machine Learning for Security

Considering the complex and constantly changing environment of UAS application regarding security threats and a large amount of data that must be analyzed by security specialists once required to identify security-related events.

Due to this constantly changing environment, the limitations of traditional tools: human-based or fixed algorithms for security events detection are highlighted for its limitations. To overcome these limitations, Machine Learning techniques can be applied to act on these events in a fast manner with high accuracy and most important when compared with legacy options, can evolve in time with the attacks attempts and evolutions.

Already applied in many domains like spam filtering or image recognition, Machine Learning is a great option to support Security Domain due to its inherently of learn and making predictions/decisions from data overcoming static algorithms.

2.4 De-authentication Attack

De-authentication is a functionality available on IEEE 802.11 protocol that composes the management packets category. This type of packet allows the access point or user/station itself to de-authenticate the user/station from access point to save resources, during highly demanded period or due to inactivity for example.

Despite its protocol functionality value, de-authentication has been used by malicious agents that impersonate a legit user and requests a de-authentication to the access point, in this case UAS acting as an Access Point (Fig. 2).

This impersonating act has none security implementation by IEEE 802.11 protocol default thus it is required an additional implementation to identify and ignore this

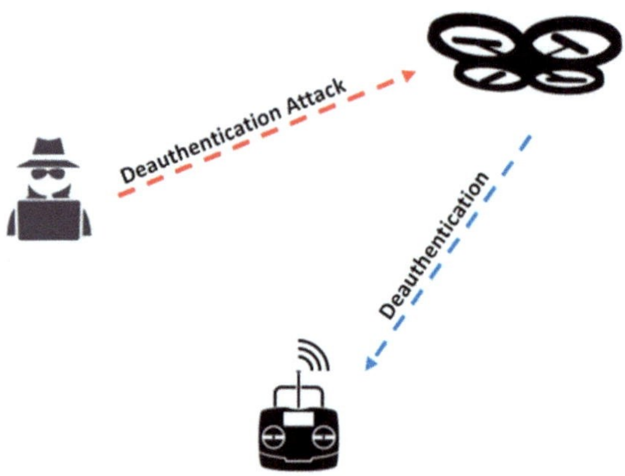

Fig. 2 De-authentication attack

Fig. 3 De-authentication attack impact

package without removing the protocol functionality but just ignoring the malicious impersonated packet.

This de-authentication attack when performed against the Parrot AR.Drone causes a catastrophic impact like crashing to obstacles during flight or crash to land due to loss of control link communication, between controller and vehicle (Fig. 3).

3 Methodology

The methodology is based on a three steps process (Fig. 4), starts with a data capturing process then a machine learning development and finally embedding the machine learning algorithm in UAS platform.

3.1 Data Capture

Data Capture, also known as Sniffing, is responsible for the Knowledge Database construction. It captures the communication frames between controller and UAS using Wireshark [11], a network analysis tool that provides a user interface that allows post-processing tasks. For this, data capturing was used a wireless hardware interface configured on monitor mode, which allows the interface to capture all frames available in the region under analysis. To avoid any unnecessary frame capture, the setup and capture routine was conducted in an isolated area with no other Wireless Access Point available or stations, only the UAS under analysis and the controller. To guarantee a maximum frame capture the computer with hardware interface on

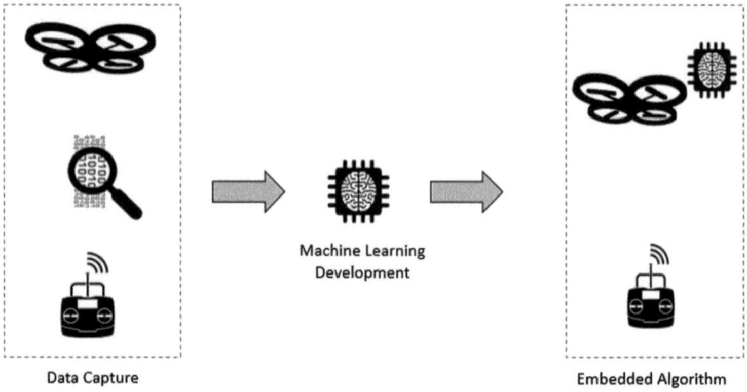

Fig. 4 Methodology overview

monitor mode was placed next to the UAS controller and in the surrounding area which UAS performed flights.

The scenarios considered for data capturing were: Connection Establishment between Controller and UAS, Take-off/Landing and Full Flight (not deterministic route) to compose the training set.

Another scenario considered for data capturing was the De-Authentication Attack. On this scenario, a third wireless interface was configured to inject malicious de-authentication frames on behalf of controller. This injection was conducted with the wireless hacking suite aircrack-ng [12].

All frames captured on these scenarios were filtered on Wireshark to uses only frames related to de-authentication.

3.2 *Machine Learning Development*

With knowledge base made in prior step, Machine Learning development process starts. The Machine Learning task is a classification problem, it aims to classify if a de-authentication frame is a legal request from the controller or if it is a malicious request to de-authenticate the controller from UAS.

On scope of this study, three classification algorithms were analyzed: neural network, decision tree and support vector machine.

These algorithms were then analyzed for accuracy, latency and false positive rate which are the major metrics for a Security Event Detection System in a UAS Application.

Based on the knowledge database obtained during the data capturing, the features considered for Machine Learning development are

- Time delta from previously captured frame (seconds)

- Frame Length (bytes)
- Duration (seconds).

3.3 *Embedding Algorithm*

The last part is embedding the algorithm in the UAS to prevent de-authentication attacks real-time.

The Parrot AR.Drone used runs on BusyBox [13] operation system that is a Linux distribution for Embedded Applications.

To embed this algorithm, it was considered IPTables [14], a firewall tool for Linux that allows filter customization. IPTables filter will be the Machine Learning Algorithm elected on prior step for this application.

4 Conclusion and Future Work

This is an on-going research project aims to correctly identify and mitigates de-authentication attacks on UAS Systems that uses IEEE 802.11 Protocol through Machine Learning algorithms.

Future works considers the achievement of de-authentication mitigation and to expand the use of Machine Learning to address other security vulnerabilities that UAS Systems using IEEE 802.11 Protocol can be exposed.

References

1. MacAskill, E.: US drones hacked by Iraqi insurgents. The Guardian. [Online]. Available: https://www.theguardian.com/world/2009/dec/17/skygrabber-american-drones-hacked (2017). Accessed 20 Sept 2017
2. Fox-Brewster, T.: Watch a very vulnerable $140 quadcopter drone get hacked out of the sky. Forbes.com. [Online]. Available: https://www.forbes.com/sites/thomasbrewster/2017/04/25/vulnerable-quadcopter-drone-hacked-by-ut-dallas-cyber-researchers (2017). Accessed 20 Sep 2017
3. Deligne, E.: ARDrone corruption. J. Comput. Virol. **8**(1–2), 15–27 (2011)
4. Vasconcelos, G., Carrijo, G., Miani, R., Souza, J., Guizilini, V.: The impact of DoS attacks on the AR.Drone 2.0. In: 2016 XIII Latin American Robotics Symposium and IV Brazilian Robotics Symposium (LARS/SBR) (2016)
5. Hooper, M., Tian, Y., Zhou, R., Cao, B., Lauf, A., Watkins, L., Robinson, W., Alexis, W.: Securing commercial WiFi-based UAVs from common security attacks. In: MILCOM 2016—2016 IEEE Military Communications Conference (2016)
6. Pleban, J., Band R., Creutzburg, R.: Hacking and securing the AR.Drone 2.0 quadcopter: investigations for improving the security of a toy. In: Mobile Devices and Multimedia: Enabling Technologies, Algorithms, and Applications 2014 (2014)

7. Agarwal, M., Pasumarthi, D., Biswas, S., Nandi, S.: Machine learning approach for detection of flooding DoS attacks in 802.11 networks and attacker localization. Int. J. Mach. Learn. Cybernet. **7**(6), 1035–1051 (2014)
8. LaRoche, P., Zincir-Heywood, A.: 802.11 de-authentication attack detection using genetic programming. Lecture Notes in Computer Science, pp. 1–12 (2006)
9. Sumra, I., Hasbullah, H., AbManan, J.: Attacks on security goals (Confidentiality, Integrity, Availability) in VANET: A Survey. In: Advances in Intelligent Systems and Computing, pp. 51–61 (2014)
10. AR. Drone 2.0 Power Edition: Parrot Store Official. [Online]. Available: https://www.parrot.com/us/drones/parrot-ardrone-20-power-edition#technicals (2017). Accessed 24 Sept 2017
11. Wireshark: Wireshark.org. [Online]. Available: http://www.wireshark.org (2017). Accessed 21 Sept 2017
12. Aircrack-ng: Aircrack-ng.org. [Online]. Available: https://www.aircrack-ng.org/ (2017). Accessed 21 Sept 2017
13. BusyBox: Busybox.net. [Online]. Available: https://busybox.net/ (2017). Accessed 21 Sept 2017
14. netfilter/iptables project homepage—The netfilter.org project: Netfilter.org. [Online]. Available: http://www.netfilter.org/ (2017). Accessed 21 Sept 2017

Detection and Visualization of Forearm Veins for Venipuncture Based on Digital Processing of Infrared Image

Kevin Espinoza, Bryan Magallanes and Guillermo Kemper

Abstract This article presents a method of detection and visualization of the map of forearm veins based on digital processing of infrared images. The proposed method employs absorption spectroscopy, a technique that uses the interaction of infrared radiation with matter (veins). This interaction is carried out under the design of a correct enclosure, allowing adequate infrared illumination and a correct capture of an image of the map of forearm veins in the infrared spectrum, for which filters were optical. Once the image is captured, we proceed to perform the appropriate digital processing for the extraction of the veins; this process consists in a rough outline of an image cut, followed by a segmentation (thresholding), a post-processing (morphological processing), a special filtering to highlight details and distinct veins for their differentiation. Results showed that the proposed method is highly reliable in terms of detecting and mapping veins and in turn, implies a lower cost of implementation, compared to its similar models. This model was tested on 90 patients for which 86.67% correct answers were obtained in terms of correct visualization of the vein map. Of the total patients, 20% exhibited difficult venous access of which it was possible to obtain a correct visualization of the vein map.

Keywords Veins map · Infrared images · Image processing · Venipuncture

K. Espinoza · B. Magallanes · G. Kemper (✉)
Universidad Peruana de Ciencias Aplicadas, Av. Prolongación Primavera 2390, Santiago de Surco, Lima, Peru
e-mail: gullermo.kemper@gmail.com

K. Espinoza
e-mail: kev.espinozac@gmail.com

B. Magallanes
e-mail: brmagasal@gmail.com

Y. Iano et al. (eds.), *Proceedings of the 3rd Brazilian Technology Symposium*,
https://doi.org/10.1007/978-3-319-93112-8_4

1 Introduction

Venipuncture, the extraction of blood from veins, is a necessary process in order to perform a series of tests which can provide important keys about overall human health. A critical part of this process, and the one of greatest concerns on the part of patients, is the insertion of the needle for extraction. This is because if a correct detection of the vein map has not been made in the area where the puncture will be performed, the procedure can cause intense pain, skin damage and even infections. Although veins map detection is performed by qualified personnel, it is not completely reliable for all patients, since many of them are difficult to access venously. Therefore, the need arises to use technology that facilitates the detection of the vein map.

At present, electronic equipment is varied which seeks to facilitate the detection and visualization of the vein map, each one of them with unique characteristics. There exists equipment which works by showing a projection of light on the forearm, using near-infrared light at 850 nm wavelength, identifying deep veins of up to 15 mm. This equipment usually has a high price, and is not always accessible to low-income medical centers [1].

Other similar equipment is characterized by its great portability at only 275gr. This machine uses an infrared laser which detects the surface of the vein and then projects onto the patient's arm. However, the price of this product is very high, and the range of vision presented is between 10 and 45 cm. Outside this range the images obtained are variable [2].

Other products that boast good portability and consist of an array of red and orange LEDs that illuminate uniformly towards the center, giving a better contrast between the vein and the fabric. In addition, the price of this product is relatively accessible. However, it presents a great disadvantage, since it is necessary the given that two people must perform the puncture, one that holds the equipment and keeps it pressed lightly against the skin; and another that performs blood extraction. Hence, the benefits of this type of technology are overshadowed by the excess labor that it requires to function [3].

All the equipment mentioned above is characterized by using a near-infrared spectrum to obtain the vein map, using mostly a projection on the forearm as the visualization medium. This implies the selecting correctly focused illumination on the infrared spectrum band where a greater contrast between veins and skin is obtained. Also, it is necessary to have a correct acquisition sensor (camera, either CCD or CMOS) accompanied by the exact optical filter that captures an image on the desired spectrum.

Algorithms developed for this purpose should achieve a great contrast between veins, skin and hair.

In [4] the images are captured through a CCD camera illuminated with a DOM system with 112 LEDs that emit a wavelength of 850 nm over the area of interest (ROI). To obtain a better image, the authors made a closed room with a system to fix the position of the arm, with the positioning of the camera and the DOM system below the wrist to achieve the smallest difference between the lens and skin. Also, optical filters were placed to obtain the infrared near-infrared band of interest.

In [5] the authors used a CMOS webcam to which the infrared filter was removed to increase spectral response. The algorithms used were performed and simulated in MATLAB and generated a VDHL code. The algorithms used were: conversion of RGB to Gray scale averaging the R, G, and B values of each pixel; "Contrast Limited Adaptive Histogram Equalization (CLAHE)" which greatly improved the contrast of each image, since it independently equalizes each block of the image and bilinearly interpolates the ends of each block to eliminate artificial borders. It also eliminates arm hairs, which are filtered in the image.

In [6] three methods of equalizing the image histogram are presented to increase image contrast and evaluate the results, using a transformation function that is derived from the product of the cumulative histogram and a factor. The types of transformation are: linear, parabolic, and logarithmic.

In [7] are presented the considerations that the LED lighting system should have for optimal performance. These considerations are based on the number of LEDs in the array, the distance between the surface and the array, and the inclination of each led.

The method proposed in this work is an alternative solution based on the digital processing of infrared images, which is obtained from a properly designed room for the location of infrared light LEDs and acquisition sensor (camera). Visualization of the vein map is on screen located above the forearm, which significantly differentiates the present equipment from the similar systems mentioned above. The results shown at the end of the work not only demonstrate the reliability of the employed method, but also the quality of the equipment developed.

2 Description of the Proposed Method

Figure 1 shows the block diagram of the proposed method. General development can be summarized as follows:

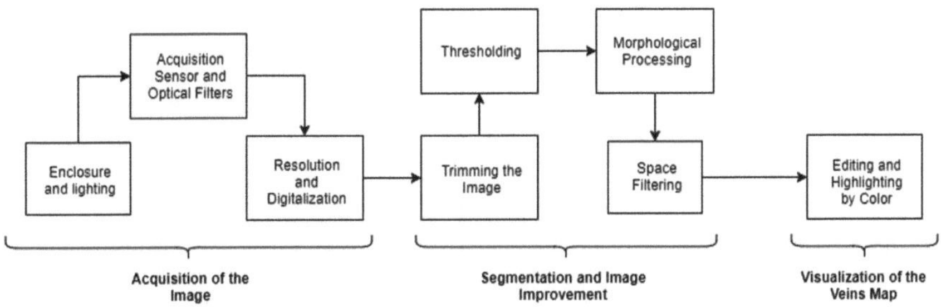

Fig. 1 Block diagram of the proposed method

The process begins with the acquisition of the image with an infrared sensor, an optical filter and an adequate illumination to achieve the highest possible contrast between veins, skin, and hair.

Once the image is captured it is digitized and transformed into gray scale format. Then, the image is cut out in order to obtain a region of interest for real-time processing.

The contrast is increased through the equalization of the obtained image segment. The result is binarized to generate a strong contrast between veins and skin. However, when surpluses produced by the thresholding are obtained, post-processing is carried out with a morphological opening operation followed by a binary AND operation between the current resulting image and that obtained in the previous acquisition. Once the filtered image is obtained, the veins are colored to differentiate them from the skin, and can thus be shown on a screen or monitor.

In the following sections, the most important functional blocks that make up the proposed method will be described in detail.

2.1 Acquisition of the Image

As mentioned above, the design of both the enclosure and the infrared lighting system is fundamental. For this, it is important to choose a suitable position for the lighting LEDs with respect to the acquisition sensor (camera). For this purpose, the radiance and irradiance were evaluated in oblique and frontal positions as presented in Fig. 2.

The oblique model was chosen because it better distributed luminous energy and avoided saturation of the camera.

The camera used was a Pi NoIR Infrared high definition with a Sony IMX219 sensor and a spatial resolution of up to 8 megapixels.

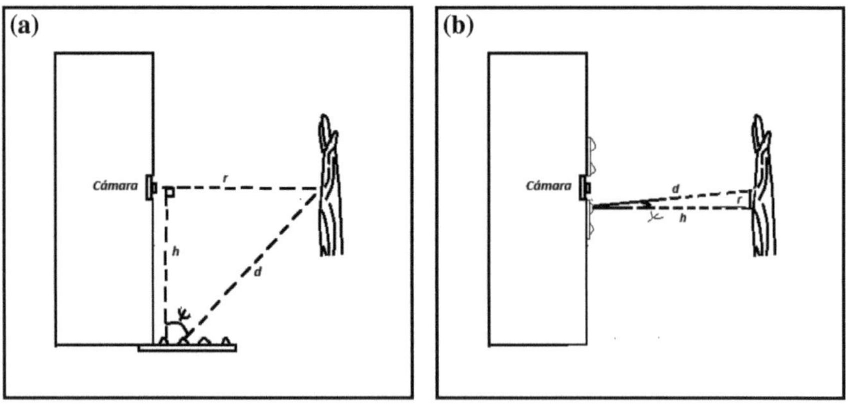

Fig. 2 **a** Arm illuminated obliquely. **b** Arm illuminated frontally

This camera was coupled with an optical band-pass filter in the 800–900 nm band with a central frequency of 850 nm. On this band, the highest intensity of reflectance and contrast of the veins against the skin and other tissues was recorded.

For real-time image digitalization, an average capture time resolution of 25 frames per second and a spatial resolution per frame of 380 × 380 pixels was configured in RGB format—24 bits (true color: Red-Green-Blue).

The image acquired in an instant of discrete time t is expressed through RGB primary components defined as $F_R^t(x, y)$, $F_G^t(x, y)$ and $F_B^t(x, y)$, where x and y constitute the spatial coordinates of the image.

2.2 Segmentation and Improvement of the Image

For processing, a region of interest (ROI) is defined as:

$$F_{1R}^t(x, y) = F_R^t(x + x_0, y + y_0) \tag{1}$$

$$F_{1G}^t(x, y) = F_G^t(x + x_0, y + y_0) \tag{2}$$

$$F_{1B}^t(x, y) = F_B^t(x + x_0, y + y_0) \tag{3}$$

For $x = 0, 1, \ldots, 189$, $y = 0, 1, \ldots, 189$, $x_0 = 95$ and $y_0 = 95$.

The green box in Fig. 3a shows the ROI and Fig. 3b the image segmented with primary components $F_{1R}^t(x, y)$, $F_{1G}^t(x, y)$ and $F_{1B}^t(x, y)$.

The extracted image segment (with dimensions of 190 × 190 pixels) is then converted to gray scale.

This conversion can be expressed as:

$$I_1^t(x, y) = 0.299F_{1R}^t(x, y) + 0.587F_{1G}^t(x, y) + 0.114F_{1B}^t(x, y) \tag{4}$$

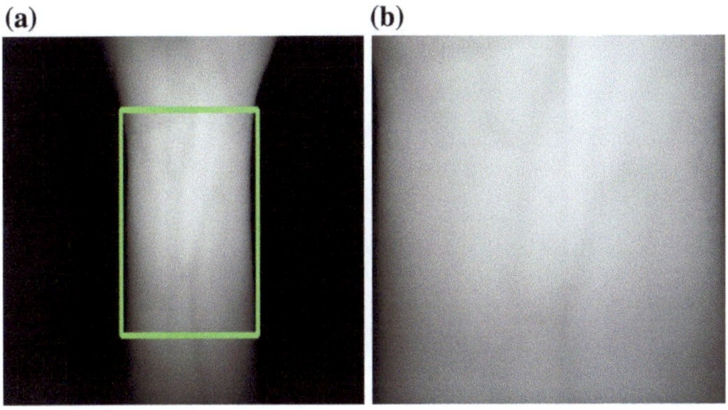

Fig. 3 **a** Original image. **b** Image of interest

To achieve a contrast enhancement between vein and skin, the image $I_1^t(x, y)$ is subjected to a Adaptive Histogram Equalization by Contrast Limitation (CLAHE), a method that starts by dividing the image into regions of equal size. Then the histogram of each region and the cutoff is calculated. This value establishes a horizontal cut of the histogram on the abscissa. Surplus is evenly distributed in the histogram [8].

Finally, an interpolation of each pixel is applied to its neighborhood [9].

The resulting image of the equalization is be denoted as $I_2^t(x, y)$ and is shown in Fig. 4.

For the segmentation of the veins, it is assumed that a vein pixel is of lesser value than a tissue pixel. In these cases, the difference between the two is negative, being that the result of the position of the pixel evaluated in the output image is equal to 1. Otherwise, it is considered tissue and the result is 0.

This procedure is applied in 4 orientations: $\theta = 0°, \theta = 45°, \theta = 90°$ and $\theta = 135°$, where θ is the angular location of the neighboring pixel in the separated Cartesian plane d pixels of the evaluated pixel. For this case it is considered as $d = 5$.

Applying the procedures indicated above with the remote pixel neighborhood method [10], the following images are obtained:

For $\theta = 0°$:

$$I^t_{2B(\theta=0°)}(x, y) = \begin{cases} 1 & , \begin{array}{l} I_2^t(x, y) - I_2^t(x + d, y) < 0 \wedge \\ I(x, y) - I(x - d, y) < 0 \end{array} \\ 0 & , \begin{array}{l} I_2^t(x, y) - I_2^t(x + d, y) \geq 0 \wedge \\ I_2^t(x, y) - I_2^t(x - d, y) \geq 0 \end{array} \end{cases} \tag{5}$$

For $\theta = 45°$:

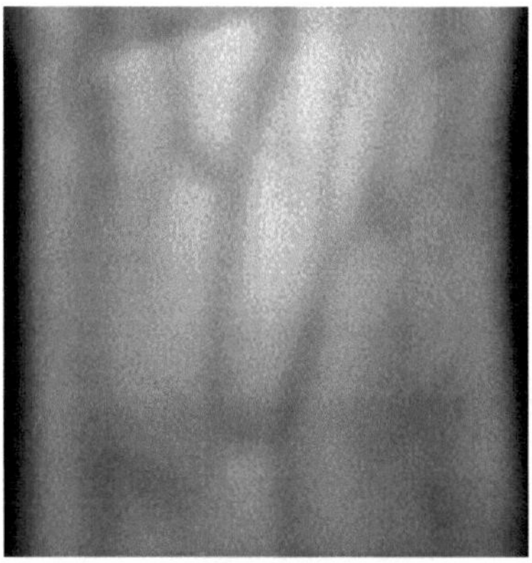

Fig. 4 Resulting image from the CLAHE equalization process

$$I^t_{2B(\theta=45^\circ)}(x,y)=\begin{cases}1, & I^t_2(x,y)-I^t_2(x+d,y+d)<0 \;\wedge\; I^t_2(x,y)-I^t_2(x-d,y-d)<0\\ 0, & I^t_2(x,y)-I^t_2(x+d,y+d)\geq 0 \;\wedge\; I^t_2(x,y)-I^t_2(x-d,y-d)\geq 0\end{cases} \quad (6)$$

For $\theta = 90^\circ$:

$$I^t_{2B(\theta=90^\circ)}(x,y)=\begin{cases}1, & I^t_2(x,y)-I^t_2(x,y+d)<0 \;\wedge\; I^t_2(x,y)-I^t_2(x,y-d)<0\\ 0, & I^t_2(x,y)-I^t_2(x,y+d)\geq 0 \;\wedge\; I^t_2(x,y)-I^t_2(x,y-d)\geq 0\end{cases} \quad (7)$$

For $\theta = 135^\circ$:

$$I^t_{2B(\theta=135^\circ)}(x,y)=\begin{cases}1, & I^t_2(x,y)-I^t_2(x-d,y+d)<0 \;\wedge\; I^t_2(x,y)-I^t_2(x+d,y-d)<0\\ 0, & I^t_2(x,y)-I^t_2(x-d,y+d)\geq 0 \;\wedge\; I^t_2(x,y)-I^t_2(x+d,y-d)\geq 0\end{cases} \quad (8)$$

The final image $\mathbf{I}^t_3$ resulting from the integration of the images of each case is expressed as:

$$\mathbf{I}^t_3 = \mathbf{I}^t_{2B(\theta=0^\circ)} \text{ OR } \mathbf{I}^t_{2B(\theta=45^\circ)} \text{ OR } \mathbf{I}^t_{2B(\theta=90^\circ)} \text{ OR } \mathbf{I}^t_{2B(\theta=135^\circ)} \quad (9)$$

The resulting image $\mathbf{I}^t_3$ of binarization by distance neighborhood of pixels is shown in Fig. 5.

Next, an opening morphological operation [11] is applied to the image $\mathbf{I}^t_3$ to eliminate the surplus produced in the previous process and maintain vein patterns. This operation consists of 2 morphological operations. First, an erosion is applied to thin the vein pixel patterns eliminating the surplus. Then, dilation is applied to return to its original thickness. The procedure is expressed as:

$$\mathbf{I}^t_4 = (\mathbf{I}^t_3 \Theta \mathbf{I}_{\text{ker nel}}) \oplus \mathbf{I}_{\text{ker nel}}$$
$$\text{for the structural element } \mathbf{I}_{\text{ker nel}} = \begin{bmatrix} 11 \\ 11 \end{bmatrix} \quad (10)$$

The resulting image of this operation $\mathbf{I}^t_4$ is shown in Fig. 6.

Fig. 5 Image of the binarization process

Fig. 6 Image with opening morphological operation

However, this above procedure is still insufficient to eliminate the residuals, so an iterative AND binary operation is applied in time that is expressed as:

$$\mathbf{I}_{\mathbf{A}}^{\mathbf{t}} = \mathbf{I}_{4}^{\mathbf{t}} \textbf{ AND } \mathbf{I}_{\mathbf{A}}^{\mathbf{t}-1} \tag{11}$$

for $\mathbf{I}_{\mathbf{A}}^{-1} = \mathbf{I}_{4}^{0}$.

Then we do

$$\mathbf{I}_{5}^{\mathbf{t}} = \mathbf{I}_{\mathbf{A}}^{\mathbf{t}} \tag{12}$$

The resulting image $\mathbf{I}_{5}^{\mathbf{t}}$ is shown in Fig. 7.

Fig. 7 Image resulting from binary operation and iterative in time

2.3 *Visualization of the Vein Map*

To highlight and visualize (in an instant of time) the vein patterns obtained in red, image is generated from its primary components RGB:

$$I_{6R}^{t}(x, y) = \begin{cases} 255, & I_{5}^{t}(x, y) > 0 \\ 0, & other\ case \end{cases} \tag{13}$$

$$I_{6G}^{t}(x, y) = 0 \tag{14}$$

$$I_{6B}^{t}(x, y) = 0 \tag{15}$$

Figure 8 shows the obtained image.

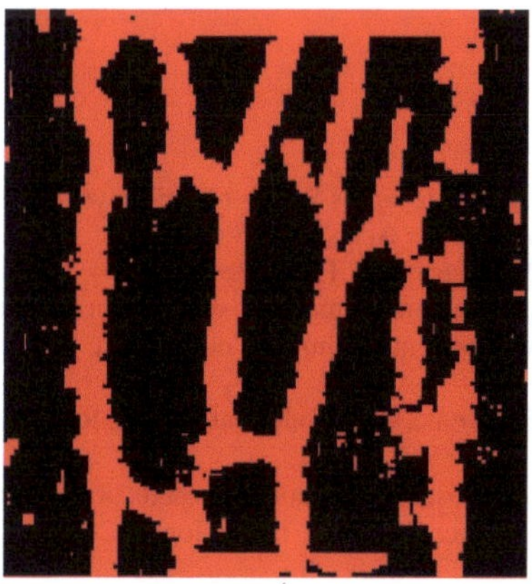

Fig. 8 Final image processed and visualized

3 Results

The validation of the proposed system was supported by the nursing department of the National Institute of Child Health, Breña, Lima, Peru.

This procedure consisted first of detecting those patients in whom the visualization of the vein map was possible with the naked eye to corroborate what was seen with developed equipment. Then, the nurses focused on the patients in whom it was no longer possible to define a vein map with the naked eye, and therefore techniques such as tourniquet (partial obstruction of the blood circulation to increase its volume in a specific area) were necessary. Finally, there were patients in whom it was very difficult to detect the vein map either with the naked eye or with a tourniquet, so only the nurses' expertise was available to locate the vein map.

A total of 90 patients between 17 and 65 years old participated in the study, for which the nurses located the vein map using the procedures described above. A total of 78 successful cases were developed, within these cases are those patients with easy venous access and those in which a tourniquet was applied, representing 86.67% effectiveness. On the other hand, 13.33% error was obtained which was reflected in the small contrast between veins and skin.

Within the margin of error, 12 patients were those with difficult to access veins. Among the 78 successful cases, 18 were those in which the patient also had difficult to access veins, yet the vein map could be correctly visualized. In total 20% of the patients had difficult venous access for which good visualization of the veins map was achieved.

Acknowledgements The authors are grateful to the Research Department of the UPC (Universidad Peruana de Ciencias Aplicadas) for their support in the development and financing of the project. Likewise, we thank the Nursing Department of the National Institute of Child Health Breña, who gave us their support for the validation of the results, with special mention to head of the Research Department, Teaching and Care in Pediatric Medicine of the aforementioned institute, Dr. Franklin Aranda Paniora.

References

1. Christie Medical Holding: Whitepapers Support VeinViewer Capabilities and Benefits
2. AccuVein: Manual del usuario de AccuVein AV400
3. TransLite LLC: Manual de usuario de TransLite LLC
4. Suárez, J.: Mecanismos de Captura y Procesado de Imágenes de Venas para identificación Personal. Tesis doctoral, Departamento tecnología electrónica, Universidad Carlos III de Madrid, Leganés, Julio (2011)
5. Sureskumar, V., et al.: Electronic Vein Finder. Int. J. Adv. Res. Comput. Commun. Eng. **4**(10) (2015)
6. Chakravorty, T., et al.: Low-cost subcutaneous vein detection system using ARM9 based single board computer. In: Electronics Computer Technology (ICECT), 3rd International Conference (2011)
7. Nundy K.K., Sanyal, S.: A low cost vein detection system using integrable mobile camera devices. In: Annual IEEE India Conference (INDICON) (2010)

8. Benitez, G., et al.: Face identification based on contrast limit adaptive histogram equalization (CLAHE). Escuela de Ingeniería Mecánica y Eléctrica, Instituto Politécnico Nacional de México. México D.F, México
9. Pizer, S., et al.: Adaptive histogram equalization and its variations. Comput. Vis. Graph. Image Process. **39**, 355–368 (1987)
10. Alejo, W., et al.: Biometric method based on the matching of dilated and skeletonized IR images of the veins map of the dorsum of the hand. IEEE Latin Am. Trans. **13**(5), 1438–1445 (2015)
11. Universidad Nacional de Quilmes: Operaciones Morfológicas en Imágenes Binarias – 2da parte. Ingeniería en Automatización y Control Industrial, Agosto (2005)

Measuring the Level of Mildew in Quinoa Plantations Based on Digital Image Processing

Gian Oré, Alexis Vásquez, Guillermo Kemper and Jonell Soto

Abstract This paper presents a computer vision system that successfully detects the level of affectation produced by Mildew present in quinoa leaves. The system consists of two subsystems in sequence: processing for the discrimination of earth with vegetation, and processing for the discrimination of vegetation with the affected leaves, which is useful for disease detection. The Lab color method and non-complex image processing techniques are used for both subsystems, such as: histogram equalization, thresholding, filtering, and tagging and connectivity algorithms. This model has been tested in different scenarios, different angles for different crops in the coastal area (where problems with the Mildew are presented with greater emphasis) during different times of the year. In addition, it has been shown that this system produces acceptable results even in difficult conditions, such as in the presence of objects outside the field, either pipes or waste.

Keywords Computer vision · Image processing · RGB · Lab · Filtering
Threshold · Subjective evaluation

G. Oré · A. Vásquez · G. Kemper (✉)
Universidad Peruana de Ciencias Aplicadas, Av. Prolongación Primavera 2390,
Santiago de Surco, Lima, Peru
e-mail: guillermo.kemper@gmail.com

G. Oré
e-mail: gc.ore.huacles@gmail.com

A. Vásquez
e-mail: vasquezgarciaalexis@gmail.com

J. Soto
Instituto Nacional de Innovación Agraria, Av. La Molina 1981, La Molina, Peru
e-mail: jonellsjeri@gmail.com

Y. Iano et al. (eds.), *Proceedings of the 3rd Brazilian Technology Symposium*,
https://doi.org/10.1007/978-3-319-93112-8_5

1 Introduction

For some years, Precision Agriculture (PA) has been taken with greater emphasis by farmers and organizations engaged in agricultural research. The purpose of PA is to record detailed information on the state of crops, making use of technological methods. In this way, problems can be identified such as pests, lack of fertilization, lack of water, among others, and thus it is possible necessary corrective measures. With the current projections of population growth and the consequent reduction of available land and natural resources, there will be a need for cheaper, more efficient and environmentally friendly agriculture [1].

Management of specific pests in the field usually includes the application herbicides in a strictly necessary dose based on the infestation of plagues and location the crop [2]. In addition, crop analysis should be as non-invasive as possible, and hence viable methods include computer vision and image processing.

There are papers related to the classification and comparison of weeds or pests in crops. Research in this field is generally mentions two steps to identify infestation in crops: (1) Segmentation of vegetation with the bottom (soil and/or sky debris) and (2) detection of vegetation pixels representing the damaged part [2].

In this context, detection of infestations requires a high level of precision and a minimum time to obtain results. In the present paper, the use of two components is highlighted: the "Lab" color space to obtain values which characterize an object less susceptible to variations in luminosity, and the two-dimensional histogram.

2 Description of the Proposed Algorithm for the Segmentation of Vegetation

An input image with primary color components $I_R(x, y)$, $I_G(x, y)$ and $I_B(x, y)$ is transformed to L*a*b* color space. Then, a two-dimensional histogram is calculated using a* and b* components previously calculated. From the two-dimensional histogram, the vegetation region is calculated, which then is segmented through limits or thresholds. The segmented image is filtered by a mean filter. The goal of this process is to reduce the object number in the segmented image. Then, an object labeling process is performed, where objects which having a size less than a threshold are removed. Figure 1 shows the block diagram of the described algorithm. The proposed segmentation algorithm is described in detail following.

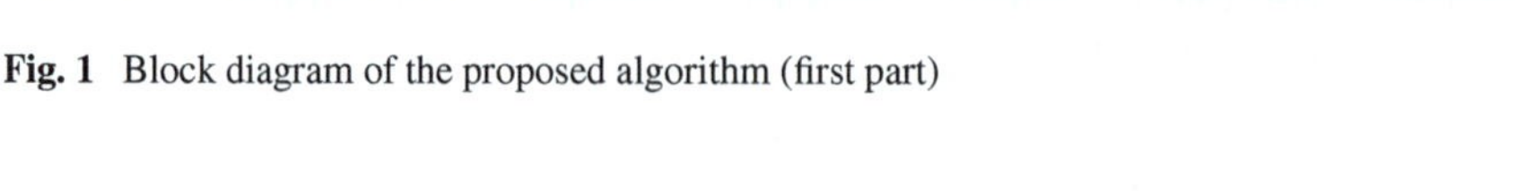

Fig. 1 Block diagram of the proposed algorithm (first part)

2.1 Conversion RGB to L*a*B*

To convert the image to L*a*b* color space, the image must be converted to an XYZ color space [3], using a white-point value of D56, to obtain the $\mathbf{I}_{L*}$, $\mathbf{I}_{a*}$ and $\mathbf{I}_{b*}$ components. This color space is chosen because luminosity is ignored in the $\mathbf{I}_{a*}$ and $\mathbf{I}_{b*}$ components. Obtaining two uncorrelated components helps to create a two-dimensional histogram for more segmentation flexibility. The $\mathbf{I}_R$, $\mathbf{I}_G$ and $\mathbf{I}_B$ variables are defined as components of the RGB color space from the system input image and X, Y and Z variables are defined as components resulting of the transformation to the XYZ color space [4]. Then, $\mathbf{I}_{L*}$, $\mathbf{I}_{a*}$ and $\mathbf{I}_{b*}$ components are obtained using the following equations:

$$\mathbf{I}_{L*} = 116 \times f(\mathbf{Y}/Y_0) - 16 \tag{1}$$

$$\mathbf{I}_{a*} = 500 \times [f(\mathbf{X}/X_0) - f(\mathbf{Y}/Y_0)] \tag{2}$$

$$\mathbf{I}_{b*} = 200 \times [f(\mathbf{Y}/Y_0) - f(\mathbf{Z}/Z_0)] \tag{3}$$

where

$$f(t) = \begin{cases} t^{1/3} & t > (\frac{6}{29})^3 \\ \frac{1}{3} \times (\frac{29}{6})^2 \times t + \frac{4}{29} & otra\ forma \end{cases} \tag{4}$$

The X_0, Y_0 and Z_0 constants have the following standard values:

$$X_0 = 0.9492483 \tag{5}$$

$$Y_0 = 1.03644 \tag{6}$$

$$Z_0 = 1.088754 \tag{7}$$

2.2 Segmentation Using Two – Dimensional Histogram

After arriving at the $\mathbf{I}_{a*}$ and $\mathbf{I}_{b*}$ components, which are values between -100 and $+100$ [5], a two-dimensional histogram is generated. An example of the two-dimensional histogram is shown in Fig. 2. The abscissa axis represents the a* component and the ordinate axis represents the b* component. Values which are related with each ordered pair is represented by the white color intensity. The greater the number of values, the greater the white intensity.

From the histogram, two easily visible regions are separated by a line. The upper region corresponds to the vegetation and the lower region corresponds to the ground. For segmentation, a line equation is used to separate both regions of the histogram. Then, a binary image is derived that it is represented by the $\mathbf{M}_b$ matrix. This process is described by the following equation:

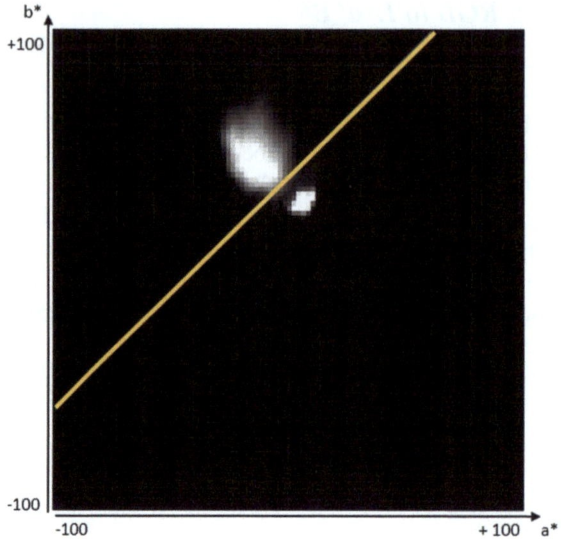

Fig. 2 Two-dimensional histogram for separation of regions (vegetation—soil)

$$M_b(x, y) = \begin{cases} 255, & [0.97 \times I_{a^*}(x, y) + I_{b^*}(x, y)] > 39.4 \\ 0, & othercase \end{cases} \tag{8}$$

2.3 *Image Filtering*

After the binary image is obtained, it is filtered by a mean filter $\mathbf{F_{p1}}$ of 11×11. The filter is described in the following expression [6]:

$$\mathbf{F_{p1}} = \frac{1}{121}\begin{bmatrix} 1\ 1 \ldots 1 \\ 1\ 1 \ldots 1 \\ \vdots\ \vdots \ldots \vdots \\ 1\ 1 \ldots 1 \end{bmatrix}_{11\times 11} \tag{9}$$

Then, the filtered image is represented by $\mathbf{M_{f1}}$ array

$$M_{f1}(x, y) = \sum_{m=0}^{10}\sum_{n=0}^{10} F_{p1}(m, n)M_b(x - m, y - n) \tag{10}$$

The next step is to binarize the image using a threshold. In this case, the proposed threshold is 80. The result is represented as $\mathbf{M_t}$ array.

$$M_t(x, y) = \begin{cases} 255, & M_f(x, y) > 80 \\ 0, & othercase \end{cases} \quad (11)$$

The objective of this method is to reduce the number of small objects after the segmentation by removing objects which are isolated and joining adjacent objects.

2.4 Segmentation of Large Objects

In this procedure, the obtained binary image is processed by a labeling and connectivity algorithm to work with each object independently and to only segment objects which have a pixel quantity greater than a threshold. Small objects often represent leaves on the ground or small segmented ground areas as vegetation. The labeling and connectivity algorithm is based on pixel 8-connectivity [7].

As a result of this block, an image is obtained from which the objects with a smaller size than the threshold is removed, which has the value of a total percentage of pixels of the image, which in this case is 0.2%. The resulting image of this block and the first part of the total proposed system is displayed in Fig. 3.

Fig. 3 Output image of large object segmentation

3 Description of the Proposed System for Segmentation of Yellow Leaves

Figure 4 shows the second part of the block diagram of the proposed algorithm. This algorithm receives as input the obtained image from the algorithm described previously.

Segmentation of the two-dimensional histogram, which is the same from the previous algorithm, is done at the same time with the segmentation in the RGB color space to finally obtain a unique binary image. After the segmentation, the resulting image is filtered and binarized to reduce the amount of objects in the image. This image is processed by a labeling and connectivity algorithm where small objects are removed. Then, the eccentricity geometric feature is calculated for each object. Objects with an eccentricity value greater than a proposed threshold are also removed. The proposed algorithm is explained in detail in the next section.

3.1 *Segmentation of Two-Dimensional Histogram and RGB Space*

Segmentation is carried considering the same histogram calculated in the previous algorithm. In this process, the histogram area for segmentation is difficult to discriminate, so the RGB space color is used to improve segmentation of the two-dimensional histogram. The two-dimensional histogram limits are fixed using a line, and simple logical conditions are used in RGB color space. Figure 5 shows a sample two-dimensional histogram and a line which is used for yellow leaf segmentation. To obtain a binary image as result of the segmentation, the line equation and logical conditions in RGB color space are used. That binary image is represented by the matrix $\mathbf{M}_{\mathbf{b2}}$. The process, where the $\mathbf{M}_{\mathbf{b2}}$ array is obtained, is described as follows:

$$M_{b2}(x, y) = \begin{cases} 255, & \begin{array}{c} (0.557 \times I_{a*}(x, y) + I_{b*}(x, y)) > 57 \\ \vee \\ I_R(x, y) > (I_G(x, y) + I_B(x, y)) \end{array} \\ 0, & othercase \end{cases} \quad (12)$$

Figure 5 shows the two-dimensional histogram used in this process.

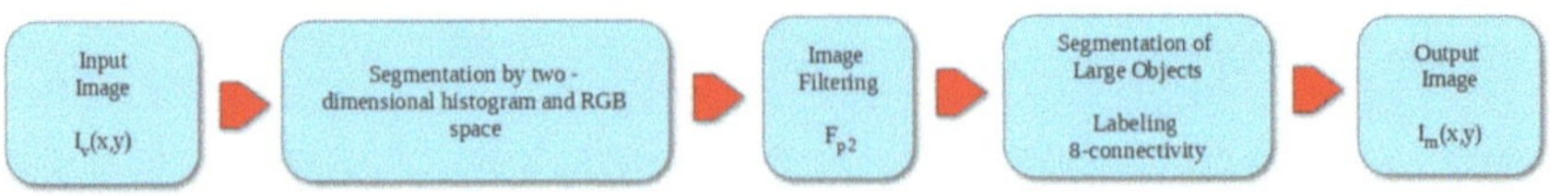

Fig. 4 Block diagram of the proposed algorithm (second part)

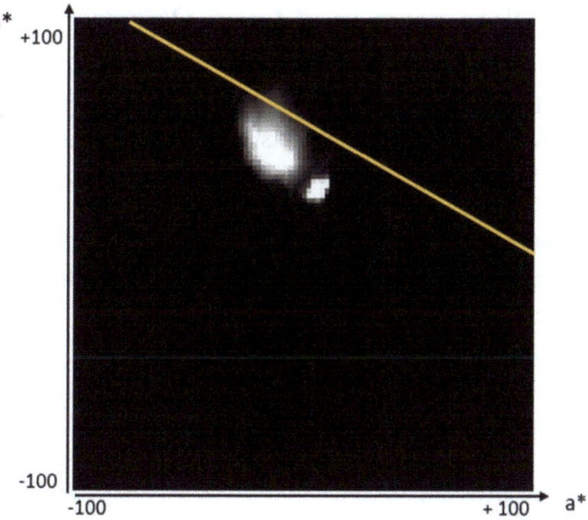

Fig. 5 Two-dimensional histogram for separation of yellow leaves—green leaves

3.2 *Image Filtering*

In this process, the obtained binary image of the segmentation is filtered by a mean filter. The objective is to reduce the number of objects to alleviate computational calculation of the next process. The equation of the mean filter of 5 × 5 [6], which is represented by the $\mathbf{F_{p2}}$ array, is shown below:

$$\mathbf{F_{p2}} = \frac{1}{25}\begin{bmatrix} 1 & 1 & 1 & 1 & 1 \\ 1 & 1 & 1 & 1 & 1 \\ 1 & 1 & 1 & 1 & 1 \\ 1 & 1 & 1 & 1 & 1 \\ 1 & 1 & 1 & 1 & 1 \end{bmatrix} \tag{13}$$

Then, the filtered image, which is represented as $\mathbf{M_{f2}}$ array, is obtained with the following expression:

$$M_{f2}(x, y) = \sum_{m=0}^{4}\sum_{n=0}^{4} F_{p2}(m, n)M_{b2}(x - m, y - n) \tag{14}$$

The next step is to binarize the image using a threshold. The proposed threshold is 80. The result is represented as $\mathbf{M_{t2}}$ array:

$$M_{t2}(x, y) = \begin{cases} 255, & M_{f2}(x, y) > 80 \\ 0, & othercase \end{cases} \tag{15}$$

3.3 Segmentation of Large Objects

In this process, a labeling and connectivity algorithm is applied to the $\mathbf{M}_{t2}$ binary image to discriminate objects which have an amount of pixels smaller than a proposed threshold. This threshold has a value of 0.002% of the total image. This process discriminates small objects which were wrongly segmented, either because of an illumination effect or small parts of the vegetation which have a yellow color. The pixel 8-connectivity concept is used to develop the labeling and connectivity algorithm.

3.4 Discrimination by Eccentricity Property

In this process, the calculus of eccentricity geometric property is done for each labeled object. Then, objects which have an eccentricity value greater than a proposed threshold are removed. To calculate the eccentricity value of each object, a math expression is used [8], as shown below

$$exc = \frac{\sqrt{MaL^2 - MiL^2}}{MaL} \tag{16}$$

where:

*Ma*L Semi-major axis of an object.
*Mi*L Semi-minor axis of an object.

Then, objects which have an *exc* value greater than the proposed threshold of 0.98 are removed. The objective of this process is to discriminate objects which have a linear shape because these objects often represent branches, stems and leaf edges.

Finally, as result of the algorithm, Fig. 6 shows an image where the background is removed, and yellow leaves are highlighted using a blue color.

4 Results

For the evaluation of the Mildew detector system, the Coefficient of Kappa of Cohen Method was used [9]. This method measures the agreement between two examiners; in this case, that of the specialist agronomist with that of the detector system. For this paper, the table is completed following the pixel pattern in a digital image. The procedure proposed by this method can be summarized as follows: First, the specialist will receive a printed image where he/she will shade with blue color the parts where he considers that Mildew is present. Then, the image returned by the agronomist will be processed so that a segmentation of shaded pixels can be made. Contrarily, the original digital image is processed with the proposed Mildew detector system. In this way, there are two resulting images, one following the opinion of the agronomist

Fig. 6 Output image of the eccentricity discrimination process

Table 1 Table of results obtained from the kappa index

Image	1	2	3	4	5	6	7	8
k	0.7462	0.8126	0.7783	0.8127	0.7616	0.7688	0.7362	0.7811
Image	9	10	11	12	13	14	15	16
k	0.7946	0.7309	0.8009	0.7934	0.8296	0.7811	0.7946	0.7309
Image	17	18	19	20	21	22	23	24
k	0.8909	0.8934	0.92996	0.8156	0.7733	0.8037	0.7620	0.7788

and another as an output of the proposed detector system. These images will serve to complete the Kappa of Cohen table and to find the final Coefficient of Kappa k [9]. The resulting k value provides the level of agreement that exists between two evaluators. This value can range between 0 and 1, where 0 means that they completely disagree and 1 that they are completely in agreement. The proposed system is based on the value of Kappa to validate its operation and verify the reliability of the results. A sample of 24 images was taken, during the day between the hours of 9:00 a.m. to 3:00 p.m. (UTC—05:00), which were captured by a single capture sensor of 8 megapixels of resolution. In Table 1, the results are presented, obtained after the analysis of the Kappa index.

In Fig. 7, the procedure performed for a sample obtained to find the kappa index k is presented graphically.

5 Conclusions

Analysis of crops through image processing involves working in uncontrolled lighting environments where the use of methods where one can quantify characteristics of objects of interest that are least susceptible to changes in lighting. Using the appropri-

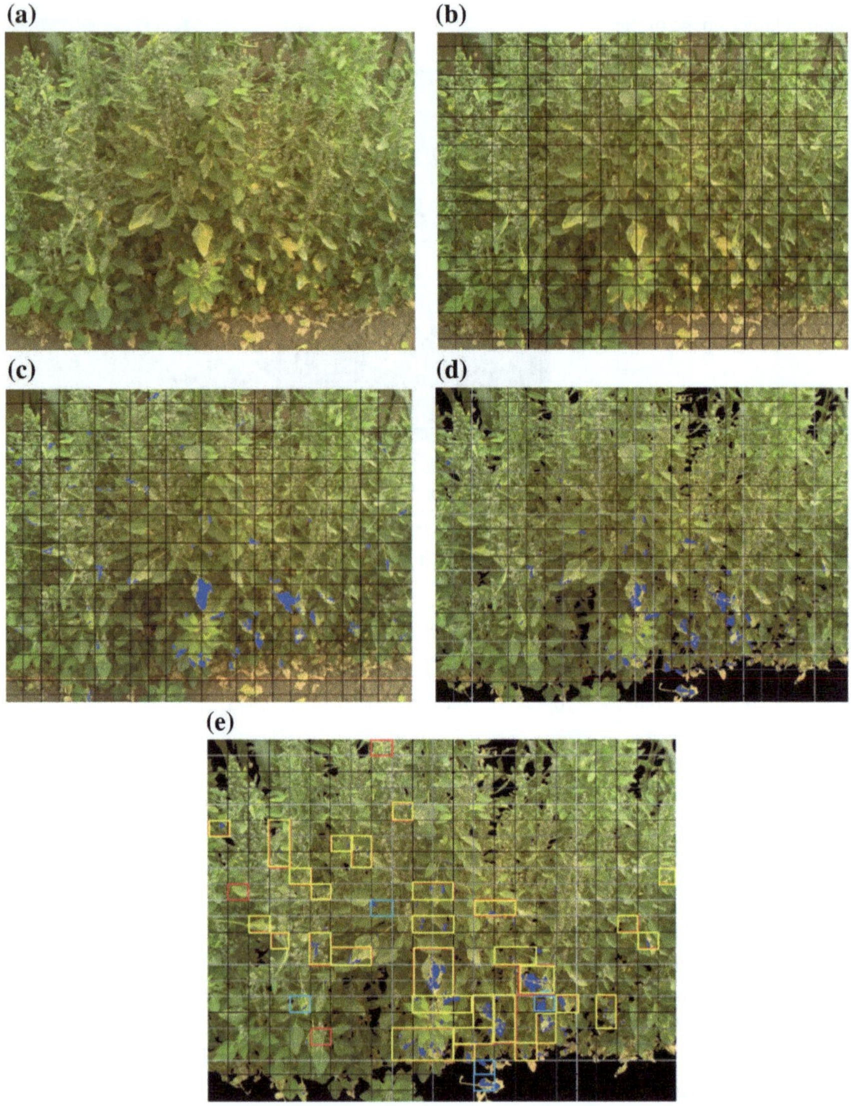

Fig. 7 **a** Original image. **b** Original sectioned image. **c** Imagen resultante tras evaluación del agrónomo. **d** Resulting image after evaluation of the agronomist. **e** Comparative image of the analysis of the agronomist versus the proposed system (Red—Agronomist, Blue—Proposed System, Yellow—Matches)

ate color space, channels can be obtained that characterize an object. Due to lighting these channels may vary to a lesser or greater extent. The proposed method, the Lab color space, was a suitable model to more effectively de-correlate lighting color. The results also indicate that the Mildew detector system against objects external to the

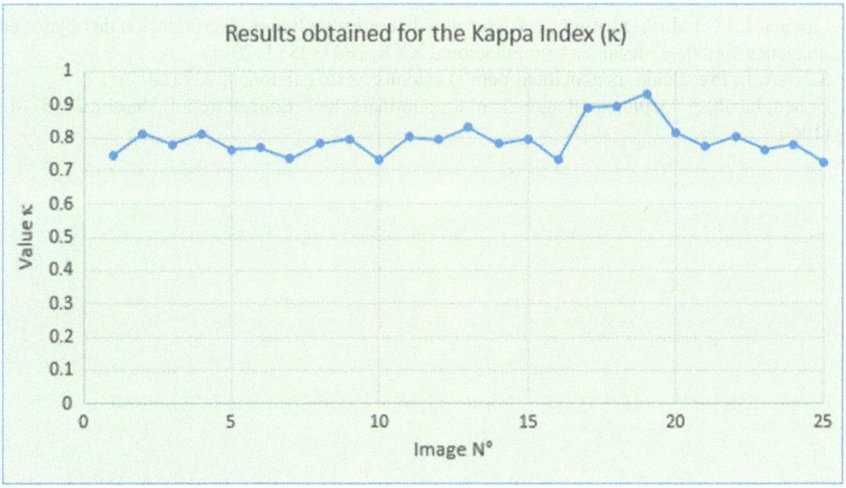

Fig. 8 Historical Data. Graph of Image N° versus Value Kappa

crop, such as pipes or waste, manages to discriminate adequately such elements. On the other hand, the employed evaluation method has gone through a certain number of specialists to provide their opinion and guidance during the testing process of the detector system. The proposed system offers reliable results, since the value of the kappa k index has a mean value of 0.7811 (from values higher than 0.7 it is considered "adequate") acceptable for field use. In Fig. 8, the historical data of the kappa values obtained according to the images captured during field tests of the proposed system is presented.

Finally, it is important to point out that the detector system also has limitations, and considers the effect of external variables preventing its operation in adverse and extreme environments.

References

1. Srinivasan, A.: Handbook of Precision Agriculture. Principles and Applications. The Haworth Press, New York, NY (2006)
2. Burgos, Xavier, Ribeiro, Angela, Guijarro, Maria, Pajares, Gonzalo: Real-time image processing for crop/weed discrimination in maize fields. Comput. Electron. Agric. **75**(2), 223–372 (2011)
3. Plataniotis, K.N., Venetsanopoulos, A.N.: Color Image Processing and Applications, Cap. 1, p. 35. Feb 2000
4. Plataniotis, K.N., Venetsanopoulos, A.N.: Color Image Processing and Applications. Cap. 16, p. 35. Feb 2000
5. X-rite: Guía para entender la comunicación del color, p. 12 (2002)
6. Boris Escalante, R.: Procesamiento Digital de Imágenes, p. 7, Oct 2016

7. Larcher, L.I., et al.: Algoritmo para detección de bordes y ulterior determinación de objetos en imágenes digitales. Mecánica Computacional **XXX**, 2841–2852 (2011)
8. Stewart, J.: Precálculo, matemáticas para el cálculo, Sexta edición, p. 737 (2012)
9. Cohen, Jacob: A coefficient of agreement for nominal scales. Educ. Psychol. Measur. **20**, 37–46 (1960)

An Architecture for Flow-Based Traffic Analysis in a Cloud Environment

Tiago Primini, Eric Baum, Leonardo Mariote, Matheus Ribeiro and Giovanni Curiel

Abstract In today's cloud environments, where thousands of virtual machines are executed simultaneously, it can be very complicated and complex to identify and trace, in real time, suspicious or malicious traffic. Performing a per-packet analysis or doing a deep packet inspection in such high volume of traffic may impact the hypervisor's performance negatively, which can render the unfeasible instantiation of a large amount of virtual machines. In order to enhance this procedure, this paper proposes an architecture that is able to collect flow-based information from all running virtual machines. In addition, its intent is to be easily pluggable to cognitive analysis tools, such as network intrusion detection systems, traffic profiling and classification, behavior prediction, and so forth.

Keywords Virtualization · Network function · Intrusion detection system

1 Introduction

Virtualization techniques have been proven effective in increasing agility and scalability while creating significant cost savings for Information Technology (IT) environments and data centers. Network Function Virtualization (NFV) [1] aims to use such techniques to virtualize all network function running on custom hardware appliances, emerging as a new paradigm bringing the benefits of virtualization to the

T. Primini (✉) · E. Baum · L. Mariote · M. Ribeiro · G. Curiel
CPqD Foundation, Campinas, SP, Brazil
e-mail: tiagopri@cpqd.com.br
URL: http://www.cpqd.com.br

E. Baum
e-mail: ebaum@cpqd.com.br

L. Mariote
e-mail: lmariote@cpqd.com.br

M. Ribeiro
e-mail: matheusr@cpqd.com.br

G. Curiel
e-mail: gcuriel@cpqd.com.br

Y. Iano et al. (eds.), *Proceedings of the 3rd Brazilian Technology Symposium*,
https://doi.org/10.1007/978-3-319-93112-8_6

telecommunication industry. Companies are increasingly suffering more pressure to reduce costs, and NFV opens up the opportunity for new commercial lines, including new service offerings and cost savings through reductions in capital expenditures (CAPEX) and operating expenses (OPEX) [2].

Cloud computing is a model for enabling ubiquitous, convenient, and on-demand network access to a shared pool of configurable computing resources (e.g., networks, servers, storage, applications, and services) that can be rapidly provisioned and released with minimal management effort or service provider interaction [3]. An architecture where software systems decrease the requirement of new, expensive, and custom hardware appliances while increasing the scalability, elasticity, measurement capacity, and robustness of computer networks, is very commercially attractive. In this context, it is not a surprise that Software-as-a-Service (SaaS) is expected to increase 20.1% in comparison with 2016, reaching US$46.3 billion in investments worldwide in 2017, according to [4].

NFV and cloud computing may work together bringing several innovation mechanisms to the telecommunication industry, as an example we can cite a NFV platform developed in [5]. However, in a public cloud environment, some physical resources (e.g., networks, servers, storage, virtual network functions (VNFs)) are shared between different customers creating security breaches, as detailed in [6]. Besides that, there are several cloud computing security threats [7], as abuse of cloud services, insecure application programming interfaces (APIs), malicious insiders, insufficient due diligence, data loss/leakage, account or service hijacking, denial-of-service attacks and shared technology vulnerabilities. In this scenarios, common techniques attempting to detect malicious traffic from nodes or external hosts might not be able to work properly due to the quality, reliability, or amount of traffic generated.

As a remediation for these vulnerabilities, the network monitoring should be comprehensive for detecting malicious packets and activities in order to alert when some suspicious behaviors are verified. Nowadays, it can be done using Intrusion Detection System (IDS) [8]. An IDS system that is monitoring a cloud environment must be able to collect data from Virtual Machines (VMs) and containers in real time. In this way, an IDS can act directly in the virtualized environments, so they can be managed, quarantined, and healed at any given time, when unexpected or unsafe situations happen, mitigating risks and avoiding attacks in order to maintain the service availability and to protect sensitive data.

2 Background and Related Works

IDS is an appliance or software responsible for monitoring, analyzing, and detecting abnormal behaviors in a cloud environment. Basically, it works by collecting logs and events, correlating them, and generating alerts about some malicious attacks, suspicious activities, or policies violations. Some efforts have been made in cloud security research, motivated by the increasing security concerns involving cloud environments, mainly focused on the development of IDS. As the use of cloud environments becomes more widespread, the number of attacks originated from within the infrastructure increases, given the ease of launching a virtual machine instance in the cloud and considering that the organizations are responsible for deploying their own access controls to grant or restrict the access level for their assets; this approach does not guarantee the appropriate assurance and protection level for a particular resource in the cloud [9]. The tendency of multi-tenancy in this environments, especially for cloud service providers, where several customers share the same physical infrastructure, turns this into an even more critical issue.

Some efforts have been done [10] for integrating an IDS into Eucalyptus [11], an open-source cloud environment, using Snort [12] as an inspection tool based on packet analysis. Using this approach, Snort is deployed at cloud controller nodes as well as on the hypervisors to detect intrusions coming from external networks. In this architecture, it is not necessary to deploy multiple instances of IDS, which could be fast and cost-effective solution. Despite the results, the packet inspection approach might demand so many computing resources that can render the solution unfeasible in some circumstances, especially when analyzing encrypted payloads. As another solution based on packet analysis, [13] proposed a multi-thread centralized IDS for detection of malicious traffic using signature comparisons and a predefined rule set, passing the alerts to a third-party monitoring service, which, in turn, would directly inform the cloud user that their system is under attack. According to this architecture, it does not compromise hypervisor performance as it acts directly on switching devices, but it might impose some restrictions to scalability due to switches limitations and also to an increased complexity when configuring these network elements.

Considering such restrictions and limitations based on packet analysis, a different tendency has been well accepted, especially for cloud environments: the use of flow statistics instead of packet inspection for anomaly detection. It gives an efficient way of providing security in a cloud environment because of the lightweight nature and scalability of this method, and also could be easily integrated with appliances installed in the market. This approach was used by [14], which proposed a flow-based analysis IDS, using a labeled dataset provided by [15], through an SVM (Support Vector Machine) algorithm to detect suspicious traffic signatures. According to its interface specification, this work could be applied directly to our proposed architecture as a cognitive plugin. Even considering flow-based inspection, neural networks are also possible to be used to detect malicious traffic [16]. A specialized type of artificial neural network implementation called Self-Organizing Map (SOM), which is an unsupervised learning technique, could also be applied to the detection of network threats like botnets in the network [17], that is, one of the most destructive cybersecurity menaces nowadays.

It is also possible to combine events from different sources in order to detect some threats, as demonstrated in [13], where an event-based system is proposed, which is composed by a node, an IDS service, an event auditor, and a storage service. The event auditor is the responsible for capturing data from different sources (logs, files, and so forth), representing a key element for the solution. The storage service holds the data for posterior analysis. By using an artificial neural network algorithm, the

IDS service is responsible for detecting an intrusion behavior pattern. According to simulations, the solution performance was satisfactory for a real cloud environment implementation, but no further details were provided.

Using an innovative approach, [18] has proposed an intrusion prevention system based on Software-Defined Network (SDN) for cloud environments virtual networks. It uses Snort [12] as the intrusion detection tool, because it is able to automatically reconfigure the network using SDN capabilities, such as the redirection of traffic, Quality of Service (QoS) adjustments, virtual traffic isolation, and port blocking.

Although relevant work has been done in this area using different approaches, as far as we know, none of them can be used as a simple extension for a cloud orchestration platform, maintaining scalability, without being a concern in performance overhead and not requiring new physical elements. This paper explores these issues and presents a distributed architecture, which is able to collect relevant data for both cognitive and rule-based intrusion detection algorithms or other security tools. The extensibility of this proposal was also a matter of attention, since most of the current initiatives could be easily combined with the proposed architecture.

3 Problem Statement

In large and complex cloud environments where thousands of VMs and containers owned by different tenants are deployed spanning a multitude of different physical servers connected through complex networks, it is extremely difficult to detect and locate the origin of a particular traffic pattern which could be classified as abnormal, suspicious, or even malicious. Besides that, nowadays, most IDS systems are heavily based on packet analysis, which is CPU intensive. However, cloud computing providers rely on selling hardware capacity, which make the CPU resources available on each of the hypervisor servers invaluable. That expose an issue when running IDS for traffic analysis locally on each of the hypervisors as it may degrade the performance that would be offered by that server, moreover, replicating the traffic and sending it to nodes dedicated to traffic analysis may incur on significant costs to the cloud infrastructure network [10].

In this context, the problem addressed by this paper is the one of providing an architecture for cloud security analysis based on network traffic in order to improve the security of the cloud environment as a whole without compromising the performance and availability of resources. This solution has its principal intent, being extensible while maintaining the isolation of per-tenant contexts, making it easier for legacy IDS systems without actual support for multi-tenancy, to be integrated as pluggable intrusion detection mechanisms.

This paper presents a model for a lightweight and distributed architecture, which is able to collect data using flow analysis to be delivered to pluggable machine learning algorithms in order to analyze complex traffic patterns, making it able to detect possible issues and failures and alert a cloud administrator or even notify automatic systems with information about which set of virtual machines are responsible or affected by the problem that was detected. The implementation of a proof of concept for this architecture was built on top of an OpenStack [19] cloud environment.

4 Architecture

Taking into account the aforementioned requirements, a novel distributed architecture is proposed, aimed at making use of already existing infrastructure at the level of an OpenStack cloud deployment compute nodes. We have chosen OpenStack [19] cloud operating system because it is recent and is under active development, it has great potential due to its community, and it is under Apache 2.0 license. As an open-source software, OpenStack has attracted a very broad community: nearly, 2500 independent developers and 150 companies are part of it, including giants as AT&T, HP, Cisco, Dell, and Intel.

The semestral OpenStack community summit, where IT leaders, cloud managers, and members of the development community meet to exchange ideas and explore new directions for the software, now attract over 1000 participants from many different countries [20]. The proposed architecture is composed of three basic elements: a flow collecting agent, an external manager node, and the detection plugins.

The collecting agent is the element responsible for collecting traffic information locally at the compute nodes, without compromising their performance. Using OpenStack, the infrastructure and networking configuration is created in a way to implement both virtual machines connectivity and traffic protection. Part of this infrastructure is a per-port Linux bridge that connects the client virtual machine to the nodes aggregation switch (Open vSwitch—OVS). On this bridge, a set of IPTables [21] rules that were defined in the virtual machine's security group configuration are added, in order to limit the network exposure and to enforce traffic constraints.

From this basic network topology that is already deployed to every compute node (the ones most impacted by CPU usage requirements), it is proposed the addition of a Conntrack [22] based flow information collector. Such collector is responsible for parsing information that is already measured by Conntrack and for making it available to a machine learning plugin implementation located in a remote manager node. The flow data that was processed by the collectors is made available to the manager nodes via the Netflow [23] protocol.

The manager node is a host that has access to the cloud's internal network and knows its topology. It is responsible for gathering the information that was collected by the agents on each compute node and providing this data to the detection plugins. If something is detected by any plugin that is running on the manager nodes, alerts can be triggered, allowing cloud administrators or automated systems to act directly on the affected hypervisors or VMs. To obtain the environment data and the network topology, the manager node is also integrated with the OpenStack controller node via its public interface.

The detection plugins are the elements which implement the algorithms responsible for detecting any anomaly or threats to the environment. The main purpose of this work is to provide all the required infrastructure for constructing cognitive plugins, through the use of machine learning algorithms having as input the network traffic history or a real-time perspective of the network. Nevertheless, signature-based techniques may also be constructed over the proposed architecture, since they are based upon traffic flows information. The relation between this different elements is represented in Fig. 1.

Looking more deeply into the compute node, it is possible to identify the integration between the collecting agent and the bridges used for VM communications. This detail of the local architecture for compute nodes is shown in Fig. 2.

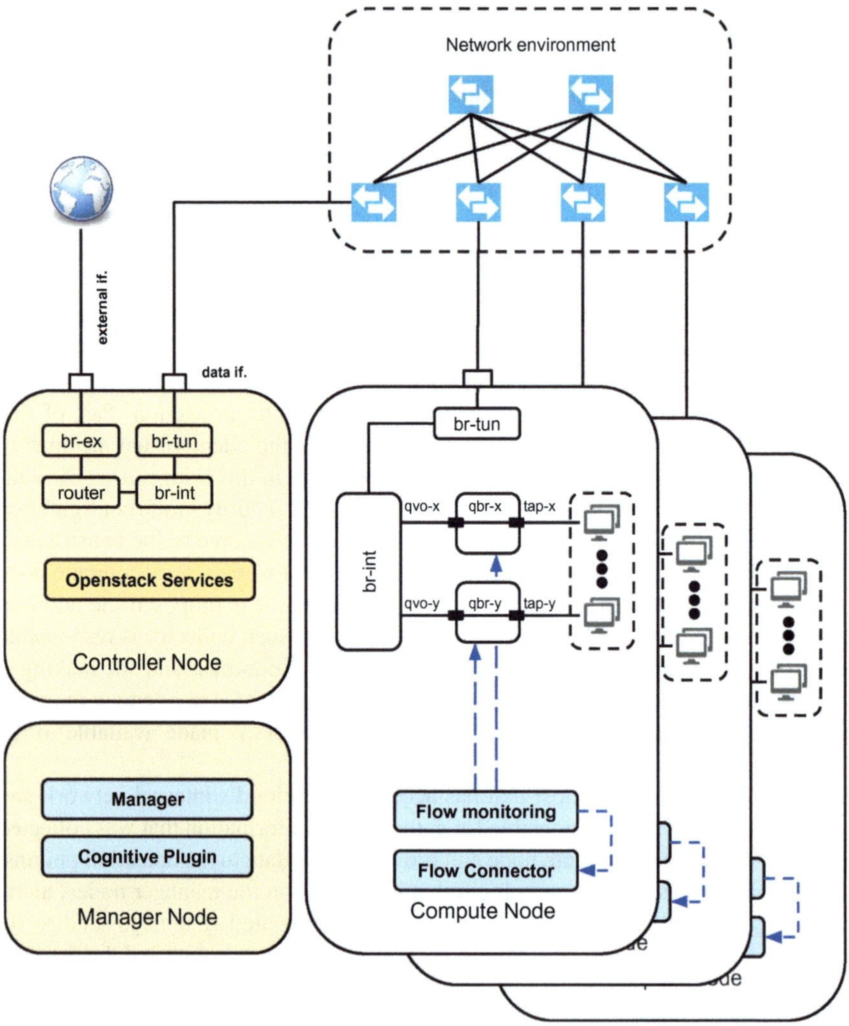

Fig. 1 Proposed architecture

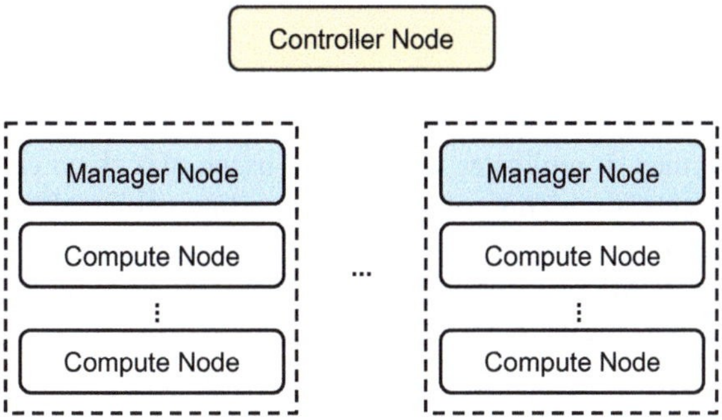

Fig. 2 Modules distribution

Besides the benefits already exposed in this section, the proposed architecture has a distinguished characteristic: the physical isolation between the collecting agents and the intrusion detection algorithm. With a big concern on impacting performance in hypervisors, this isolation makes it safer from a computing overhead point of view to install any new plugin, since its impact could never affect the hypervisors, once they are deployed in different nodes.

5 Implementation

In order to validate the proposed architecture, a proof of concept was implemented with two basic objectives:

- check that it works as described in the previous section;
- analyze the processing overhead imposed by the solution.

The implementation of the flow information collector is twofold. First, an instrumented version of neutron agent was created allowing automatic management of a set of IPTables rules which segregate, at Conntrack level, the hosts traffic from both the nodes, infrastructure communication (i.e., OpenStack agents intercommunication) and the other virtual instances allocated on the hypervisor. Such traffic segregation is accomplished by creating a Conntrack zone per virtual host on the hypervisors IPTables. Second, a new collecting agent is installed on each one of the compute nodes to monitor, parse, and expose the metered flows from the Conntrack tables to the manager node.

As this is a proof of concept, it is not expected that it should have the best performance possible to perform its task. Instead, it provides useful data so that future works might improve it. Also, the analysis done in this paper considers only the "vanilla" version of the solution, i.e., without plugins. Additional features could be included and analyzed with such functionality, which could be addressed in future works.

6 Tests

In order to validate the proposed architecture, a set of tests was designed and performed. It replicates the network usage of a cloud customer which serves external clients, simulating people accessing Internet in a day-to-day basis. It contains most common traffic profiles, such as TCP and UDP traffic, focused on flow-based scenarios. ICMP was not included in these tests because its traffic is significantly smaller than its TCP and UDP counterparts.

All the servers running on OpenStack Cloud are analyzed, and their CPU usage values are gathered in the hypervisor. This parameter is retrieved from a single compute node, as measuring it is a local procedure to the evaluated node. This procedure is repeated for different numbers of connections/second (from 10 to 500) and for different numbers of running servers (from 1 to 5) within the same hypervisor.

The tests were performed in two scenarios: with and without the collector agent running on the hypervisor. Figure 3 depicts the baseline CPU usage of the compute node in an OpenStack environment without the flow collector agent. The highest CPU usage in this scenario is around 7.5% for 500 flows per second and 5 servers. For Fig. 4, the flow collector agent was used—the highest CPU usage is around 13.5% (with 500 flows per second and 5 servers). There is no significant difference of CPU usage for flow rates lower than 100/second, regardless of number servers and presence of the flow collector.

The graph shown in Fig. 5 presents the overhead added to the baseline usage by the flow collector agent. As preliminary results, considering flow collector agent, the variation of flows per second, and the number of running servers in the environment, we have an overhead of near to 5% of CPU usage.

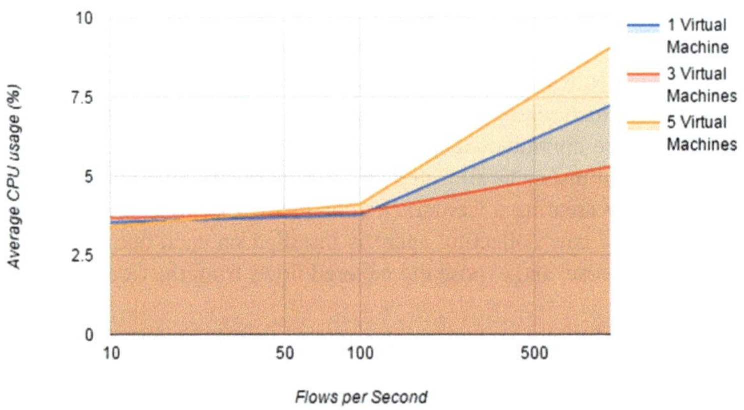

Fig. 3 Results with vanilla OpenStack

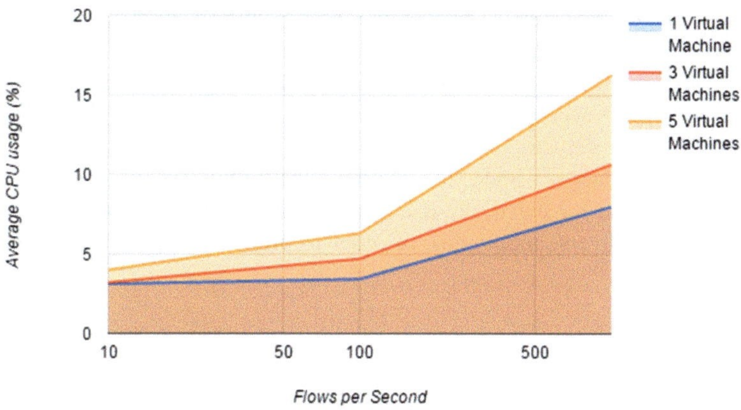

Fig. 4 Results with extended solution

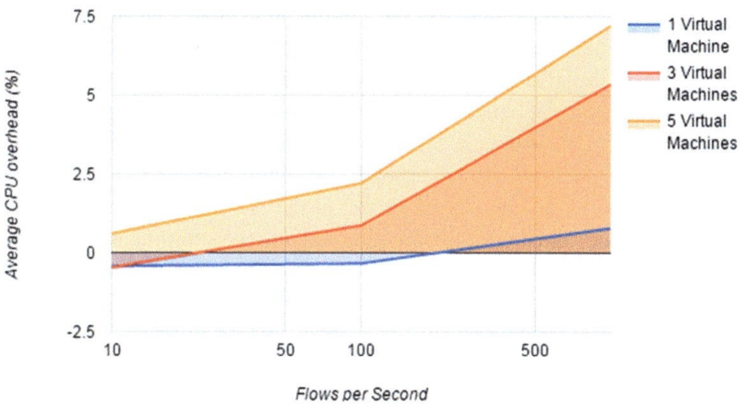

Fig. 5 Processing overhead

7 Future Works

With a good preliminary validation provided by the proof of concept tests, the solution should be submitted to a deeper test scenario, including more than one cognitive plugin. In this test, the communication overhead should be analyzed and the usage of hypervisors resources isolation should be demonstrated. For a better performance, the collecting agent should also receive some improvements, as for example, deploying it at kernel level, reducing the overhead added by running it on user level.

Regarding the extensibility, some machine learning plugins have to be constructed in an integrated test scenario, so a variety of simultaneous analysis techniques can be validated, obtaining their accuracy when running isolated or even when their results are combined, increasing the spectrum of methods that can be used when looking for network security failures.

In the manager node side, we expect two focus points on optimizations for future developments, the messaging between the manager and the collector agents and the data serialization methods that are used when constructing such messages.

By achieving this accomplishment, we expect to improve the overall performance and also reduce the network overhead. A possible extension is the use of dedicated interfaces, on the manager node side, isolating the traffic generated by the collecting agents from the regular traffic from the virtual machines. It is also necessary to run tests with diverse topologies, in order to properly measure this bandwidth consumption.

8 Conclusion

This paper proposes an extensible distributed architecture for intrusion detection in cloud environments using OpenStack. With lightweight agents installed locally in the hypervisors, the solution can concentrate a rich, per-tenant data, and is able to provide that information to specific algorithms running as pluggable modules. Those plugins are orchestrated by a manager using a flow-based traffic information in order to detect malicious and suspicious behavior. This approach simplifies the traffic data collection and also minimizes the total CPU overhead, especially in contrast with a per-packet approach. Other distinguished characteristic is the isolation between performance impact on hypervisors and any algorithm plugged into the solution, making the whole architecture safer. The preliminary tests demonstrated good performance results in a controlled environment. Further work includes the development or using of existent cognitive algorithms, which is based on anomaly detection on flow information in order to enhance the current architecture. This way we can validate this architecture as a whole and look for improvements in the accuracy of intrusion detection techniques.

References

1. European Telecommunications Standards Institute (ETSI), Network Function Virtualisation. Available http://www.etsi.org/technologies-clusters/technologies/nfv. Retrieved 22 Mar 2017
2. Paper, N.W.: Network functions virtualisation: an introduction, benefits, enablers, challenges & call for action. Issue 1 Oct 2012
3. Mell, P., Grance, T.: The NIST definition of cloud computing. National Institute of Standards and Technology (NIST) (Oct 2009)
4. Forbes, Roundup Of Cloud Computing Forecasts - 2017. Available https://www.forbes.com/sites/louiscolumbus/2017/04/29/roundup-of-cloud-computing-forecasts-2017/. Retrieved 29 May 2017
5. Nascimento, M., Primini, T., Baum, E., Martucci, P., Cabelo, F., Mariote, L.: Acceleration mechanism for high throughput and low latency in NFV environments. To appear in 18th IEEE International Conference on High Performance Switching and Routing (HPSR) (June 2017)
6. Srinivasamurthy, S., Liu, D.Q., Vasilakos, A.V., Xiong, N.: Security and privacy in cloud computing: a survey. Parallel Cloud Comput. **2**(4), 126 (2013)

7. Kazim, M., Zhu, S.Y.: A survey on top security threats in cloud computing. Int. J. Adv. Comput. Sci. Appl. (IJACSA) (2015)
8. Roschke, S., Cheng, F., Meinel, C.: Intrusion detection in the cloud. In: Eighth IEEE International Conference on Dependable, Autonomic and Secure Computing, DASC'09, pp. 729–734. IEEE, New York (2009)
9. Ashfaq, R.A.R., Wang, X.Z., Huang, J.Z., Abbas, H., He, Y.L.: Fuzziness based semi-supervised learning approach for intrusion detection system. Inf. Sci. **378**, 484–497 (2017)
10. Mazzariello, C., Bifulco, R., Canonico, R.: Integrating a network IDS into an open source cloud computing environment. In: 2010 Sixth International Conference on Information Assurance and Security (IAS), pp. 265–270. IEEE, New York (2010)
11. Github. Available https://github.com/eucalyptus/eucalyptus. Retrieved 31 May 2017
12. Snort: Snort Project. Available https://www.snort.org/
13. Shelke, M.P.K., Sontakke, M.S., Gawande, A.: Intrusion detection system for cloud computing. Int. J. Sci. Technol. Res. **1**(4), 67–71 (2012)
14. Winter, P., Hermann, E., Zeilinger, M.: Inductive intrusion detection in flow-based network data using one-class support vector machines. In: NTMS, pp. 1–5 (2011)
15. Sperotto, A., Sadre, R., Van Vliet, F., Pras, A.: A labeled data set for flow-based intrusion detection. In: International Workshop on IP Operations and Management, pp. 39–50. Springer, Berlin (2009)
16. Jadidi, Z., Muthukkumarasamy, V., Sithirasenan, E., Sheikhan, M.: Flow-based anomaly detection using neural network optimized with GSA algorithm. In: 2013 IEEE 33rd International Conference on Distributed Computing Systems Workshops (ICDCSW), pp. 76–81. IEEE, New York (2013)
17. Le, D.C., Zincir-Heywood, A.N., Heywood, M.I.: Data analytics on network traffic flows for botnet behaviour detection. In: 2016 IEEE Symposium Series on Computational Intelligence (SSCI), pp. 1–7. IEEE, New York (2016)
18. Xing, T., Xiong, Z., Huang, D., Medhi, D.: Sdnips: Enabling software-defined networking based intrusion prevention system in clouds. In: 2014 10th International Conference on Network and Service Management (CNSM), pp. 308–311. IEEE, New York (2014)
19. OpenStack. Available https://www.openstack.org/. Retrieved 30 May 2017
20. NASA. Available https://www.nasa.gov/offices/oct/40-years-of-nasa-spinoff/openstack-cloud-computing-platform. Retrieved 30 May 2017
21. Netfilter.org: Iptables project www.netfilter.org/projects/iptables/index.html (2014)
22. conntrack-tools.netfilter.org
23. Claise, B.: Cisco systems netflow services export version 9 (2004)

Development of Residential Automation Modules for Performing Low Complexity Activities Using IoT

Lahis G. de Almeida, Rachel B. de Lima and Edgard Luciano O. da Silva

Abstract The integration and popularization of Residential Automation at all levels of society is still a challenge due to high prices gain this technology. This work proposes to unite Residential Automation and Internet of Things to develop four modules that automate, at a low cost, simple tasks from day to day, allowing more people to have access to these types of solutions. The paper also performs a price comparison between the proposed method and other preexisting in the market.

Keywords IoT · Residential automation · Esp8266 · MQTT

1 Introduction

Residential Automation has grown since the 1980s. Advances in Electronics and Informatics have enabled emerge residential automation projects around the world [1]. However, problems such as the lack of standardization of communication protocols between devices, high implementation cost and the difficulty of including old equipment to this technology, still make it difficult to disseminate and popularize [2]. Simple tasks such as turning lights on or off, opening or locking doors, monitoring water levels or garbage, require unnecessary time that could be availed in other activities if these tasks were automated.

In the Internet of Things (IoT) paradigm, many of the objects that surround the population are networked. Information and communication systems are invisible to their users. To provide this scenario, Internet of Things joins Cloud Computing

L. G. de Almeida (✉) · R. B. de Lima · E. L. O. da Silva
Amazonas State University, Manaus, AM 69050-010, Brazil
e-mail: lga.eng@uea.edu.br

R. B. de Lima
e-mail: rbdl.eng@uea.edu.br

E. L. O. da Silva
e-mail: elsilva@uea.edu.br

Y. Iano et al. (eds.), *Proceedings of the 3rd Brazilian Technology Symposium*,
https://doi.org/10.1007/978-3-319-93112-8_7

in search of a virtual infrastructure, which integrates monitoring devices, storage devices and analysis tools, enabling its services in real time from any place [3].

In this context, the main contributions of this research are: to develop an efficient and low cost residential automation approach, using IoT concepts and protocols, and compare and analyze it with preexisting solutions.

The remaining part of the paper is organized: Sect. 2 presents preliminary concepts needed to understand this study, Sect. 3 points out the main researches related to this article, Sect. 4 describes the proposed residential automation approach, Sect. 5 exposes the experiments and results got and Sect. 6 presents the conclusions and possibilities of future work.

2 Theoretical Foundation

2.1 Residential Automation

A set of connected services that perform monitoring and control tasks in the home branch characterize Residential Automation, also known as Domotics [4]. These systems can detect events such as individuals' presence and variations of physical quantities such as temperature, humidity and pressure; and controlling devices such as light bulbs, air conditioners, fans and doors.

Despite the facilities that domotics systems offer their users, the price to use this technology is still high, making it difficult for a large part of the population to adhered [4]. In this context, the search for low cost solutions is essential to include in the routine of citizens this technological area, making common scenarios such as access control or monitoring of household appliances in homes.

2.2 Internet of Things

The Internet of Things (IoT) is an innovative paradigm that is exploding in the scenario of modern wireless networks of Telecommunications. The concept of the pervasive (unnoticed) presence of a diversity of things or objects is its base, which surround the population and are connected to each other, such as sensors, actuators and smartphones [5]. Thus, solutions that link residential automation to IoT will allow greater mobility to its users, since the monitoring and control of the residence can be done anytime and anywhere.

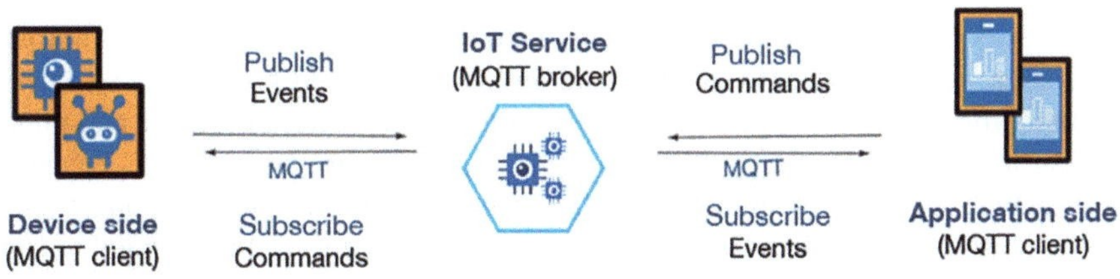

Fig. 1 MQTT protocol overview [11]

2.3 *MQTT*

Message Queuing Telemetry Transport (MQTT) is an open protocol widely used in embedded systems and presents a simple and light architecture [6]. TCP/IP communication protocol is its base and default message exchange is *Publish/Subscribe*.

In MQTT, topics identify messages. Network elements can post topics or select the topics they want to subscribe to Broker, known as publications and subscriptions manager (see Fig. 1). When a client aims to get information, it executes a request for Broker (subscribe). Elements that wish to publish data also use this manager, sending the information they have to him [7].

2.4 *ESP8266*

The ESP8266 is a microcontroller that has a 32-bit Tensilica L106 microprocessor core [5]. It is a dedicated device to create solutions that need a Wi-Fi connection. The board main features are: an internal flash memory for storing files and another for the Firmware, built-in antenna, GPIO ports, low power consumption and operating in a 3.3 V range [3].

3 Related Work

In this new paradigm that aims to combine IoT with Domotics, Egidio et al. [8] developed a work to implement improvements to the people's life quality highlighting the low cost with an Arduino Uno board and an Ethernet module ENC28J60 to carry out activities such as control of lights, fans and radio, enabling time gain of the through simple automations.

As exemplified in the paper by Wanzeler et al. [9], the use of IoT with the help of the Arduino Uno microcontroller, an Ethernet Shield board and temperature, light and movement sensors can implement various functionalities such as: dimerized lighting systems, temperature monitoring and alarm system residential.

The work of Kumar [10] features a Smart Home system that, through an Android application, an Arduino MEGA 2560 and an Ethernet Shield boards, allows users to control the lights, check the ambient temperature, presence of intruders and fires; sending the collected data by e-mail. In this paper, Representative State Transfer (REST) Web services communicate the hardware components and the application.

The present work differs from those mentioned by the fact of proposing develop IoT modules of residential automation with lower cost, allowing a greater integration of this technology in the life of the population. The research still performs a comparative of prices and technologies used with other approaches.

4 Proposed Approach

The proposed approach seeks, through embedded modules of Residential Automation and IoT, to reduce the time to carry out activities and to allow the control and monitoring thereof remotely. Four tasks (Fig. 2) were chosen to be automated: lamp control, garbage level monitoring and presence and fire detection. These tasks have been chosen because they are inserted into the population routine and to show that activities like these can also integrate Domotics solutions. The four modules are

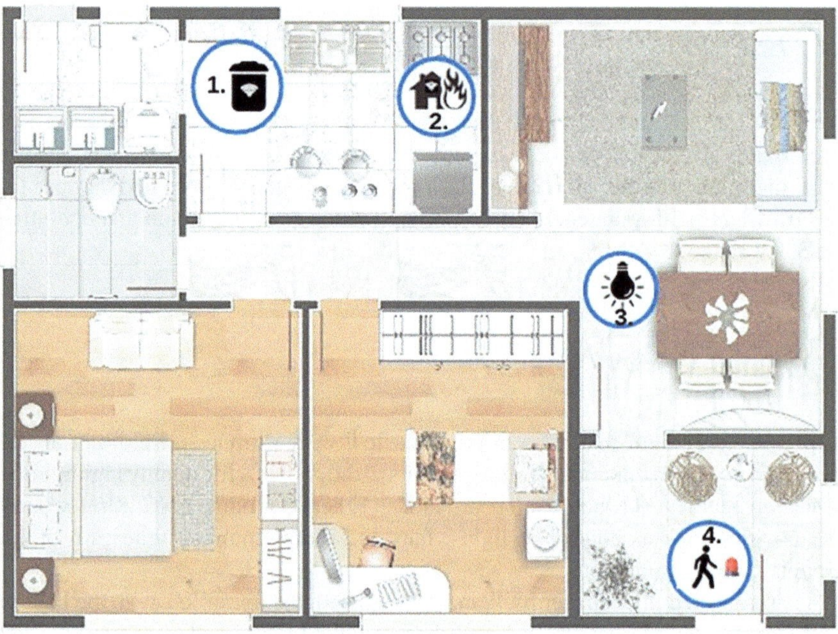

Fig. 2 Disposition of residential automation modules [12]

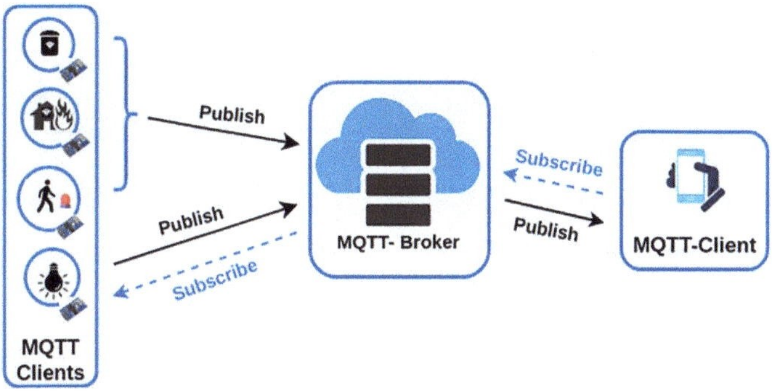

Fig. 3 Network communication between modules and users

1. **Smart Trash**: this module uses an ultrasonic sensor to measure the amount of waste.
2. **Smart Fire Alarm**: this module check fire occurs through a flame sensor.
3. **Smart Lamp**: this module controls a lamp (on/off) by activating a relay.
4. **Smart Invader Alarm**: this module check with a presence sensor if there is any unauthorized person in the environment.

In each automation prototype there is a Wi-Fi module (ESP8266), responsible for reading the sensors, controlling devices and connecting to the Internet. It is possible collected data to be sent to the Broker and passed to the clients. As well as, Broker receive commands which clients sent (see Fig. 3).

5 Experiments and Results

In the experiments phase, each module circuit was assembled. The ESP8266-01 was chosen for its low cost and the fact it has few doors, avoiding the waste them, because the sensors and actuators chosen to use a maximum of two GPIOs. For the programs to be recorded on each automation module, an FTDI converter was used, allowing serial communication between computer and ESP8266. Figure 4 shows the prototype of the modules.

The next step was communicated network elements. Modules that have sensors, publish the data collected, allowing the user to monitor unauthorized persons, fire and garbage level in their residence. The Smart Lamp subscribes messages for user commands to take the lamp on or off. Figure 5 shows the resident's monitoring and residence control through the application. You can see that in the Subscribe tab of the app, all the messages sent by the sensors are attached; while in Publish, the components responsible for equipment control are set.

Fig. 4 Prototypes of residential automation modules

The MQTT service used was open platform Eclipse IoT that hosts the Broker Mosquitto. On the user side, the tool chosen for message exchange was the MQTT-DashBoard application, as it is free and easy to use.

5.1 Results and Discussion

The modules circuits were assembled and the communication programming between the network elements was done, it was possible to simulate the proposed residential automation approach in real time. As a result of the experiments, was verified that simple tasks, such as controlling lamps and monitoring level of garbage, intruders' presence and even occurrence of fire, can be automated in a simple way.

The experiments showed that Domotics allied with IoT brings benefits such as check of network elements and low cost in the implementation, being possible anywhere and at any time to monitor and act on the devices of the residence.

A weakness found in this approach was the use of the Eclipse IoT Service because it does not accept cryptography, allowing anyone who knows the MQTT topics of the residence and who uses this service to have access to the monitoring and control of the equipment. However, since Eclipse IoT was only used for the prototype and verification of viability of the method, more robust MQTT services, such as Amazon's, would solve this problem.

The solution developed in this work proved to be much more accessible than the other low-cost solutions found in the market. Table 1 presents a comparison of

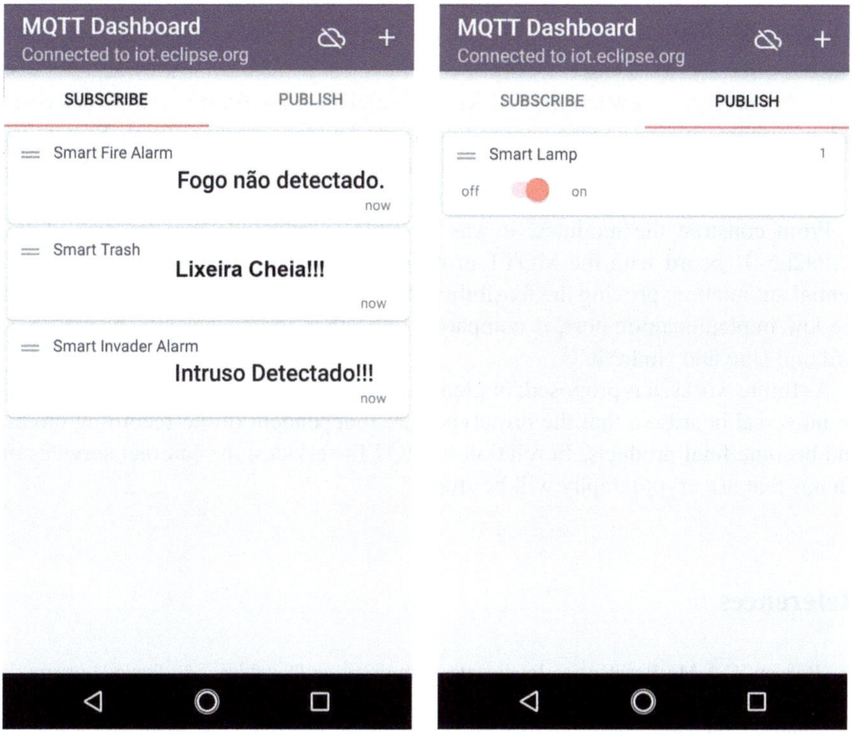

Fig. 5 MQTT-dashboard application screens [13]

Table 1 Residential automation solutions

Embedded boards	Auxiliary boards	Total price (R$)
Raspberry Pi 3	–	120,000
Arduino Uno	Ethernet Shield × 4	39,900
NodeMCU	–	19,960
ESP8266-01	FTDI × 1	1196

the prices of the most common home automation solutions. Prices were taken from FilipeFlop site, reference in the boarding market.

According to the table, is possible to realize that the user would save up to R$ 1000.00 if opted for the approach developed in this article instead of modules controlled by boards like Raspberry Pi. In relation to other solutions such as Arduino Uno and NodeMCU, the savings would be approximately 66.73 and 83.36% respectively. However, for modules that require the use of more GPIO ports, analog ports, and even image processing capability, the solution adopted in this research become ineffective because ESP8266-01 does not have these features.

6 Conclusion

This work presented develops modules of residential automation to perform activities of low complexity using IoT. Four modules were developed: Smart Trash, Smart Fire Alarm, Smart Lamp and Smart Invader Alarm, each one making use of an ESP8266-01 for communication with the MQTT protocol.

From construe the modules, it was possible to conclude that the use of the ESP8266-01 board with the MQTT protocol satisfactorily met the proposed residential automation, proving the feasibility of managing the modules in real time and the low implementation cost, if compared with other approaches (Raspberry Pi 3, Arduino Uno and NodeMCU).

As future works, it is proposed: implement battery-powered modules and mounted on universal boards so that the prototypes are independent of the recording circuit and become final products. In relation to MQTTs services, the Internet services of Things that use cryptography will be studied.

References

1. Bolzani, C.A.M.: Residências Inteligentes: um curso de Domótica. São Paulo: Livraria da Física (2004)
2. Mainardi, E., Banzi, S., Bondiè, M. e Beghelli, S.: A low-cost Home Automation System based on Power-Line Communication Links. In: 22nd International Symposium on Automation and Robotics in Construction, ISARC, Ferrara, Itália (2005)
3. Gubbi, J., Buyya, R., Marusic, S., Palaniswami, M.: Internet of Things (IoT): A Vision, Architectural Elements, and Future Directions. Future Generation Computer Systems, No. 7, pp. 1645–1660. Elsevier, Amsterdam (2016)
4. Guerra, F.H.M.: Automação residencial de baixo custo com protocolo X10 e ESP8266, (2016)
5. Souza, M.V. d.: Domótica de baixo custo usando princípios de IoT, M.S. Thesis, Brasil (2016)
6. (2014, Nov). MQTT[Online]. Available: http://mqtt.org/
7. Barros, M.:. MQTT - Protocolos para IoT[Online]. Available: https://www.embarcados.com.br/mqtt-protocolos-para-iot/ (2015, June)
8. Egidio, L., Ukei, T.: Internet das Coisas (IoT): Uma análise de aplicabilidade, WSEE (2015)
9. Wanzeler, T., Fulber, H., Merlin, B.: Desenvolvimento de um sistema de automação residencial de baixo custo aliado ao conceito de Internet das Coisas (IoT), SBrT (2016)
10. Kumar, S.: Ubiquitous smart home system using android application. arXiv preprint arXiv:1402.2114 (2014)
11. Tang, C.B.: Explore o MQTT e o serviço de Internet of Things no IBM Bluemix[Online]. Available: goo.gl/H2ba4 J (2015, Dez)
12. Plantas de Casas Prontas Grátis [Online]. Available: goo.gl/7noFsY (2016, Jun)
13. IoT MQTT DashBoard [Online]. Available: http://bit.ly/mqttdash (2016, Set)

Internet of Things: An Overview of Architecture, Models, Technologies, Protocols and Applications

J. R. Emiliano Leite, Paulo S. Martins and Edson L. Ursini

Abstract The Internet has established itself as a network and international communication medium, thus allowing people, computers and machines to communicate. The reduction in costs and size of interfaces that access the Internet allowed common devices (e.g. appliances, air conditioners, alarm systems and lighting) the access to a wide network. In a short time, the Internet of Things emerged, allowing the connection of an increasing number of intelligent devices from diverse fields of the economy. The growing application of IoT must be well understood, due to its potential to increase and use data in real-time, and the abundance of information available from sensors. Within this context, this work contributes with an updated overview of this new network environment in relation to its architecture, models, technologies, protocols and applications.

Keywords Wireless sensor network · RFID · NFC · Bluetooth · ZigBee 6LoWPAN · MANET

1 Introduction

The Internet was introduced in the sixties with the goal of connecting people and computers at the universities and research environments. Its success led to its use in the business field, and consequently, to commercial areas. Its use was then widespread with both the World Wide Web and social networks, leading to a global network. The advance of the field of microelectronics allowed the reduction of size and price of network interfaces as well as the access of new devices (or objects and things) to

J. R. E. Leite · P. S. Martins (✉) · E. L. Ursini
University of Campinas (UNICAMP), Limeira, SP 13484-332, Brazil
e-mail: paulo@ft.unicamp.br

J. R. E. Leite
e-mail: j750465@dac.unicamp.br
URL: http://www.ft.unicamp.br

E. L. Ursini
e-mail: ursini@ft.unicamp.br

Y. Iano et al. (eds.), *Proceedings of the 3rd Brazilian Technology Symposium*,
https://doi.org/10.1007/978-3-319-93112-8_8

the network. Research on RFID (Radio-Frequency IDentification) tags in 1999, and the use of sensors and actuators, enabled the increase of intelligence in diverse fields of the economy, including smart home, buildings, cities, grid, logistics, industry, hospitals, health and commercial automation (retail and wholesale) among others. The intelligence and automation were a result from the processing, memory and communication of the involved objects. This scenario triggered the emergence of the Internet of Things (IoT), i.e. the network of physical devices that enables objects to connect and exchange data. Examples of such objects include but are not limited to vehicles, home appliances and other items embedded with electronics, software, sensors, and actuators.

The IoT promotes a new digital transformation by connecting devices, increasing the value of business, redefining enterprises and opening new possibilities and frontiers. Examples of international open standards for the interconnection of heterogeneous systems independently of type, model and manufacturer are ITU-T IoT-Reference Model (Y4000)/ISO/IEC, IoT-A (IoT standardization), EPCglobal (RFID standardization), IPSO (Standardization of Intelligent Objects), IEEE, 3GPP (mobile phones), and IETF (standardization of Internet).

Despite the body of existing literature on IoT (Sect. 2), the main contribution of this overview paper is that it summarizes and updates in a generic way current IoT architecture and models, technologies, protocols and potential applications. None of the work reviewed presents an updated and brief but comprehensive view of these aspects. Whereas there are a number of propositions and views of the IoT trend, special emphasis was given to the approach by ITU-T/ISO/IEC standards.

The remainder of this paper is organized as follows: Sect. 2 presents related work. Section 3 addresses the IoT architecture and models, Sect. 4 shows the technologies that compose IoT as well as the IoT protocols. Section 5 covers a number of applications and Sect. 6 presents the conclusions.

2 Related Work

The ITU-T [4] introduced the IoT architecture (including the management and information security layers) and its reference model on a collaborative basis with ISO and IEC. It also presented sensor and network technologies, protocols and several case studies.

Bassi et al. [1] presents the IoT architecture according to the European technical committee IoT-A, including the IoT reference model, technologies and case studies, e.g. in transport/logistics, smart home, smart cities, smart factory, retail, E-health, and smart energy (grid).

The work by Xu et al. [10] introduces a survey of the IoT application in industrial sets. It shows the technologies, a five-layer architecture (sensing, accessing, networking, middleware and application). The author also describes the Service-Oriented Architecture (SOA), QoS aspects, and the following challenges: (1) interoperability of heterogeneous networks, (2) service discovery methods, (3) interconnection with legacy systems, (4) mediators and big data, (5) information security, protection and privacy, (6) evolution of RFID to handle sensors, (7) integration with social networks, (8) connection to the cloud, (9) green IoT technologies, (10) artificial intelligence techniques for self-configuration,optimization and healing.

The work by Li et al. [5] presents IoT definitions, the architecture, basic technologies, open issues, challenges and research work. It shows the IoT evolution from RFID sensors (1999-) to wireless sensor networks (WSN, 2005-), smart things (2012-) and IoT (2017-). IoT applications such as business, social networks, healthcare,infrastructure, and security/surveillance are also introduced and discussed.

Farooq and Waseem [2] present an overview of the IoT landscape,including technologies and sensor networks, a six-layer architecture (business, applications, middleware, network,perception, coding) as well as the challenges for its evolution.

Fersi et al. [3] present the Middleware Software Layer,between the Technology and the Application layers. This layer provides common services and functions to the applications allowing them to abstract away the implementation details. This facilitates the development and deployment of complex applications. Their work explores the main challenges in a middleware, such as interoperability, heterogeneity of devices, scalability, mobility, security and privacy. It also introduces the main functions such as publish/subscribe, SOA, semantic Web/Ontology/Context-aware.

Madakam et al. [6] present an IoT literature review, where they show the concepts, academic and industry research work, and discussion with experts in the field. Some definitions, requirements and IoT features are addressed, as well as the architecture, technologies and case studies. The architecture is presented following the ITU-T standard as well as the IoT-A European committee.

The paper by Miraz et al. [7] discusses the difference between IoT, IoE and IoNT. It shows future scenarios and applications. It presents the state-of-the-art for these technologies and the applications, as well as relevant literature work. The work by Weber [8, 9] highlights the regulatory and legal aspects of the new technology. The main issue in privacy is the large number of collected data. However, the key idea of the new technology is to reuse the network and security concepts already in place such as HTTPS, IPsec and cryptography. Other considerations about security are pointed out, such as the need for the Open Interconnect Consortium (OIC), which is now responsible for the regulation of key security issues related to large data sets [9].

3 Architecture and Models

The IoT-Reference model [4] consists of the following sub-models with different abstractions:

- *Domain*, which is responsible for the identification and grouping of systems.
- *Information*, which addresses the transport of information generated by sensors and RFID.
- *Functional*, responsible for the aggregation of existing functions in the system.

– *Communication*, which deals with the separation of the communication in seven layers according to the ISO/OSI model. Examples of functions are QoS, identification, addressing, discovery, publisher-subscribe API.
– *Trust, security and privacy*, which include functions for information security, privacy and certification.

The IoT architecture consists of the following layers (Fig. 1): Application Layer, which contemplates all the target IoT applications. Service and Application Support Layer, which includes, as the name implies, all the support services for the application layer, e.g. remote file transfer; Network Layer addresses the basic functions of transport and routing. The Device Layer specifies the functions for the connectivity of sensors, actuators, RFID and gateways; clearly, the Security Layer and the Management Layer cover all the aspects of security and management of devices, respectively. The IoT architecture is based on the ISO/OSI seven-layer model (Table 2), where the functionality is split into seven layers due to its complexity. Nevertheless, for the sake of efficiency, the IoT (along with the Internet and the Web) followed the TCP/IP architecture, which opted for grouping the three upper layers into one.

The RFID architecture was also accepted by IoT and it consists in the layers of Identification,Capture, and sharing of information.The Identification is responsible for identifying RFID tags using a unique umber. The Capture layers defines the collection of information received from RFID tags. The Sharing layer specifies the distribution of information collected from the tags. Due to the specific nature of the IoT applications, IoT adopts the Service-Oriented Architecture (SOA), where the functionality implemented by the applications are made available in the form of services [10].

4 IoT Technologies and Protocols

The IoT architecture (Fig. 1) adopted open, simple, and well established technologies such as Ethernet, Wi-Fi, WiMax, ZigBee, Bluetooth, RFID, NFC, Barcode, WSS (Wireless Signal Solutions), sensors, actuators, wireless sensor networks (WSNs) and Ad-hoc networks. Clearly, Ethernet/Wired solutions, GPS, 3G, 4G, WEB, Social Nets, cloud computing and future ones (Wireless USB, 5G and ultra-broadband) are to be used as well.

IoT has selected and adopted standards and internationally used technologies. Therefore, much of the new IoT environment is similar to the one already used in residential and enterprise Internet. One general guiding principle driving the IoT development is that *the IoT is not meant to reinvent the wheel*. In order to ensure flexibility and scope, most technologies used are from radio and wireless (radio access technologies), under the IEEE 802.11, IEEE 802.15 and IEEE 802.16 standards.

Connectivity allows the exchange of data between participants within a functional domain, in a system and between systems. The data may include sensor updates, telemetry data, events, alarms, protocols, files, logs, change of status, control com-

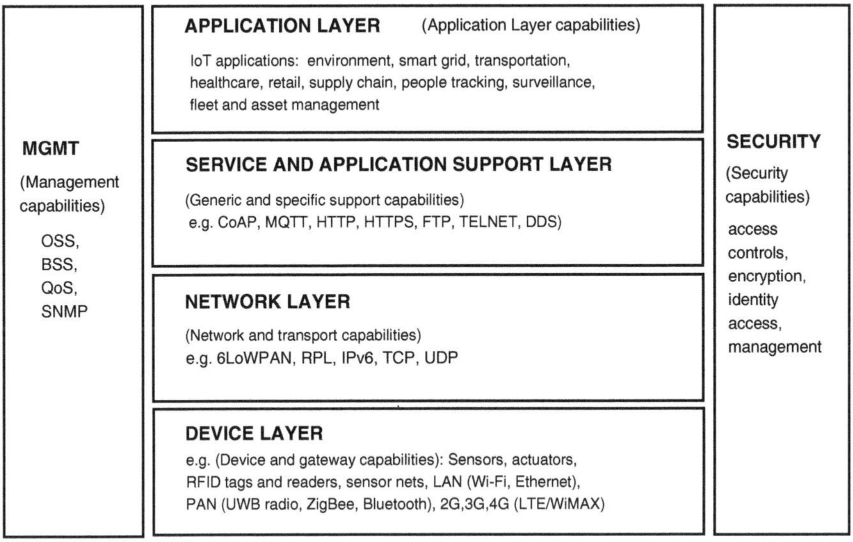

Fig. 1 IoT reference model and main protocols

mands, failure data and configuration updates. Table 1 summarizes the main access network technologies for the IoT and their protocols. Table 2 shows the IoT and the Web services protocol stack in relation to the OSI reference model.

The focus of IoT is on the transported information (application data) generated from RFIDs and sensors. The remaining layers (Physical, Link, Network ID, End to End) import their functionality from the ISO/OSI layers. Due to the large number of connected devices, the volume of data may be large. The web services protocol stack (web stack) is also considered in IoT in order to search for information in the cloud (e.g. from a smartphone). In short, the TCP/IP stack may be completely used. Since protocols from layers 1 to 4 are relatively well-known due to the popularity of the TCP/IP architecture, the relatively newer protocols are 6LoWPAN (IPv6 over Low-Power Wireless Personal Area Networks), CoAP, MQTT, AMQP, and CBOR. CoAP (Constrained Application Protocol) is a specialized Web Transfer Protocol for constrained nodes and devices of the IoT. MQTT is a machine-to-machine (M2M)/IoT connectivity protocol. It was designed as an extremely lightweight publish/subscribe messaging transport. MQTT is AMQP or JMS for the resource constrained environment of IoT. CBOR (Concise Binary Object Representation) is a data format designed with the goal of providing IoT with tiny message sizes. It is based on the JSON. While JSON uses text encoding, CBOR uses binary encoding thus resulting in more compact message sizes. These new protocols were created for constrained applications (low processing/memory/power/bandwidth and high packet loss). 6LoWPAN fragments large IPv6 packets into 802.15.4 frames (function: fragment, reassembly, and header compression).

Table 1 Comparison of access technologies and their protocols

Technology	Protocols	Rate (bps)	Power	Range (m)	Battery	Topology	# nodes
RFID	ISO/IEC	800 to 960M	1 W (reader)	5 max.	Not applicable	Star	Diverse
	18,000	2.45 or 5.8G					
NFC	ISO/IEC	424 k	100+ mW	0.01–0.1	Months/years	P2P	2
	18,092	125k, 13.56M	(reader) –		–		
Bluetooth	IEEE 802.15.1	1 M	49m/0.2 mA	1–10+	Days	Star	7
ZigBee	IEEE 802.15.4	20–250k	30m/0.356 mA	100+	Months/years	P2P/star/mesh	254 to 64,516
Wi-Fi	IEEE 802.11b	54M	400/20m mA	1–30+	H	star	32+ cabled
6LoWPAN	IEEE 802.15.4	250k	1 mW	100+	Months/years	Star and tree	100+

Table 2 IoT and the web services protocol stack

IoT Stack	OSI	Web Stack
DATA/CoAP	7. Application	Web Applications
MQTT/XMPP, AMQP	6. Presentation	HTML, XML, JSON
CBOR, JSON	5. Session	HTTP,HTTPS, DHCP, DNS, TLS/SSL
TCP/UDP/DTLS	4. Transport	TCP, UDP
RPL/IPv6/IP routing/6LowPAN	3. Network	IPv6, IPv4, IPsec
IEEE 802.15.4 MAC	2.Data Link	Ethernet, DSL, ISDN, WLAN, Wi-Fi
IEEE 802.15.4 PHY radio	1. Physical	Physical

5 IoT Applications

IoT selected several applications that are key to the world economy:

- Consumer and Home (Infrastructure, Awareness and Safety, Comfort and Convenience);
- Buildings (commercial/institutional);
- Industrial/Supply Chain/Logistic (resource automation, fluid processes and distribution);
- Monitoring (environment, security, healthcare, traffic, industrial, buildings, construction);
- Energy (supply, demand, alternative, water, oil and gas);
- Retail (store, hospitality and specialty);
- Transportation (Transport Systems, vehicles, non-vehicular);
- Security and Public Safety (emergency services, public infrastructure, tracking, equipment, surveillance);
- IT and Networks (Enterprise and Public);
- Devices: servers, storage, PCs, routers, switches, PBXs, etc. This sector includes carrier, ITData, office, mobile nets, services, E-Commerce, data-centers, fixed carriers, ISPs, etc.

Table 3 presents some examples of smart applications. Ideally, there should be a reference model for each application which supports its understanding and implementation. We now highlight some applications.

1. *Smart Home*: This application aims at the automation and remote control of houses through remote access using smart phones or tablets. Operations include switching lights, TV sets and A/C on and off, acquisition of battery lifetime, opening and closing doors, reception of runout groceries in a refrigerator, reception of information via smart TV (including videos), warning against intruders among others.

Table 3 IoT smart applications

Smart applications	Environment	Functions (monitoring of)	Remote access	Stake-holders	State-of-the-art
Home	Residential	Handle lighting, A/C, Windows, alarms appliances, doors	Smart phone	Family	Commercial solutions
Building	Civil construction	Power efficiency, water heating, solar, battery	Sensors, actuators	Condominiums	Standards, pilot
Cities	Autos, traffic lights bridges, buildings, transit, tunnels	Congestion, air quality parking, lighting, noise, etc.	Smart phones GPS	Population, police, hall	Standards, pilot
Grid	Utilities energy	Power generation transmission, distr.	Energy meters, mgmt. systems	Energy companies	Standardized, under implementation
Healthcare	Residential	Patient monitoring, medicine, tags	Smart-phone tablet, laptop	Patients, doctors, nurses, health plan	Pilot implementation
Manufactoring	Industrial	Robots, PLCs	Management	Factories automation devices	Proprietary implementations
Farming	Farms cattle	Tractors, trucks, harvesters, drones	Mgmt. systems	Farms silos, co-ops	Proprietary implementations
Transportation, Logistics	Roads, fleets vehicles	Vehicle downtime, performance	GPS, handhelds, mgmt. systems	Drivers, transport authorities, etc.	Proprietary implementations

2. *Smart Building*: The automation of a building is the key concern of this application. The goal is to address green issues (e.g. power efficiency) in regard to electric power, water, solar heating, and battery systems. High-consumption equipment (air conditioning and ambient lighting) would be dynamically adjusted as a function of the internal and external temperature/humidity/illumination of a building. Battery systems may be monitored in order to anticipate a needed change of battery.
3. *Smart Cities*: This application deals with the exchange of information between automobiles, buildings, smart-phones, and traffic lights. The goal is to improve traffic conditions in urban areas. A smart city encompasses the following services: structural health of buildings; waste management; air-quality monitoring; noise monitoring;traffic congestion;mobility;city energy consumption;smart parking, and smart lighting.
4. *Smart Grid (Energy and Utilities)*: This application addresses automation of the electric power sector. It provides residential smart meters, which transmit the information of power consumption directly to the providers, which in turn use the network to monitor and control these in view of faults and failures. A specific network is used for the electric segment, due to reliability and response time requirements. The main components are: (1) Smart Power Generation; (2) Smart Transmission Grid Application; (3) Distribution Automation, (4) Customer Home Automation and Demand Response (advanced metering infrastructure (AMI), electric cars and home-power management). Some examples of devices included are power stations, generators, power conversion stations, wind and solar generators, UPS, batteries, and hydrogen cells. Similarly, there is the application for monitoring the water and gas supply network.
5. *Healthcare (Remote Medical Assistance)*: This application enables the monitoring and follow up of remote patients in their homes through the collection of information from sensors attached to the patient, such as temperature, pressure, glucose, oxygenation, cardiac activity, as wells as sensors for detection of movement, ambient temperature, RFIDs for identification of patient, medicaments, equipment among others. This information is sent to the cloud where it can be shared by all the interested and authorized parties such as doctors, nurses, hospitals and family. The doctor (or medical team) may be able to have access to the patients information via his or her smartphone and request that the patient be relocated to a hospital if his or her prognostic indicates a critical scenario. This application also enables a new business model to the health plan providers, where the patient physically goes to the hospital only in emergency or critical cases. An interesting variant to this application is the so-called *Smart Lifestyle*.
6. *Smart Manufacturing (Industry)*: This application tackles the control of devices and instrumentation with the automation of manufacturing processes (computer numeric controls, PLC's, robots etc.). It requires short reaction/response times, often with a deadline attached, as well as small jitter, delay, and low traffic volume. Intermediate systems such as adapters, gateways and bridges are required due to the large number of application islands, each having their own proprietary protocols.

7. *Smart Farming*: This application deals with farm automation, along with its interconnection with the outside world. It allows the interconnection of combine harvesters, tractors, drones, farm workers, RFID tags and the identification of animals, the farm control and supervisory application, silos and livestock, as well as other agricultural entities. It is necessary that the farm have a small radio base station that allows the connection of these components, which ultimately enables the implementation of an operational online farm.
8. *Transportation and Logistics*: Three major aspects are included in this class of applications: the optimization of (1) vehicle performance, (2) fleet operations, and (3) traffic management. The inclusion of IoT in vehicles will enable new capabilities such as the prediction and monitoring of vehicle performance while minimizing downtime. Vehicle Area Networks (VANETs) play a crucial role in integrating the objects/devices that support vehicle functionality. Furthermore, the use of sensor, communication and processing capabilities in vehicles will allow them to be monitored and tracked anywhere on a street, road or parking stall. This feature will permit the prediction of their future location. On the other hand, the adoption of IoT in fleets will permit their tracking, the streamlining of logistics and real-time data, the monitoring of performance and the reduction of downtime and delays are they occur. The integration of IoT in the engineering of traffic (Intelligent Transportation Systems) may also largely benefit from IoT, for example in the assessment of road conditions, management of transportation infrastructures, reduction of congestion, coordination of transport networks as well as the provision of innovative services.
9. *Other applications*: Intelligent parking, commercial automation (wholesale and retail), connected construction sites, monitoring of industrial tanks, oil and gas pipeline monitoring, smart metering, public security, and catastrophe warning among others. These new applications will require substantial standardization effort, specifications, development and tests. Each one is expected to have its own reference model.

6 Conclusion

The IoT is an evolution of the traditional Internet and it enables the communication of devices/objects/things as a new technological wave. It is internationally standardized and it uses the best technology and protocols that are currently available. It creates several smart applications, which will enable the automation of several sectors of the economy. The IoT leads to a new digital transformation by increasing businesses, adding value to processes, redefining organizations and generating new opportunities. Clearly, it will require substantial research efforts on global networks and their applications.

Additionally, international standards will need to include new applications with their respective reference models. It is expected that many pilot projects will turn into commercial solutions and new pilot experiences will appear as new applications.

Finally, it is worth emphasizing that the monitoring of objects in real-time may impose critical timing constraints and further challenges to the more resource-constrained applications of the IoT.

References

1. Bassi, A., Bauer, M., Fiedler, M., Kramp, T., van Kranenburg, R., Lange, S., Meissner, S.: Enabling things to talk: designing IoT solutions with the IoT architectural reference model (2013)
2. Farooq, M.U., Waseem, M.: A review on internet of things (IoT). Int. J. Comput. Appl. (0975 8887) **113**(1), 1–7 (2015)
3. Fersi, G.: Middleware for internet of things: A study. In: Proceedings - IEEE International Conference on Distributed Computing in Sensor Systems, DCOSS 2015, pp. 230–235 (2015)
4. ITU-T (2016) Unleashing the potential of the internet of things. newblock Technical report, International Telecommunications Union, Place des Nations, CH-1211, Geneva 20, Switzerland
5. Li, S., Xu, L.D., Zhao, S.: The internet of things: a survey. Inf. Syst. Frontiers **17**(2), 243–259 (2015)
6. Madakam, S., Ramaswamy, R., Tripathi, S.: Internet of things (IoT): a literature review. J. Comput. Commun. **03**(05), 164–173 (2015)
7. Miraz, M.H., Ali, M., Excell, P.S., Picking, R.: A review on Internet of Things (IoT), Internet of Everything (IoE) and Internet of Nano Things (IoNT). In: 2015 Internet Technologies and Applications, ITA 2015 - Proceedings of the 6th International Conference, pp. 219–224 (2015)
8. Weber, R.H.: Internet of things - new security and privacy challenges. Comput. Law Secur. Rev. **26**(1), 23–30 (2010)
9. Weber, R.H.: Internet of things: privacy issues revisited. Comput. Law Secur. Rev. **31**(5), 618–627 (2015)
10. Xu, L.D., He, W., Li, S.: Internet of things in industries: A survey (2014)

Evaluation of Traffic Delays and Utilization of IoT Networks Considering Mobility

J. R. Emiliano Leite, Edson L. Ursini and Paulo S. Martins

Abstract The performance analysis and dimensioning of an IoT network is presented, using discrete-event simulation and considering node mobility. Specifically, the processor utilization on each node and the mean-time delay of the network are estimated. Additionally, given that the lower the transmission power the larger may be the packet loss, and that the tolerance to the amount of packet loss depends on the application, the simulated IoT network may provide the designer with a guideline to specify the minimum required transmission power as a function of node distance and admissible packet loss. This lower threshold on transmission power is critical, since in wireless and battery-powered systems, the transmission power has a direct impact on its expected lifespan.

Keywords IoT network · Sensor network · Mobility · Mean delay time

1 Introduction

The Internet of Things (IoT) consists of the networking of physical objects through the use of a number devices including embedded sensors, actuators, RFIDs, and other elements that collect and transmit information about the objects. The data gathered from these devices may then be analyzed to optimize services, products, and operations.

J. R. E. Leite · E. L. Ursini · P. S. Martins (✉)
University of Campinas (UNICAMP), Limeira, SP 13484-332, Brazil
e-mail: paulo@ft.unicamp.br

J. R. E. Leite
e-mail: j750465@dac.unicamp.br
URL: http://www.ft.unicamp.br

E. L. Ursini
e-mail: ursini@ft.unicamp.br

Y. Iano et al. (eds.), *Proceedings of the 3rd Brazilian Technology Symposium*,
https://doi.org/10.1007/978-3-319-93112-8_9

With this new potential from the IoT comes new challenges in terms of engineering these systems. In particular, in this paper, we identify the issue of planning and dimensioning these systems from a traffic engineering perspective. These systems generate traffic that has new features such as the diversity and large number of requests and the low-volume associated with each request. Furthermore, the traffic can be redundant and must be filtered, and new devices such as the mediator come into play, which also require dimensioning of their internal queues and processing capacity.

If the traffic and dimensioning is not properly accounted for, the development and implementation of new IoT applications may not be feasible, even when considering all the most recent advances in technology.

One of the goals of dimensioning is to identify the performance bottlenecks and reconfigure the network accordingly to remove their effects. By means of dimensioning, it is possible to determine the capacity of processor as well as link capacities. It is also possible to place an upper bound on the number of devices that may be connected to a given link. One of the best indicators of network performance is the delay in the network and processor utilization on each node. For certain applications, the delay or variance in delay may be prohibitive.

This work concerns with the dimensioning of node capacity investigating through a case study the mean queuing time and the CPU utilization for each node for a given probability of connectivity of the nodes resulting from the mobility. The approach used is the so-called discrete-event network simulation model, which allows an approximate placement of the sensor nodes and RFID in a cluster, as well as their approximate traffic load, but it does not express their mobility. The mobility of nodes within a cluster is simulated through the Random WayPoint (RWP) algorithm. A degradation in connectivity for some nodes may cause the overall reduction of the traffic in the upper layers.

There are a number of papers using the approach to improve the efficiency of IoT. For example, in the area of simulation, Capone et al. [1] present the modeling and simulation for improvement of the power efficiency of the IEEE 802.15.4e DSME standard, targeting wireless sensor networks. The implementation of the simulation model is carried out in OPNET. The authors analyze the power consumption and they propose a set of improvements for low-power applications, which allows significant power reduction. Chunxiao et al. present a Middleware (mediator) solution for RFID tags and ZIGBEE Sensor networks [2], and it offers API interfaces so that the applications do not need to deal with the sensors level of the network. The work by Samaniego et al. [7] concerns with the management of heterogeneous IoT resources and uses the model of virtual resources mapped onto physical resources. However, none of these works covered in the literature tackle the traffic aspects of the network such as partial and global delays and CPU utilization. Instead, in this work, we tackle the performance of the queues, CPUs, and traffic in an IoT network, including the mediator, in addition to the clusters of sensors in the ad hoc network. The remainder of this paper is organized as follows: The network model is discussed in Sect. 2. A case study illustrating the application of the model along with the results is shown in Sect. 3. We summarize and present our conclusions in Sect. 4.

2 Network Model

The network model is a hierarchy consisting of clusters which contain nodes, which in turn have multiple CPUs, thus allowing several parallel connections. Inherent in each queue is the waiting delay before a TCP/IP packet can be processed by a server. Clearly, both queuing and processing times are subject to statistical distributions. Therefore, a network cluster may be regarded as a set of internal queues (each one associated with an outbound link).

The network components are the mediator, gateway, and endpoints. The endpoints can be RFID and sensors for different applications. Figure 1 shows the network model with its inputs (packets) and outputs (packets) to each cluster. The upper part of the model consists of the RFID network, the mediator and the Internet. There are 34 CPUs in the model to handle traffic for clusters and applications. More details are found in [4]. The model adds four representative applications that process and consume the information leaving the mediator: (1) RFID, (2) Sensor, (3) SNMP management, and (4) Smart cities. In IoT, the databases may be reached by mobile or cell phone applications through the secure HTTPS protocol.

3 Case Study and Results

To evaluate each node independently, a MATLAB routine generates random positions for the ten nodes within the cluster, every one sec (in our case). This case used the Random WayPoint Mobility Model to simulate the performance of the network. The mobility determines the location of each node that selects a random destination and travels toward it in a straight line at a randomly chosen uniform speed. The ad hoc subnet (sensors) is spread across a $1000 \times 1000\,\text{m}^2$ area. It has 10 mobile nodes and one static node in the rightmost position. For the crossover distance, $d_{crossover}$, there is a threshold ($d > d_{crossover}$) above which the two-ray propagation model is better than the free space model, and $d_{crossover} = (4\pi h_t h_r)/\lambda = 547\,\text{m}$.[1]

The input parameters were the same from Leite et al. [5], except that here transmission power was assigned with three values (i.e., two additional ones, 5, 10, and 15 dBm). Two basic propagation models (FS = Free Space and TR = Two-Ray ground propagation model) were considered, which are described by the following equation regarding distances d_{FS} and d_{TR}:

$$d_{FS} = \sqrt{\frac{P_t G_t G_r \lambda^2}{(4\pi)^2 P_{r,FS} L}} \text{ and } d_{TR} = \left(\frac{P_t G_t G_r h_t^2 h_r^2}{P_{r,TR} L}\right)^{1/4} \quad (1)$$

If the transmission power is set to 15 dBm (31.6 mW), the transmission range is 582 m for the free space model and 564 m for the Two-Ray. We set the range

[1] https://www.isi.edu/nsnam/ns/doc/node217.html, access Feb/04/2018.

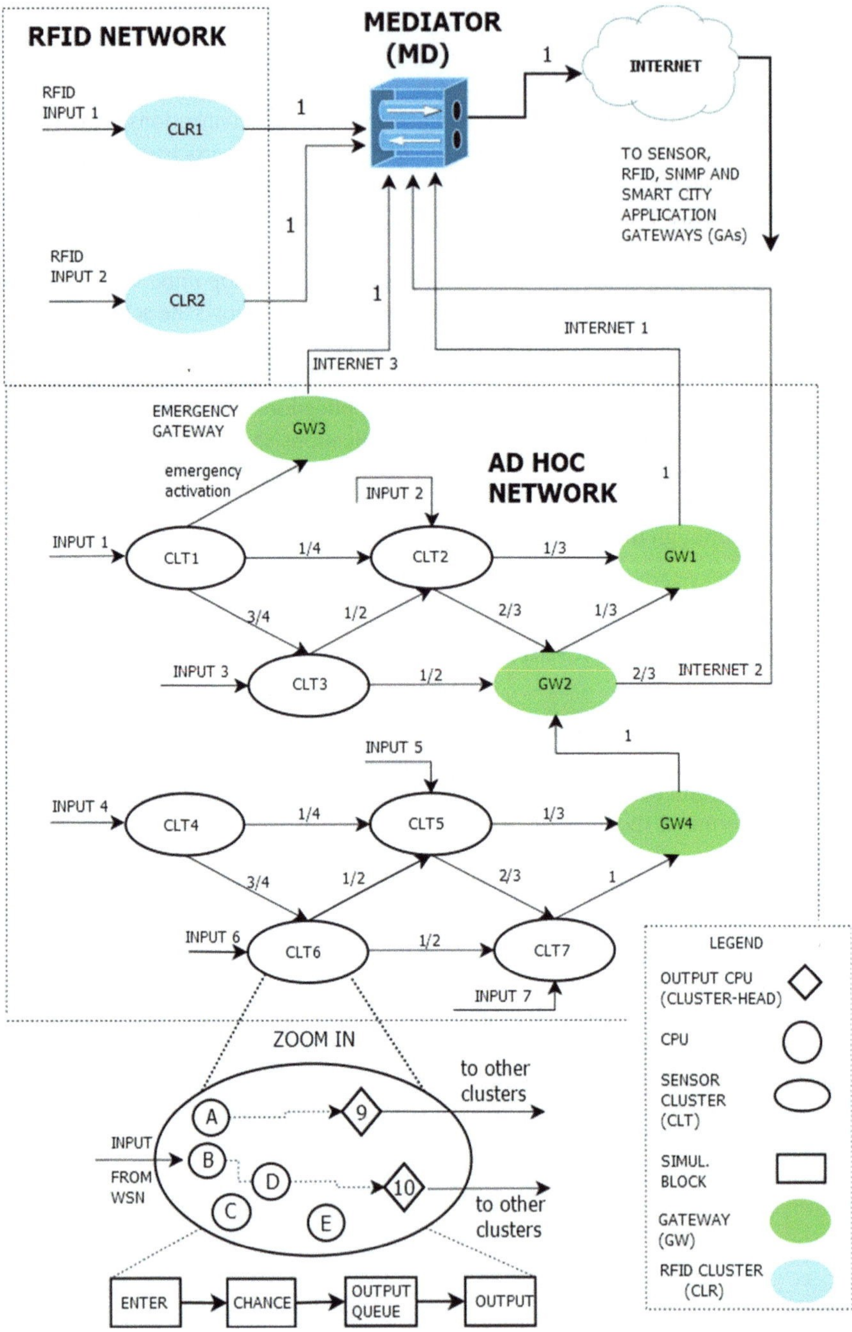

Fig. 1 IoT network model

to the more conservative value of 500 m. For the evaluation of losses, we consider that the power irradiates from the center of the mobile node, as shown in Fig. 2. Following the same assumption, the remaining values were calculated using the free space propagation model, as they present a distance that is less than 500 m. For 10 dBm (10 mW), d = 327 m, and we adopted 300 m; for 5 dBm (3.16 mW), d = 184 m, and we adopted 170 m.

Signal loss was 87.5 % for 5 dBm; 10.5% for 10 dBm; and 4.95% considering 15 dBm (accounting for nodes without connectivity in a set of runs including mobility). Due to the fact that 87.5% (5 dBm) of losses is not acceptable from a pragmatic view, we run the simulation with 20.5 and 49.5% blocking, knowing that the values correspond to a power transmission of 5 dBm$< P_t <$ 10 dBm, as shown in Fig. 2. The adopted mobility model is the one presented by Leite et al. [5, 6].

To illustrate this issue, Fig. 3 shows distance as a function of transmitted power. In this scenario, we also used the Jackson's network analytical model [3], which was calculated as a Markov process to validate the simulation model under the exponential distribution for both arrival and service distributions. Therefore, note that the simulation model is not limited to the use of the Poissonian distribution

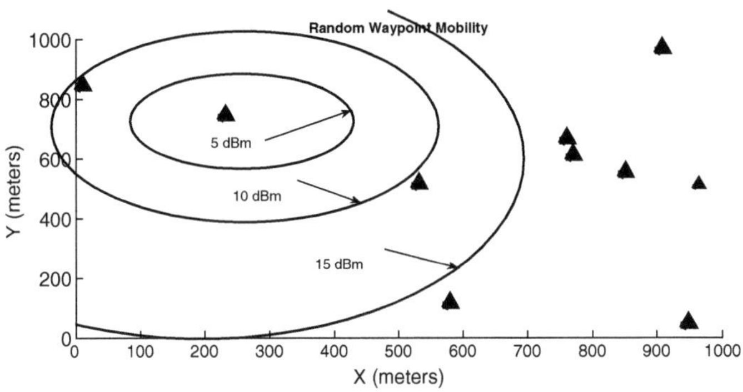

Fig. 2 Power range for 5, 10, and 15 dBm (ten mobile nodes)

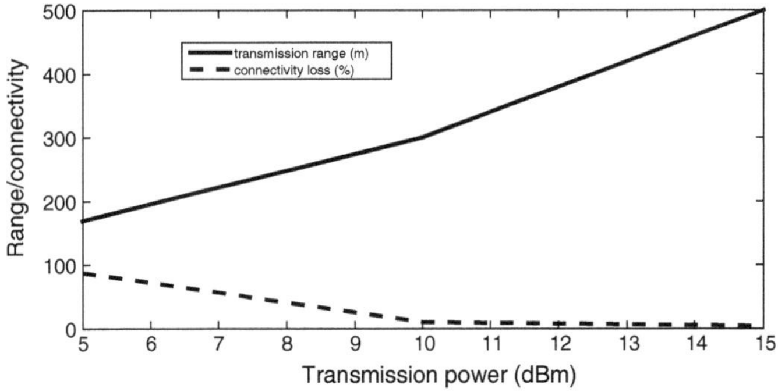

Fig. 3 Connectivity and distance as a function of power range

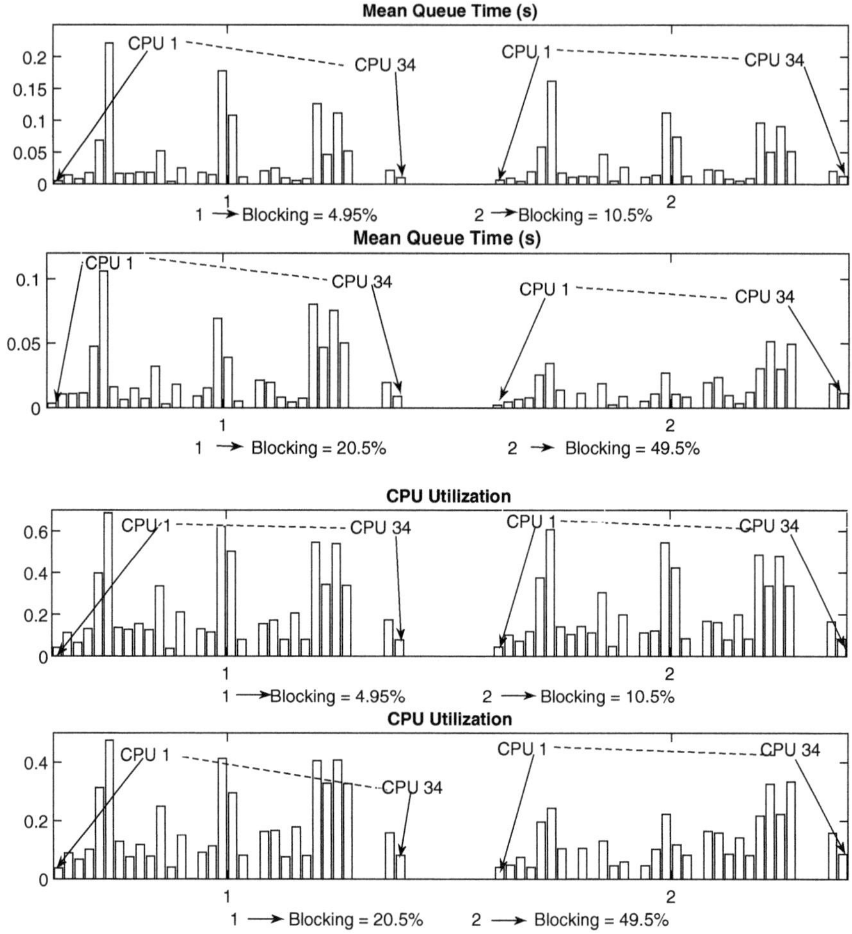

Fig. 4 CPU mean queuing time and utilization: connectivity loss (blocking) of 4.95, 10.5, 20.5, and 49.5%

initially assumed in this work. In Fig. 4, we see the mean queue time and utilization for each considered CPU (1-34). Using the Jackson's arrival rates and considering the simulated values and the ones obtained using the M/M/1 model [4] (CPUs arrival rates are 1/.6 = 1.67 bps and service rates are 1/.1 = 10 bps; the mediator has 5 bps for arrival rate and 50 bps for service rate), the mean network delay is $\overline{W_a} = 233.2$ ms (analytic model) and $\overline{W_s} = 247.3$ ms (simulation model). However, the actual values will be lower due to connection loss issues.

For a given level of transmission power, it is possible to verify a corresponding level of packet loss (Fig. 2). Due to the intrinsic redundancy of information in the network, there is an acceptable level of loss in the system that does not affect the application from a functional perspective. The tolerance to the amount of packet loss depends on the application. Nevertheless, with the simulation model presented, it is possible to analyze such trade-off quantitatively, which provides the designer with a guideline to configure transmission power. This is a critical design parameter since, in wireless and battery-powered devices, transmission power has a direct impact on system lifetime.

4 Conclusion

The proposed simulation model captured a number of features of complex systems, including the mediator, gateways, emergency nodes, Internet and IoT traffic, and most important the traffic generated by RFID tags and sensors (WSN). It was based on both discrete-event and random waypoint simulation for an ad hoc network that accommodates clusters and the effects of node mobility.

Through the model, it was possible to estimate the incoming and outgoing traffic for each cluster and the IoT mediator. We analyzed through a case study the mean queueing time and the output CPU utilization for each node for a given probability of connectivity of the nodes resulting from the mobility.

The mobility model used in this work assumed for illustration purposes a lower mobility of the devices. In future work, we contemplate exploring this issue by increasing the mobility of the nodes in the network to a point that it models a VANET. The model could also be used to find out the allowable tolerance of packet loss that does not compromise the performance of the system.

The presented model was scalable, meaning that it accommodates the inclusion of other types and number of components, by using the concept of clusters. Therefore, we argue that the proposed model can provide a useful tool for dimensioning the system for other services, applications, and scenarios.

References

1. Capone, S., Brama, R., Ricciato, F., Boggia, G., Malvasi, A.: Modeling and simulation of energy efficient enhancements for IEEE 802.15.4e DSME. In: 2014 Wireless Telecommunications Symposium, pp. 1–6 (2014)
2. Fan, C., Wen, Z., Wang, F., Wu, Y.: A middleware of Internet of Things (IoT) based on zigbee and RFID. In: IET International Conference on Communication Technology and Application (ICCTA 2011), pp. 732–736 (2011)
3. Jackson, J.R.: Networks of waiting lines. Oper. Res. **5**(4), 518–521 (1957)
4. Leite, J.R.E., Ursini, E.L., Martins, P.S.: A proposal for performance analysis and dimensioning of IoT networks. In: Brazilian Technology Symposium (BTSym 2017) (2017a)
5. Leite, J.R.E., Ursini, E.L., Martins, P.S.: Simulation of ad hoc networks including clustering and mobility. In: 16th International Conference on adhoc Networks (2017b)
6. Leite, J.R.E., Ursini, E.L., Martins, P.S.: Performance analysis of a multi-mode adhoc wireless network via hybrid simulation. In: Brazilian Telecommunication Symposium (SBrT 2017) (2017c)
7. Samaniego, M., Deters, R.: Management and internet of things. Procedia Comput. Sci. **94**, 137–143 (2016)

Blocking of the Cell Overflow Traffic in Heterogeneous Networks

Loreno M. Silveira, Paulo S. Martins and Edson L. Ursini

Abstract The fifth generation of wireless networks, or 5G as it is known, will allow a radical re-evaluation on the performance of mobile telecommunication services as well as it will support new and important applications such as enhanced mobile broadband, high mobility cases (planes, trains, etc.), Internet of things, smart cities, drones, robotics, etc. The objective of this work is to research, by means of simulation experiments, the network performance of the handover procedures under traffic variation and at different deployment scenarios of wireless applications, aiming to derive core network planning instructions. The used methodology is based on the principle of incremental validation, in which, in a first step, simplified analytical and/or simulation models are enhanced and validated by comparison with real situations or analytical solutions. Then, by means of increments in the complexity of the distributions and/or new functionalities, and with the increase of the knowledge of the problem, the dimensioning solution is refined.

Keywords Handover · Heterogeneous networks · Simulation

1 Introduction

The International Telecommunications Union Radiocommunication Sector (ITU-R) is the body responsible for the emission of radiofrequency standards and related service requirements. The ITU-R has defined the objectives, process, and timeline for the development of IMT-2020 (or Fifth Generation Systems). ITU-R also published in September/2015 at [1], its vision concerning applicable requirements on new

L. M. Silveira (✉) · P. S. Martins · E. L. Ursini
School of Technology UNICAMP, Limeira, Brazil
e-mail: loreno.silveira@pos.ft.unicamp.br

P. S. Martins
e-mail: pmartins@ft.unicamp.br

E. L. Ursini
e-mail: ursini@ft.com.br

Y. Iano et al. (eds.), *Proceedings of the 3rd Brazilian Technology Symposium*,
https://doi.org/10.1007/978-3-319-93112-8_10

mobile wireless systems. Three different usage scenarios or main applications were envisaged:

- massive Machine-Type of Communications (mMTC): Applications where a very large number of devices are interconnected exchanging a low traffic of non-delay sensitive data. Examples include Internet of Things (IoT), e-health, e-Farm, smart city, smart home etc;
- enhanced Mobile Broadband (eMBB): The demand for traditional human-centric mobile broadband communication does not stop to increase, leading to performance and seamless improvements, and sometimes aiming at very different applications like hotspots and wide area coverage. Expected new functionalities are Virtual Reality, Augmented Reality, Holograms, High Mobility cases (Planes, trains), etc;
- Ultra-Reliable and Low Latency Communications (URLLC): Critical applications, with very stringent Quality of Service (QoS) requirements. Examples include industrial control processes, smart grid distribution automation, Drones & Robotics, transportation security, Driving support systems, autonomous cars, etc.

The technology vendors' assessment, and the commercial launch are expected to happen before 2021, however the traffic performance of heterogeneous access networks (which combine several types of base stations with diverse cell sizes to provide coverage of an area) is still an issue, specially concerning three or more levels of cell splitting, as described in the following.

One of the main aspects addressed has been the question of hyperdense and heterogeneous networks. The situation has been a commonplace since the 2G and 3G generations but has become extremely relevant with the introduction of small cells with low power base stations (especially home stations, remote or retransmission stations). In this sense, some authors have investigated by means of simulation tools, the adequacy of the Poisson spatial distribution to model the location of base stations in practical scenarios of cell splitting network deployment (Martin-Vega et al. [2] and Chen et al. [3]).

The traffic offered to a given microcell can be partly transferred to a macrocell through the handover procedure. Glabowski et al. [4] presents a new analytical method for the calculation of overflow systems that does not require the calculation of the mean and variance of the overflow traffic, requiring only the knowledge of the value of the traffic offered to the system and the capacity of the direct resources and alternatives. However, it did not handle heterogeneous networks with three or more cell sizes. Cominardi et al. in [5] addresses the issue of handover procedure after establishing the communication with the destination data network. The new procedure releases the previous gateway and forwards the packets directly to the destination, avoiding the internal retransmission of packets in the 5G core network.

Another fundamental question concerns the improvement of performance parameters. Özçevik et al. [6] presents and compares three end-to-end optimization solutions to avoid bottlenecking of UDP traffic by TCP traffic in 5G systems, and Mahmood et al. [7] studies the potential throughput gains for full duplex communication in conventional half-duplex transmission in a small cell network and asymmetric traffic.

Finally [8] and [9] propose different models of simulation, from which several parameters can be derived. Faraci and Lombardo in [9] propose a simulation tool for 5G networks, which is capable of capturing delay statistics due to the CPU load and transmission link congestion in the virtual presence points of an operator, applying a study of case.

The contribution of this work is to further investigate the real or total blocking probability when deploying heterogeneous networks aimed to eMBB applications, particularly, models of multiple cell layers and dynamic assignment of resources at base stations are addressed.

2 Methodology

After the definition of the problem, implementation and validation of the corresponding conceptual model, the experimental procedure included the following steps [10]:

- Exercising the simulation model to verify its validity, using simulation or analytical methods, and comparing with known real data;
- Submitting the model to the conditions established in the research hypothesis;
- Analyzing the results and refining the model.

The validation of the conceptual and simulation model makes use of the principles of incremental validation, in which, in a first step, simplified analytical and/or simulation models are enhanced and validated by comparison with real situations. Then, by means of increments in the complexity of the parameters that reflect the variables of entry or inclusion of new functionalities, and with the increase of knowledge of the problem, the model is being refined. The validation will be performed by comparison with the real situations and known facts or by means of analytical models for which we already know that they reflect real conditions.

Two network scenarios are initially designed, both dealing only with enhanced Mobile Broadband applications, considered a well-known area: a two layers cell environment and a three layers cell environment. The objectives are mostly the same, to analyze the effects of traffic or bandwidth increase in the network final blocking, regarding cells configurations and capacities.

3 Results

3.1 Case Study 1: Overflow from Multiple Cells

The objective of this Case Study is to evaluate the total blocking when traffic from several microcells overflows to a single macro cell, considering a steady increase in the microcell traffic and a fixed macrocell original traffic. This could correspond to a dense urban situation where the user experienced data rate could reach 350 Mbps, whereas in the macro urban, it is kept below 75 Mbps, during busy hour.

The total blocking, which reflects que user perception, is considered the main indicator to evaluate the impact of traffic increase over the network performance.

For compliance purposes, the studied scenarios meet the frequencies and bands specified so far for 5G New Radio [11] with the following additional assumptions:

- Larger cells occupy the same geographical area of smaller cells cluster;
- Call generation rate: Holding Time/Traffic per cell
- Holding time: A time parameter in the range 100–200 s
- Microcell carrier bandwidth: 100 MHz
- Macro cell carrier bandwidth: 25–50 MHz
- Cell number of channels according to the frequency band and cluster size n (n = 3, 4, 7, and 12)
- Traffic per cell defined by the number of channels and Service Grade B = 2%.

Results showed minor changes in the total blocking, up to microcell nominal traffic, followed by a significant rise before converging to 100%, but never surpassing microcell blocking. Figure 1 illustrates one possible configuration and Fig. 2 presents the achieved results in this case.

Complementary, the Equivalent Random Traffic Method, e.g., as described at Iversen [12], allows to compute the real traffic blocking probability of 0.6% when microcells and macrocells have a blocking probability of 2%.

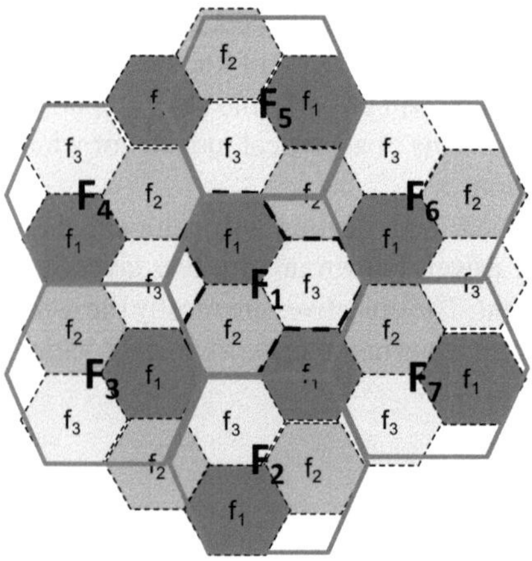

	Macrocell	Microcell
Frequencies	Fj (j=1,2,..7)	fi (i=1,2,3)
Cluster size	7	3
Number of channels	16	10
Nominal traffic	9.5 Erl (fixed)	5 Erl

Fig. 1 Example Case Study 1

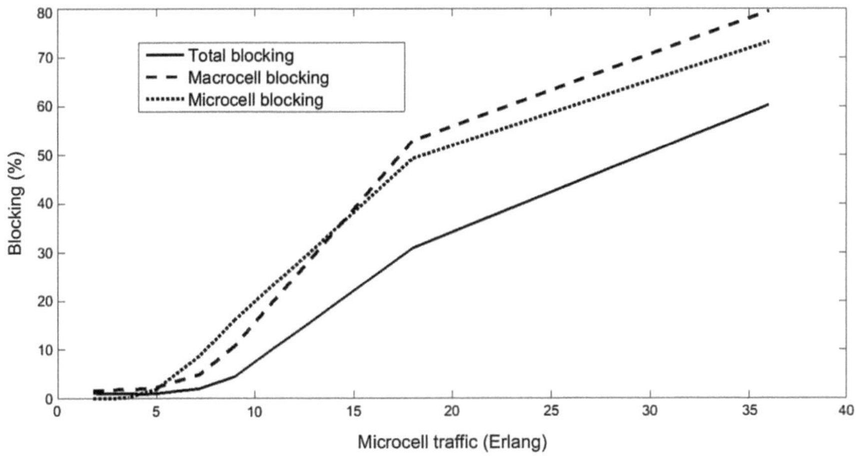

Fig. 2 Blocking probability and traffic in the microcell

3.2 *Case Study 2: Blocking Probability at a Three-Layer Cell Scenario*

The objective of this Case Study is to evaluate the total blocking when traffic from several picocells overflows to a micro cell, whose traffic overflows to a macrocell whenever the microcell is congested. Again, picocell traffic is variable, but the original microcell and macrocell traffic are kept constant.

Figure 3 depicts resulting Blocking Probabilities, when the number of channels is $N_{picocell} = 10$, $N_{microcell} = 8$ and $N_{macrocell} = 16$. In this case, on purpose, microcell and macrocell capacities were chosen below recommended values.

3.3 *Case Study 3: Blocking Probability with Dynamic Bandwidth Assignment*

The objective of this Case Study is to evaluate the sensitivity of the total blocking to the number or bandwidth channel increase in a Three-layer cell environment. Although this is a non-probable scenario, the Case Study aims to investigate possible strategies to expand the network capacity.

Figure 4 illustrates the results when the three cells have the same number of channels N = 10, and traffic a = 18 Erlangs. The number of channels of each layer is increased while in the others it remains constant.

It can be noticed that the worst performance case is obtained when the increase of channels happens at the picocell level, and the best at the macrocell level. In other

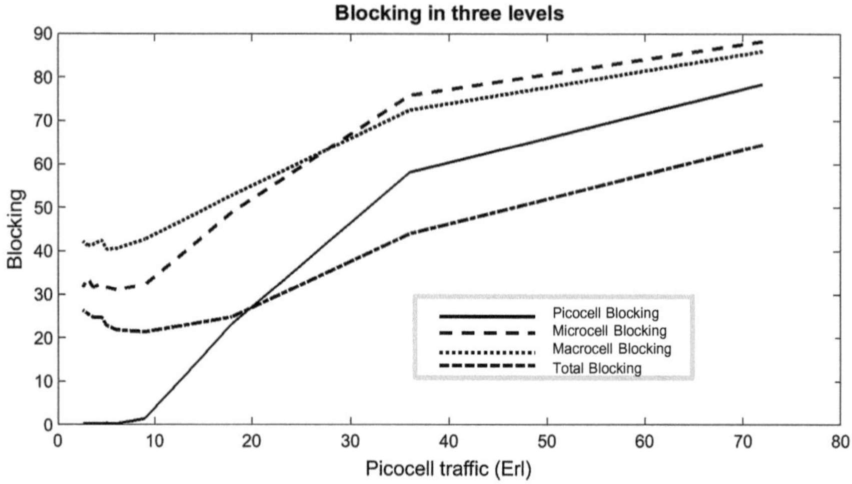

Fig. 3 Three-layer heterogeneous network blocking probability

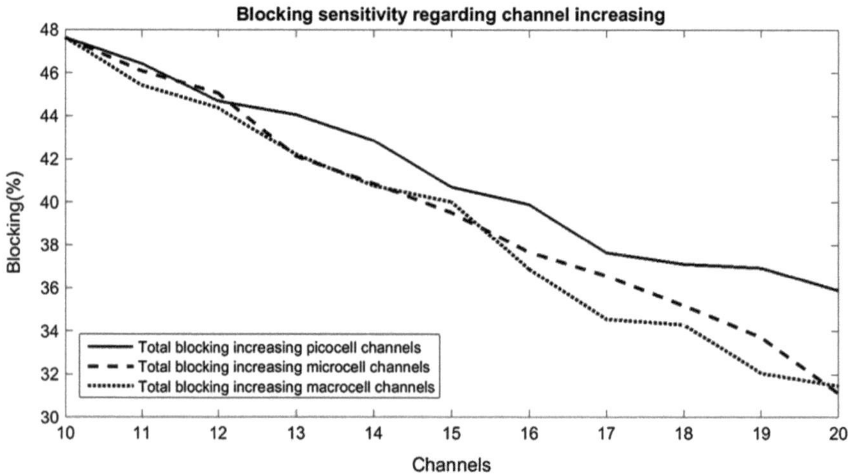

Fig. 4 Blocking sensitivity to channel expansion

words, in this scenario, if capacity expansion should occur, the best level to start is the macrocell layer.

4 Conclusions

Three Study Cases were visited aiming to derive core network planning guidelines. Despite the simplicity of the models, interesting conclusions have arrived from the simulation data, showing both the feasibility and difficulty while dimensioning a heterogeneous network, as it will be required in 5G scenarios. Only one category of 5G applications was analyzed, the enhanced Mobile Broadband. Dimension rules in this case are well known and were applied although not explicitly in the paper.

Other applications and other Study cases apart from the handover procedure should be analyzed as well.

References

1. ITU-R, Recommendation M.2083-0: IMT Vision—Framework and overall objectives of the future development of IMT for 2020 and beyond (2015)
2. Martin-Vega, F.J., Di Renzo, M., Aguayo-Torres, M.C., Gomez, G., Duong, T.Q., Paris, G.: Stochastic geometry modeling and analysis of backhaul-constrained hyper-dense heterogeneous cellular networks. In: International Conference on Transparent Optical Networks (ICTON), Budapest (2015)
3. Chen, S., Ji, X., Xing, C., Fei, Z., Wang, H.: System-level performance evaluation of ultra-dense networks for 5G. In: IEEE Region 10 Annual International Conference, Proceedings/TENCON 2016, vol. 2016, pp. 0–3. Singapore (2016)
4. Głąbowski, M., Hanczewski, S., Stasiak, M.: Modelling of cellular networks with traffic overflow. Hindawi Math. Probl. Eng. **2015** (2015)
5. Cominardi, L., Giust, F., Bernardos, C.J., De La Oliva, A.: Distributed mobility management solutions for next mobile network architectures. Comput. Netw. **121**, 124–136 (2017)
6. Özçevik, M.E., Canberk, B., Duong, T.Q.: End to end delay modeling of heterogeneous traffic flows in software defined 5G networks. Ad Hoc Netw. **60**, 26–39 (2017)
7. Mahmood, N.H., Sarret, M.G., Berardinelli, G., Mogensen, P.: Full duplex communications in 5G small cells. In: 2017 13th International Wireless Communications and Mobile Computing Conference Proceedings, pp. 1665–1670 (2017)
8. Bąkowski, K., Roadziewicz, M., Sroka, P., Rodziewicz, M., Sroca, P.: System-level simulations of selected aspects of 5G cellular networks. In: Proceedings of the International Symposium on Wireless Communication Systems ISWCS 2015, pp. 711–715 (2015)
9. Faraci, G., Lombardo, A.: A simulative model of a 5G telco operator network. Procedia Comput. Sci. **110**, 344–351 (2017)
10. Sargent, R.G.: Validation and verification of simulation models. In: Winter Simulation Conference, pp. 13–24 (2004)
11. 3GPP TS 38.104, Radio Access Network: NR—Base Station (BS) radio transmission and reception, V.15.0.0 (2017)
12. Iversen V.B.: Teletraffic engineering and network planning Course. In: Technical University of Denmark (DTU), available http://www.dtu.dk/english (2015). Last accessed 20 Apr 2017

Digital Image Processing with Data Storage for Security Applications

Daniel Izario, Yuzo Iano, Bruno Izario and Diego Castro

Abstract This article discusses a web tool oriented for computational vision applications, considering digital image processing techniques such as pattern recognition and filtering. It provides to the user a real-time location monitoring; the captured images pass through a filtering process to extract relevant features of the scenes and the image generated is compressed using the properties of the discrete cosine transform. All the inherent applications of the developed tool favor to the area of security and monitoring, mitigating possible human errors.

Keywords Computational vision · DCT/IDCT · Digital image processing
Surveillance · Security applications · License plate recognition

1 Introduction

The development of security-oriented applications is growing due to the advanced multimedia information retrieval techniques proposed in the last years. Computer vision allows public agencies and security companies the possibility to extract relevant information of public interest. This information can be used for many purposes, such as the recognition of people, objects, and texts. In the most general way, the extraction of patterns that could serve as the country security.

To capture the multimedia information, the surveillance cameras (IP cameras) are often used, fixed on lighting poles to give a broader view of the area to be monitored.

D. Izario (✉) · Y. Iano · B. Izario · D. Castro
University of Campinas, Campinas, SP, Brazil
e-mail: daniel_izario@hotmail.com

Y. Iano
e-mail: yuzo@decom.fee.unicamp.br

B. Izario
e-mail: bruno_izario@hotmail.com

D. Castro
e-mail: diego.pajuelo.castro@gmail.com

Y. Iano et al. (eds.), *Proceedings of the 3rd Brazilian Technology Symposium*,
https://doi.org/10.1007/978-3-319-93112-8_11

People usually monitor the video captured from the cameras and give technical suggestion to the public entities to take some kind of action, subject to internal procedures used by each agency. This process is known as naked eye monitoring. The addition of computational vision techniques to replace the current operation of the security system reduces the operational cost and mitigates possible human errors. The computational vision techniques require a large-scale image and more computational resources to be able to mitigate failures when testing in production. The proposed solution is an intelligent tool capable of recognizing patterns using IP cameras. A website contains what is needed to ensure the integrity of the information to be taken from the captured images.

Section 2 explains the website development. In Sect. 3, the model proposed based on the Discrete Cosine Transform (DCT) and the inverse Discrete Cosine Transform (IDCT) is detailed. Section 4 presents the way of storing all information obtained. Finally, the conclusion and future work are presented in Sect. 5.

2 Website

The developed website [1] serves as a tool to provide real-time monitoring for security purposes. This application monitors several areas of a city at the same time, using different video channels in a centralized system as shown in Fig. 1. The system stores images by using the DCT/IDCT technique, which is explained in Sect. 3.

Fig. 1 Monitoring process with 4 IP cameras

After analyzing and validating the multimedia information, the system returns to the website the information about vehicle plates, people faces, etc.

2.1 Real-Time Video

As shown in Fig. 1, real-time video monitoring uses as many cameras as required to cover the entire region. It is necessary to reposition the channels on the screen to allocate each channel to a window or use the automatic channel framing option on the website.

The monitoring model is based on research works carried out by the failures performed by employees throughout the process of data analysis. There are several types of degradations and distortions, inherent to the processes of acquisition, transmission, and visualization of images, which contribute, directly or indirectly, to limit the capacity of the human eye. Considering this, it is possible to analyze in Fig. 2 how the computational vision approach takes into account the treatment of the image through digital filters, digital image processing techniques, and guaranteeing satisfactory results for analyzing.

This tool stores and displays the videos in real-time. To extract relevant information, in some cases, is necessary to apply a filter mask. If the license plate recognition feature is enabled, the vehicles with deleted license plates will have their extraction process difficult. It could be not possible to know if the vehicle is part of the database of stolen cars, with fines or taxes.

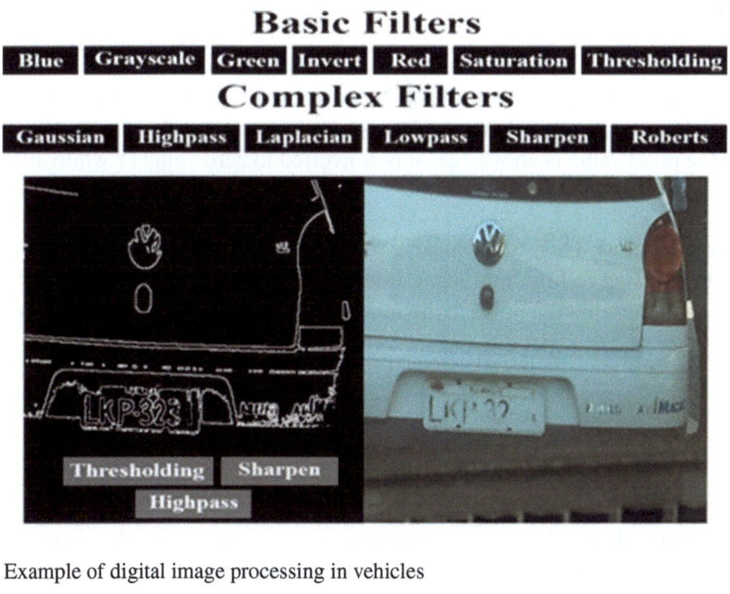

Fig. 2 Example of digital image processing in vehicles

2.2 Digital Image Processing

To search accurate information as shown in Fig. 2, digital image processing techniques are considered and are part of the website tool. The application has several edge detection filters and techniques for improving image quality.

Every filter is coded in JavaScript, using an Application Programming Interface (API) for the HTML CANVAS [2] element. The filters manipulate the image in many different ways and are categorized depending on the application. The first category restores an image to minimize the distortions and degradations inherent to the image. The second category emphasizes the visibility of a scene for visual interpretation.

The first step, before filtering, is to develop a code capable of applying a convolution process, which applies a mask to all pixels of the image, creating a new adjusted value for the image used. The main problem is the high computational complexity because it spends much time in reading pixel by pixel. Thus, the main idea was to develop a fast algorithm using simple matrix operations between all pixels of the RGB domain. It is only necessary, the creation of a function that returns the pixel value of the image by applying the masking filters using transformation matrices. These basic filters, as a grayscale filter, inverse filter, image saturation techniques, sepia, and thresholding methods, are often needed to be used prior to the convolution process.

The digital processing methods used to recognize the license plate of a vehicle in Fig. 2 were the thresholding, high-pass filtering, and sharpening methods as explained below:

Thresholding Method The luma sample is calculated as follows (1):

$$0.2126\mathrm{R} + 0.7152\mathrm{G} + 0.0722\mathrm{B}. \tag{1}$$

This component represents the brightness or the grayscale domain of an image. The thresholding or grayscale method consists of inserting a threshold value to each pixel. If the pixel value is above 128, it is converted to zero value or white, otherwise a value one or black.

High-pass Method This filter makes the transitions between different regions sharper, being used to highlight some image characteristics, such as borders, lines, curves or smudges, but the noise in the image is also emphasized. To apply this filter, it is necessary to use a specific matrix, as (2).

$$\text{operator} = \begin{matrix} -1 & -1 & -1 \\ -1 & 8 & -1 \\ -1 & -1 & -1 \end{matrix} \tag{2}$$

Sharpen Method The filter aims to accentuate the contours of the image, providing a detailed and sharpness image, whenever there is an outline of two distinct areas. It can be between light and dark areas or between different regions of colors, especially between complementary color models: Cyan/Red, Magenta/Green, and Yellow/Purple. To apply this filter, it is necessary to use a specific matrix, as (3).

$$\text{operator} = \begin{matrix} 0 & -0.2 & 0 \\ -0.2 & 1.8 & -0.2 \\ 0 & -0.2 & 0 \end{matrix} \quad (3)$$

3 Dct/Idct

The Discrete Cosine Transform (DCT) [3] makes use of real numbers and is widely used in digital image processing and data compression. It expresses a finite sequence of data points in terms of a sum of cosine functions oscillating at different frequencies, where the high-frequency components may be discarded with no visual impact.

The human eye is more sensitive to changes in brightness than color changes. For this reason, the RGB color model is transformed to the YCbCr color model, which consists of one brightness component Y' and two-chrominance components, Cb' and Cr', which can be subsample in a 4:2:0 format.

There are several image formats and in some cases are tailored to specific applications. For storing purposes, the Joint Photographic Experts Group (JPEG) image format is widely used, but Portable Network Graphics (PNG) is more versatile and recommended for applications-oriented for quality.

Considering these facts, the PNG format for the HTML CANVAS was chosen, and the JPEG for storing. Before applying the DCT transform to an image, is important to split a frame into blocks of sizes, 4×4, 8×8 *or* 16×16. In some case, sizes of 32×32 or 64×64 blocks are used in order to optimize the computational complexity and improve the coding efficiency. The JPEG standard uses 8×8 blocks. The block size has an impact on the compression ratio, and the larger the blocks, the lower the compression ratio.

The proposed method performed a compression process and coded in the JavaScript programming language. This algorithm divides the luma component and the two-chrominance components matrices into several matrices, each with a size of 8×8 pixels. The first matrix generated, with the original values of the first 64 pixels. The DCT transform is applied to these matrices, whose values are close to zero. This process transforms the time domain to the frequency domain.

The $N\times M$ DCT, is presented in (4)

$$F(\mu, \vartheta) = \left(\frac{2}{N}\right)^{\frac{1}{2}} \left(\frac{2}{M}\right)^{\frac{1}{2}} \sum_{i=0}^{N-1} \sum_{j=0}^{M-1} \Delta(i) \times \Delta(j) \times \cos\left[\frac{\pi \times \mu}{2 \times N}(2i+1)\right] \cos\left[\frac{\pi \times \vartheta}{2 \times M}(2j+1)\right] \times f(i, j), \quad (4)$$

where

$F(\mu, \vartheta)$ coefficient in the field of processing;
μ horizontal axis in the transform domain;
ϑ vertical axis in the transform domain;

$$\Delta(x) = \frac{1}{\sqrt{2}},\ for\ x = 0;$$

$$\Delta(x) = 1,\ \text{for}\ x \neq 0;$$

$$x = \mu|\vartheta;$$

$f(i, j)$ amplitude in the time domain;
i horizontal axis in the time domain;
j vertical axis in the time domain;

The IDCT, is calculated as follows (5):

$$F^{-1}(u, v). \tag{5}$$

Because of the DCT coefficients are real numbers instead of integer numbers, a quantization process is applied to reduce the number of bits required to store a value by reducing the precision of an integer. This process removes the spatial redundancy and is known as lossy compression.

After the quantization process, the image is stored in the database with a reduced size, being able to return to the website when it is necessary to apply the digital image processing, using the IDCT. Upon applying (5), it is seen that when the pixels of the image were recovered, the values were very close to the original values, proving that the losses are negligible, and only the high frequencies were discarded.

4 Storage

For the storing system is used the Structured Query Language (SQL) [4] commands implemented in conjunction with JavaScript, with a PHP (Hypertext Preprocessor) [5] script language.

The first step is to create a class instance and specify the driver, in this case, the MySQL [5], which is a Database Management System (DBMS), which uses the SQL language as an interface, the database name, the username, and password as primary information. Is necessary a try-catch code design to alert about possible errors that may occur during storing. An identifier (Id) and a unique locator are required for each stored image.

5 Conclusion

This article discusses a simple tool, easy to apply and implement and of high importance for military and civil applications. The use of digital image processing techniques allows a flexible tool because the relevant extracted features can be used to adjust and correct images before their analysis, these features are essential for viewing hidden areas or low-resolution images, which makes more efficient and accurate the operation.

As a future work, the proposed system will incorporate the recognition of people through artificial intelligence using similarity of punctual information obtained from the data. This facial recognition from the captured images along with a broad face analysis using standardized lines will expand the applicability and use of the recognition system.

References

1. Flanagan, D.: JavaScript: The Definitive Guide, 6th edn. O'Reilly Media, (2011)
2. Fulton, S., Fulton, J.: HTML5 Canvas: Native Interactivity and Animation for the Web, 2nd edn. O'Reilly Media, (2013)
3. Ahmed, N., Natarajan, T., Rao, K.R.: Discrete cosine transform. IEEE Trans Comput v. C-23, i.(1), 90–93 (1974)
4. ClydeBank Technology, SQL QuickStart Guide: The Simplified Beginner's Guide to SQL, 1st edn. ClydeBank Media LLC, (2015)
5. Thomson, L., Welling, L.: PHP and MySQL Web Development, 5th edn. Addison-Wesley Professional, (2016)

Development of a Digital Image Processing Web Tool for a Monitoring System Relying on an Unmanned Ground Vehicle

Daniel Izario, Yuzo Iano, Bruno Izario, Letícia Magalhães and Diego Castro

Abstract This paper describes the development of a remote control system of the unmanned ground vehicle. Also, the proposed system provides remote sensing video and gives the user a real-time view of the monitored location. As a part of the development, the study of different image filters, as well as its implementation is detailed throughout the work. These kinds of tools favor the border security area, agricultural sector, and video surveillance.

Keywords Digital video processing · Remote sensing · Unmanned ground vehicle · Surveillance · Web tool

1 Introduction

Unmanned ground vehicle (UGV) is a vehicle that operates while in contact with the ground and without an onboard human presence [1]. It has been developed for applications that involve limitations on communications. For instance, when exploring places with restricted local area networks, communication is fundamental, for this reason, the UGV's provides a secure communication channel with high autonomy.

D. Izario (✉) · Y. Iano · B. Izario · D. Castro
University of Campinas, Campinas, SP, Brazil
e-mail: daniel_izario@hotmail.com

Y. Iano
e-mail: yuzo@decom.fee.unicamp.br

B. Izario
e-mail: bruno_izario@hotmail.com

D. Castro
e-mail: diego.pajuelo.castro@gmail.com

L. Magalhães
National Institute of Telecommunications, Santa Rita do Sapucaí, MG, Brazil
e-mail: leticiapaiva@get.inatel.br

Y. Iano et al. (eds.), *Proceedings of the 3rd Brazilian Technology Symposium*,
https://doi.org/10.1007/978-3-319-93112-8_12

Especially, when the vehicle travels long distances and sends telemetry signals to a command center for large periods of time.

UGV's control often uses a radio frequency (RF) channel to transmit the commands and the video signal. The video camera is installed in front of the vehicle to register all scenes of the excursion. This signal is transmitted on-line to the command center. With the help of digital image processing techniques, such as the image filters, is possible to identify and extract useful information from the recorded scenes.

Section 2 explains the development stages of unmanned ground vehicles and the communication system used. Section 3 presents the web application and its functionalities. Section 4 explains the possible applications of the tool. Finally, the conclusion and future work are presented in Sect. 5.

2 Unmanned Ground Vehicle

The main parameters to consider in the construction of the vehicle are weight and robustness. A good trade-off between these two parameters allows the vehicle to remain stable on uneven ground and its battery autonomy can last a long time. At least, to finish the scheduled mission.

The chassis of the constructed vehicle is made of aluminum because it is a lightweight and sturdy material. The vehicle includes motors of 6 volts, prepared to support the vehicle and change the speeds. Each wheel is made of durable polymer, where the rubber tires used are for mixed use. All motors are connected via a Shield L293d—Driver H-Bridge, which controls the vehicle's speed and the camera's bracket.

The rotation of each motor allows the vehicle to control the speeds and directions: go forward and reverse, turn to the right and left, or turn to 30 grades with respect to the central axis. For greater battery autonomy, this work installed a rechargeable battery of 12 V/7A in the vehicle.

In Fig. 1, a complete diagram of all the parties involved in the project is being presented [2].

3 Digital Video Processing

To monitor the remote video signal, a web application was developed for capturing and processing the digital video in real time. Noise, which is inherent to the communication systems, distorted the video signal. This can occur due to climatic factors, equipment life and interferences in the transmission channels.

There are two forms of noise in digital video processing, additive and multiplicative noise. This work considers only the additive, which is one of the most used in practice, both in the spatial and in the frequency domain.

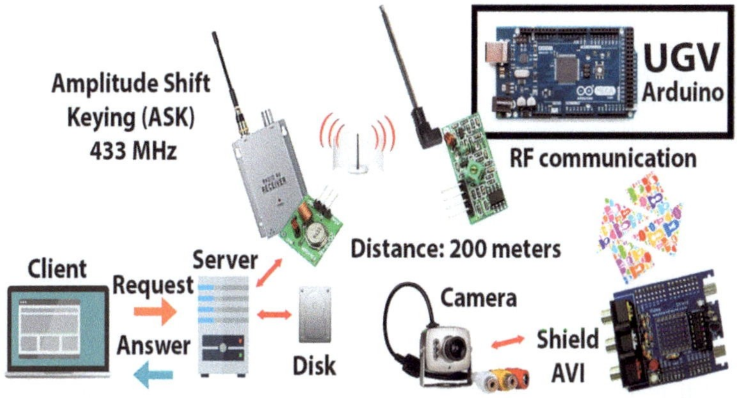

Fig. 1 UGV complete diagram

The simulated additive noise is close to the real one, that is, it models the sum of several small sources of noise. The function that represents the Gaussian probability density function of a random variable x is given in (1), where x represents the gray level present by filtering, μ is the mean of the gray values and α is its standard deviation.

$$\rho(x) = \frac{1}{\sqrt{2\pi\alpha^2}} e^{\frac{-(x-\mu)^2}{2\alpha^2}}. \tag{1}$$

If we apply (1), for example, in blur filtering, the graph shown in Fig. 2, proves the relation of the real parameter (Lr) versus the estimated parameter (Le) in an image matrix, performing calculations within a threshold for noisy pixels.

Figure 2 allows analyzing the real and estimated values of a captured image, focusing on those distorted pixels caused by noise. Thus, this tool extracts relevant information with a large relative error rate, this error model measures the difference between the actual parameter and the estimated parameter and divides this difference by the actual parameter value, as shown in Fig. 3.

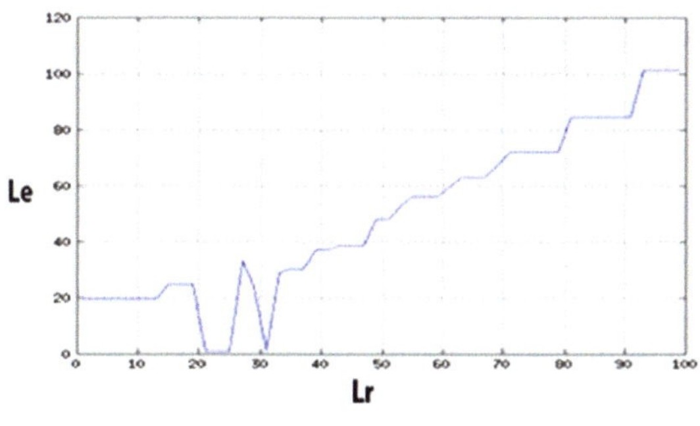

Fig. 2 Relationship between the real parameter (Lr) versus the estimated parameter (Le)

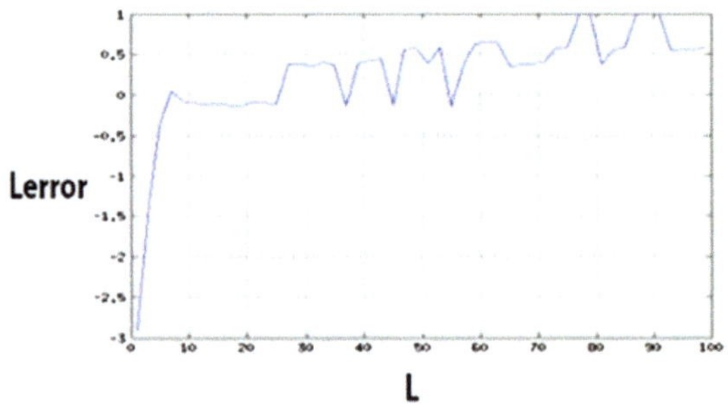

Fig. 3 The relationship between the parameters (L) and the relative error (Lerror) without correction in the application

For relative error rates greater than 1, the application is very sensitive to noise. Due to this, a restoration process is necessary to reduce the noise at the capturing process. This technique removes noise with a relative error rate close to zero, as shown in Fig. 4.

The application of the inverse filtering process results in the approximation of the original image by multiplying the noise matrix against the inverse of function $H(u, v)$ as shown in (2) which represents the inversion of each value of a matrix instead of the inverse of the entire matrix.

$$F(u, v) = \frac{1}{H(u, v)} G(u, v). \quad (2)$$

After the reverse filtering process, the tool allows that the user select different image filters that can be applied to the remote video signal. This work uses the bank of image filters, enabled in CCS [3] and HTML [4] language and integrated via JavaScript [5] in the final development. These filters can be divided into two models, which are: Scalable Vector Graphics (SVG) [6] and graphic filters. In the following statement, each filter will be described:

Blur method: This filter is based on the value applied to the deviation radius of the Gauss curve, which can vary from 0 to 20. This causes a blur effect in the video. Negative values are invalid [7, 8].

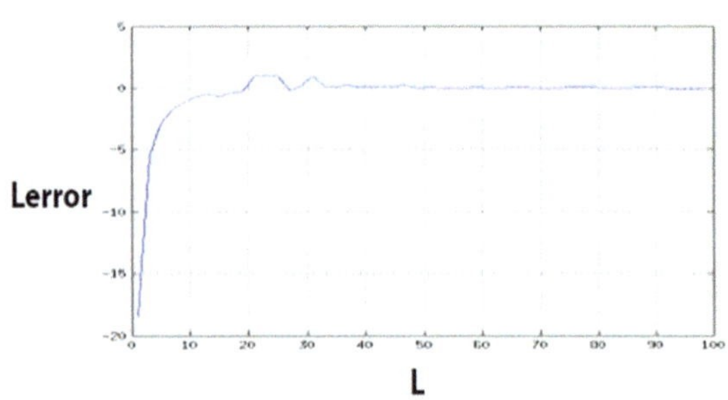

Fig. 4 The relation between the parameters (L) and the relative error (Lerror) with correction in the application

Saturation method: This filter saturates the color of a video, the values for the parameter of this function are a rate of change between 0 and 5. Negative values are invalid. The value 0 results in an unsaturated video and the value 2.5 is the default value, that is, the original video. Values greater than 2.5 result in supersaturation of the video [7, 8].

Hue method: This filter applies a function capable of converting the colors of the video to tones according to the angle of the Hue property of the color. The values for this parameter range from 0 to 360. Negative values are invalid. The value 0 is the default, that is, the original video [7, 8].

Brightness method: The filter controls the brightness of the video. Values range are between 0 and 1, for this parameter a step of plus or minus 0.05 is added to each drive. Negative values are invalid. The value 0 results in a completely dark video and the value 0.5 is the default value, that is, the original video. Values larger than 0.5 are allowed but result in high brightness videos compared to the original one [7, 8].

Contrast method: The filter controls the contrast of the video. Values range are between 0 and 5, for this parameter a step of plus or minus 0.05 is added to each drive. Negative values are invalid. The value 0 results in a video with maximum contrast and the value 2.5 is the default value, that is, the original video. Values larger than 2.5 are allowed but result in low contrast videos compared to the original [7, 8].

Invert method: This filter reverses the colors of the video. Values range are between 0 and 1, for this parameter a step of plus or minus 0.05 is added to each drive. The value of 1 completely reverses the colors, and the value 0 is the default, that is, the original video [7, 8].

Sepia method: This filter converts the colors of the video to sepia tones. Values range between 0 and 1, for this parameter a step of plus or minus 0.05 is added to each drive. The value of 1 results in a completely sepia video and the value 0 is the default, that is, the original video [7, 8].

4 Possible Applications

This tool can monitor zones considered at risk in a remote command center. For instance, to detect illegal commerce in the borders of countries, such as drug trafficking and commerce of prohibited and/or restricted items. The remote-controlled vehicle can be sanded to the strategic points, defined by the armed forces of each country, in order to monitor and send images to the command center. This information is useful to prepare early action measures.

In the civilian area, this tool can become relevant for an agricultural application. For instance, the vehicle could monitor large plantations and extract meaningful information about the land situation. Moreover, this tool can prevent any kind of pest and excess of fertilizer products or mineral salts over ideal conditions [9].

Some extra applications in the civilian area could be monitoring residences (condominiums), musical shows and sporting events.

5 Conclusion

The growth of UGV-based monitoring applications has allowed exploring places with restricted local area networks. The innovative applications in this area can be applied for various purposes, either military or civil.

The tool, developed by this work, improves remote video-based systems because applied different digital image processing techniques to extract relevant information about the recorded scene in order to guarantee satisfactory results. The tool enables to visualize hidden areas or images, which makes the operation more efficient and accurate.

As a future work, artificial intelligence-based pattern recognition techniques using the similarity of punctual marks obtained with facial recognition are at the stage of study and will be added to the system.

References

1. Wikipedia: Unmanned ground vehicle. Wikipedia. Available in: https://en.wikipedia.org/wiki/Unmanned_ground_vehicle. Access in 03 Oct 2017
2. Ribeiro, J.A.J.: Engenharia de Antenas: Fundamentos, Projetos e Aplicações, 1st edn, Érica (2012)
3. Cederholm, D.: CSS3 for Web Designers, 2nd edn, A Book Apart (2015)
4. Keith, J., Andrew, R.: HTML5 for Web Designers, 2nd edn, A Book Apart (2016)
5. Flanagan, D.: JavaScript: The Definitive Guide, 6th edn, O'Reilly Media (2011)
6. Cagle, K.: SVG Programming: The Graphical Web, 1st edn, Apress (2002)
7. Coyier, C.: CSS Filters. CSS-Tricks. Available in: https://css-tricks.com/almanac/properties/f/filter/. Access in 02 Sept 2017
8. Meehan, J.: Creative Optical & Digital Filter Techniques, 2nd edn, Lark Books (2010)
9. Medeiros, F.A., Alonço, A.S., Balestra, M.R.G., Dias, V.O., Júnior, M.L.L.: On the use of unmanned aerial vehicle for acquisition of georrefecend image data. Revista do Centro de Ciências Rurais **38** (2008)

Edge-Detection Noise-Smoothing Image Filter Techniques

Daniel Izario, Yuzo Iano, Bruno Izario, Diego Castro and Carlos Nazareth

Abstract In recent years, applications such as image filtering are becoming increasingly popular in social media and technological apps. For this reason, it is important to develop image-filtering techniques that can deal with inherent noise when taking photos. This type of filters can serve in the process of retouching photos. This paper proposes an edge-detection noise-smoothing image filter based on partial differential equations. According to the results, this technique improves recovering of a distorted image against to conventional techniques.

Keywords Image filter · Partial differential equation · Smoothing filters

1 Introduction

An image means the visual representation of a scene where objects or persons can appear. At the acquisition process, noise appears which results in a deterioration of the image. The two most frequent types of noises are the Gaussian and Impulsive [1]. To balance this degradation, image filters are used to reduce the impact of noise. These filters are classified between linear and nonlinear. The linear filter smooth and enhances image detail and minimize noise effect, and the nonlinear applies transformation process.

D. Izario (✉) · Y. Iano · B. Izario · D. Castro
University of Campinas, Campinas, SP, Brazil
e-mail: daniel_izario@hotmail.com

Y. Iano
e-mail: yuzo@decom.fee.unicamp.br

B. Izario
e-mail: bruno_izario@hotmail.com

C. Nazareth
National Institute of Telecommunications, Santa Rita do Sapucaí/MG, Brazil
e-mail: carlosn@inatel.br

Y. Iano et al. (eds.), *Proceedings of the 3rd Brazilian Technology Symposium*,
https://doi.org/10.1007/978-3-319-93112-8_13

The main filters available in the marker use the convolution of matrix masks for noise correction, such as median and Gaussian techniques. Edge-detection filters are based on differential operators such as Sobel, Roberts, Laplacian, and Prewitt, which can eliminate some noises present in images [2].

2 Proposed Method

The use of partial differential equations [3] in digital image processing improves the restoring process of the image by doing a mathematical modeling. This is an alternative to conventional filtering techniques. The results obtained improving the quality of the filtered images with minimum loss of resolution [4, 5].

The mathematical model involves an image function $\vartheta\left(x_i, y_j\right)$, where, "*i*" and "*j*" are the rows and columns of the image matrix. To obtain the differences that approximate the partial derivatives (the finite difference approximation method) require the following expressions:

$$\vartheta_x = \frac{\vartheta_{i+j} - \vartheta_{i-j}}{2}, \tag{1}$$

$$\vartheta_y = \frac{\vartheta_{ij+1} - \vartheta_{ij-1}}{2}, \tag{2}$$

$$\vartheta_{xx} = \frac{\vartheta_{i+2j} - 2\vartheta_{ij} + \vartheta_{i-2j}}{4}, \tag{3}$$

$$\vartheta_{yy} = \frac{\vartheta_{ij+2} - 2\vartheta_{ij} + \vartheta_{ij-2}}{4}, \text{ and} \tag{4}$$

$$\vartheta_{xy} = \frac{\vartheta_{i+j+1} - \vartheta_{i+j-1} - \vartheta_{i-j+1} + \vartheta_{i-j-1}}{4}. \tag{5}$$

The solution of the partial differential equation is obtained using the time scale, $t_n = n\Delta t$, where "*n*" represents the course of time in the application. The new function is

$$\vartheta\left(x_i, y_j, t_n\right). \tag{6}$$

Finally, to obtain the resulting image (I_{result}), it is necessary to apply (7), thus generating in the application an image filtered by the method of partial differential equations according to

$$I_{result} = \delta|\nabla\vartheta|div\left(\frac{\nabla\vartheta}{|\nabla\vartheta|}\right) + \sigma(1-\delta)\left(\vartheta - I_{original}\right). \tag{7}$$

Table 1 PSNR comparison of the proposed method with the mean filter

Image	Mean filter (dB)	Proposed method (dB)
New York City 4 K	30.85	44.12
Lamborghini Centenario Roadster	31.57	43.90
Average	31.21	44.01

3 Experimental Results

The application was developed in HTML [6] and JavaScript [7]. The first process is to convert from RGB to XYZ domain. The histogram is generated to prove the change made in the process, and then an edge-detection noise-smoothing image filter is applied to minimize noise. The main difference is the use of the partial differential equation technique, because of all the images are treated in the continuous domain.

The filter used as a reference is the mean filter [1] because reduces reducing the variance between neighboring pixels, which helps to reduce noise in images. The kernel often used in mean filtering as follows:

$$\mathrm{F}_{\text{Média}} = \frac{1}{9}\begin{vmatrix} 1 & 1 & 1 \\ 1 & 1 & 1 \\ 1 & 1 & 1 \end{vmatrix}. \tag{8}$$

Figures 1 and 2 show the two images that are part of the experimental tests. According to the Table 1, when applying the mean filtering technique, the average Peak Signal-to-Noise Ratio (PSNR) value reached was 31.21 dB but when applying the technique based on the partial differential equation; the PSNR value reached was 44.01 dB.

The RGB histogram comparison between the original image, the image after passes through the filter based on the partial differential equation and the image after passes through the mean filter is shown in Fig. 3. As can be seen, lower values of red color have the peakiest pixel values of the image. The histogram referenced to the proposed method looks very similar to the histogram of the original image, compared to the histogram that refers to the mean filter. The color more affected with respect to the histogram of the original image is the blue. In both cases, the mean value decays, increasing the local variance. Finally, the green component remains almost the same.

4 Conclusion

The use of existing filters for image processing using smoothing techniques for noise elimination and edge detection in images has proved to be efficient for a simple analysis of RGB images. However, the use of the technique presented in the article

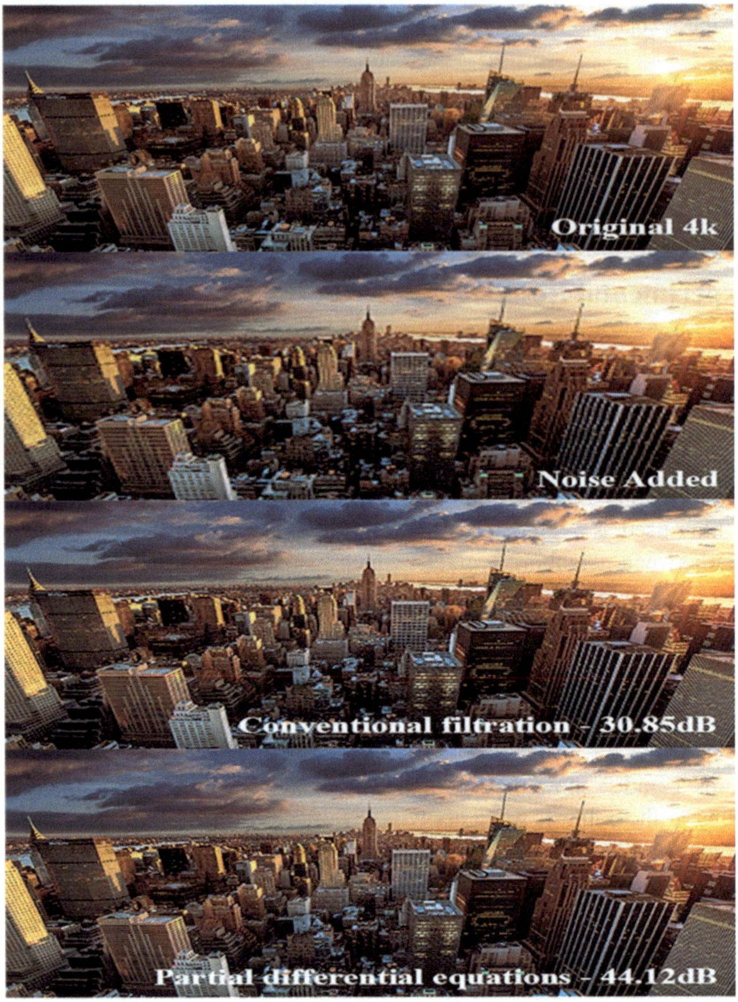

Fig. 1 "New York City 4 K" original image and the image with added noise and PSNR comparison

has a great advantage over these filters because it increases the accuracy of the color numerical process using a step-by-step conversion. To verify the results, the RGB histograms for each image were obtained and analyzed. On the average, there is a gain of 12.8 dB regarding objective quality.

In the future, it will be possible to improve the technique by including new algorithms that can have better results with edge detection and noise reducing. This will improve the acquisition of images from satellites, Unmanned Ground Vehicle (UGV), among other equipment.

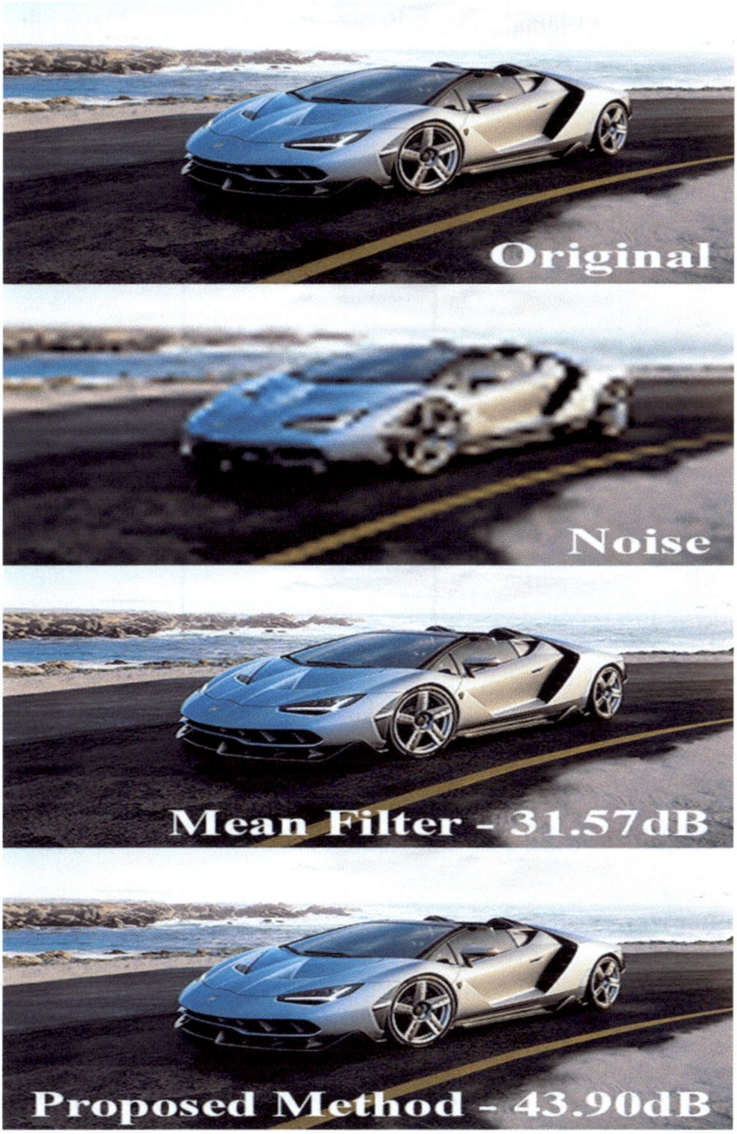

Fig. 2 "Lamborghini Centenario Roadster" original image and the image with added noise and PSNR comparison

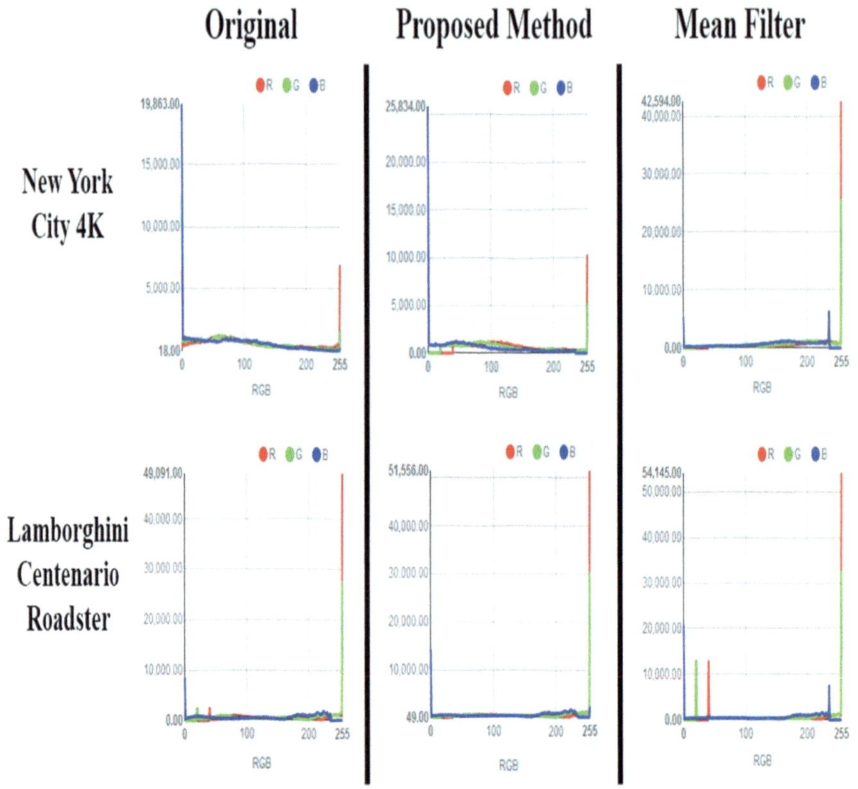

Fig. 3 RGB histogram comparison

References

1. Baxes, G.A.: Digital Image Processing: Principles and Applications, 1st edn. Wiley, (1994)
2. Gonzalez, R.C., Woods, R.E., Eddins, S.L.: Digital Image Processing Using Matlab, 2nd edn. Gatesmark Publishing, (2009)
3. Strauss, W.A.: Partial Differential Equations: An Introduction, 2nd edn. Wiley, (2008)
4. Pedrosa, G.V., Barcelos, C.A.Z.: Noise eliminations in images using partial differential equations. VIII Encontro Interno/XII Seminário de Iniciação Científica da Universidade Federal de Uberlândia, (2008)
5. Barcelos, C.A.Z., Boaventura, M., Júnior, E.C.S.: A well-balanced flow equation for noise removal and edge detection. IEEE Trans. Image Process, (2003)
6. Flanagan, D.: JavaScript: The Definitive Guide. 6th edn. O'Reilly Media, (2011)
7. Keith, J., Andrew, J.: HTML5 for Web Designers, 2nd edn. A Book Apart, (2016)

Conception of an Electric Vehicle's Robotic Platform Developed for Applications on CTS

Willian Gomes de Almeida, Juliano de Almeida Monte-Mor, Rafael Francisco dos Santos, Eben-Ezer Prates da Silveira, Sandro Carvalho Izidoro, Tarcísio Gonçalves de Brito, Natália Cosse Batista and Giovani Bernardes Vitor

Abstract Significant efforts have been made in last decades concerning intelligent systems, particularly in vehicular applications. Despite several solutions proposed for intelligent systems on the electrical vehicle platform, this work aims to demonstrate a novel approach for hardware and software interaction that allows the development of different modes of operation. Some capabilities include cooperative mode to fully autonomous mode and vice versa. Furthermore, this proposed embedded system

W. G. de Almeida (✉) · J. de Almeida Monte-Mor · R. F. dos Santos · E.-E. P. da Silveira · S. C. Izidoro · T. G. de Brito · N. C. Batista · G. B. Vitor
Laboratório Robótica, Sistemas Inteligentes e Complexos, UNIFEI, Itabira, MG, Brazil
e-mail: will.almeida@unifei.edu.br

J. de Almeida Monte-Mor
e-mail: jmontemor@unifei.edu.br

R. F. dos Santos
e-mail: rsantos@unifei.edu.br

E.-E. P. da Silveira
e-mail: eben@unifei.edu.br

S. C. Izidoro
e-mail: sandroizidoro@unifei.edu.br

T. G. de Brito
e-mail: tgbrito@unifei.edu.br

N. C. Batista
e-mail: nataliabatista@decom.cefetmg.br

W. G. de Almeida · J. de Almeida Monte-Mor · R. F. dos Santos · S. C. Izidoro · G. B. Vitor (✉)
Engenharia da Computação, UNIFEI, Itabira, MG, Brazil
e-mail: giovanibernardes@unifei.edu.br

E.-E. P. da Silveira
Engenharia Elétrica, UNIFEI, Itabira, MG, Brazil

T. G. de Brito
Engenharia Mecânica, UNIFEI, Itabira, MG, Brazil

N. C. Batista
Departamento de Computação, CEFET, Belo Horizonte, MG, Brazil

Y. Iano et al. (eds.), *Proceedings of the 3rd Brazilian Technology Symposium*,
https://doi.org/10.1007/978-3-319-93112-8_14

provides the basis for higher module integration from different levels of application. The proposed work reports real experiments carried out in a local urban environment using a medium electric vehicle platform which includes an embedded system with some embedded sensors to illustrate the validity of such platform in the context of the Cybernetic Transportation System (CTS).

Keywords Embedded systems · Hardware and software integration Instrumentation

1 Introduction

The increasing use of vehicles in urban environments has generated several problems. These include severe traffic congestions, oil dependency, pollution (environmental and noise), safety (accidents with fatalities), and general deterioration of the quality of life. Although the public transport system is a viable alternative, they still have drawbacks such as overcrowding, routes that do not serve certain regions, poor quality of services, etc. Therefore, the individual transport system (private car) remains the main option for many people in different places around the world.

A recent approach to mobility, which emerges as an alternative to the private car, is based on intelligent and automatic electric vehicles. The purpose is to provide the same flexibility with less inconvenience. These electric vehicles can be a solution for public transportation systems in specific areas and can also complement mass transit and non-motorized transportation by providing passenger service to any location and at any time. This new transport model is known as Cybernetic Transportation System (CTS) [1].

The CTS consists of a fleet of fully automated road vehicles called cybercars. These vehicles can carry passengers or goods in various ways, for example from station to station the shuttle service or in on-demand mode. All cybercars in a CTS are under control of a supervision system which manages traffic and transportation requests in order to improve people displacements. They can be used on private grounds, university camps, touristic sites exploitation, resorts, industrial parks, research platforms, among others [2]. The proposal of this work is to demonstrate a novel approach to hardware and software integration that allows the development and interaction of several modules in different application levels.

2 Related Work

Although this concept is not recent (in the past decade the first CTS system was put into practice in the Netherlands), some systems have now been put into operation [2, 3] presents an approach to ensure the safety of passengers and experiments for motion evaluation using a cybercar called robuRide, which is a fully autonomous electric vehicle that is able to transport up to 30 people. It has a control system based on three PC boards and contains a laser sensor. Localization is obtained using the gyroscope, GPS and odometric data. A public demonstration prototype CTS in Chinese cities, named CyberC3 is described by [4]. The CyberC3 includes three vehicles for up to six people. In addition, the system is centrally controlled (there is no communication between vehicles and stations) and it contains five stations, on-road monitoring cameras, vehicle camera, laser radar and ultrasonic sensors.

In Europe, different initiatives focus to boost research in this type of vehicle, trying to attain higher levels of autonomy. The main projects are CyberCar and CyberCar2 [5], CyberMove and CityMobil [6].

In Brazil, the LRM (Mobile Robotics Laboratory) group has researches involving CTS with the CARINA I vehicle [7]. This vehicle was instrumented with different sensors and actuators of stepping, acceleration and braking, not being mentioned what the system embarked on the accomplishment of these low-level controls.

3 Hardware Platform Description

It was tuned up a golf cart as the development platform of this cybernetic vehicle project, calling it as "Golfinho". It involved the adaptation of an E-Z-GO RXV electric vehicle, with dimensions of 240, 119 and 174 cm, respectively, for length, width and height. This RXV vehicle operates from a 48 VDC battery, with a continuous horsepower of 4.4 hp and a programmable speed of 13–24 kph. Figure 1 shows the proposal of our cybercar platform.

The vehicle steering has been automated with a fitting of an Akiyama NEMA 34 stepper motor into the vehicle steering column. This supports 100.0 kgf.cm, controlled by a driver Leadshine DMA860E. The RXV golf cart has electronic control

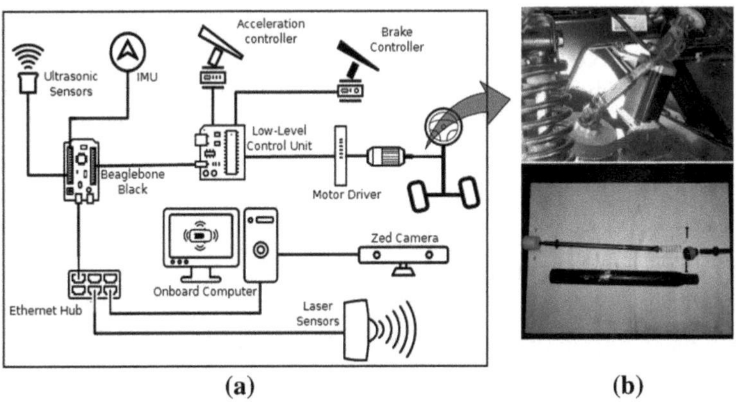

Fig. 1 Hardware architecture. **a** Golfinho hardware platform. **b** Mechanical steering system

of acceleration and brake, not requiring a mechanical adaptation of these components. The motor has been fitted to the outer front of the vehicle in which all the controls for the first steering system and pivot axis are transmitted as shown in Fig. 1b.

One important feature of this project is the possibility of the cooperative control of the vehicle. In order to achieve so, an low-level control unit was designed to allow the main central processing unit (CPU) commands and the human controls (e.g., steering wheel, brake pedals, throttle pedal, forward switch, backward switch and auto/manual switch) to communicate, independently or simultaneously, with the standard vehicle controller. This system can operate in three different ways, manual, automatic, and cooperative.

The interface is based on a TM4C123G Launchpad board, which the main features are: 80 MHz CPU ARM Cortex-M4, 32 KB RAM, and 256 KB flash. The human controls signals are monitored by the TM4C123G board, except by steering wheel, since it is mechanically connected to the steering gear box, only the feedback position signal is monitored by the interface board. Additional signal conditioning modules are necessary to meet the requirements of the analog and digital inputs of the micro-controller (limited to 3.3 V dc). Finally, the output pins of the micro-controller are connected to an output board to convert the 0–3.3 V signals to the levels required by the vehicle controller.

The board communication to the main CPU is by a serial interface to a BeagleBone Black (BBB) board (AM335x 1 GHz CPU ARM Cortex-A8, 512 MB RAM, 4 GB flash storage), with a customized protocol. BBB also controls electronic devices like ultrasonic sensors and Inertial Measurement Unit (IMU).

As shown in Fig. 2a, it was designed the following sensors to navigate avoiding obstacles: a 3D camera, five Sick laser sensors, and eight ultrasonic sensors. It may be used to perform data fusion, in order to increase the precision on autonomous navigation tasks. The images are captured from a ZED stereo camera. Further, it is drafted eight Positron PS210 car ultrasonic sensors to be installed (two per side), with a range of approximately 1.5 m each one.

To perform image processing and to manage and synchronize communication between sensors and actuators, it was inserted an onboard computer with Intel Core i7-4790 processor, 3.6 GHz, 16 GB RAM, and NVIDIA GeForce GTX 745 video card. An ethernet hub was used to provide communication between the BBB, onboard computer and laser sensors. Figure 2b shows the sensors scanning range (in meters), where the proposed platform is at the origin (0,0).

4 Experimental Results

The experimental results of this research work have been carried out in the dependencies of the Federal University of Itajub´a, advanced campus of Itabira-MG. Several experimental tests have been done to evaluate both platform design and embedded system architecture. Here, it is presented a test using a teleoperation module allowing human operators to control the electrical vehicle platform, called "Golfinho", through

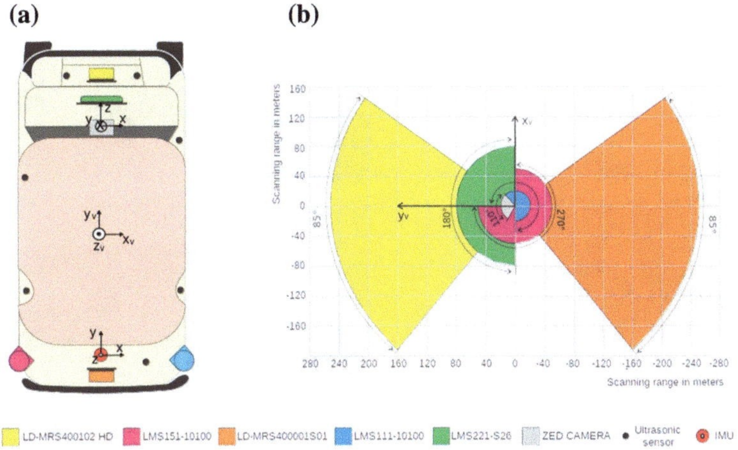

Fig. 2 Golfinho sensors diagram and scanning range

the proposed embedded system. The teleoperation module was designed using the mobile app development conception, integrating these devices into the proposed architecture. Its intention is to show the modularity and flexibility of such system. The communication management is realized by the ROS, under a local network.

In Fig. 3, it is demonstrated the path that was realized by the Golfinho‘s experimentation. This test had a duration of approximately 20 min, traveling 385 m from start point up to stop point. In this experiment, in addition to the steering and velocity command sent by the teleoperation module, a data acquisition was performed to demonstrate the sensors that were embedded in the platform at the present moment. In this Figure, it is illustrated the data obtained from the zed camera and the Sick LD-MRS laser to three distinct points on the traveled path. Further, it was added two grabbed images of the external environment of the Golfinho from a video sequence realized together with the test, at A and B points. As can be seen, the mechanical adaptation, the electronic board, and the embedded system are working satisfactory, conform planned and developed.

From data acquisition, Fig. 4 shows four signals that were saved along with the traveled path. The linear acceleration, angular velocity, and orientation signals illustrated in Fig. 4a–c were obtained from IMU sensor and the last one, Fig. 4d is the steering command sent by the teleoperation module. It should be mentioned that this signal is different from the real steering response of the adapted mechanical steering system. It is only the desired steering angle. Regarding these signals, some features may be highlighted, although the data are noisy. The existent gap observed in Fig. 4d, at time interval 778–863 s, was related to the communication loss between the mobile device and the platform. Thus, these actions reflected the signals obtained from the IMU sensor. At the time interval 109–255 s, the slight curves observed in Fig. 4a, c, from y-axis and x-axis respectively, are related to the Golfinho‘s inclination traveling by a sloping road. It is clear to see in Fig. 4c that the blue signal presents the various

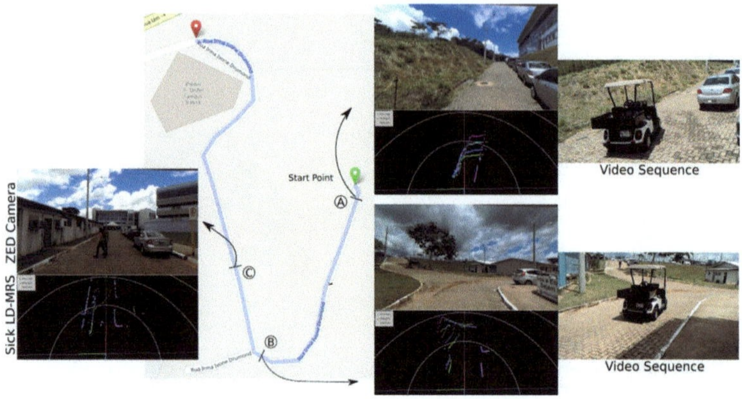

Fig. 3 The traveled path from the proposed platform. The Golfinho traveled a distance of 385 m with a duration of approximately 20 min

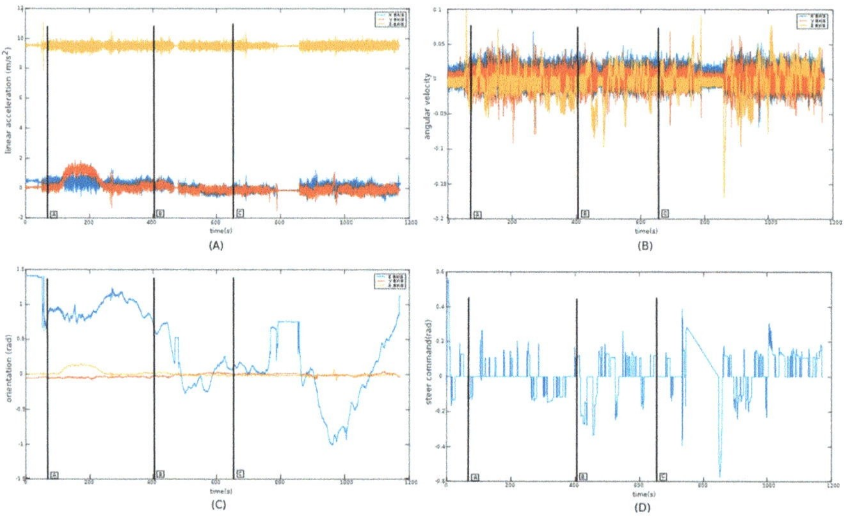

Fig. 4 Data acquisition from proposed architecture along the path. **a** Linear Acceleration, **b** Angular Velocity, **c** Orientation and **d** Steering command from teleoperation module

orientation changes that were reflected on the z-axis along the path, ensuring that the desired steering command sent by the teleoperation module was acting on the platform. At the last, the vertical black lines showed in Fig. 4 highlight the time instant that the camera and laser sensors from Fig. 3 was acquired. Demonstrating the A, B and C points.

5 Conclusion

This paper presented a novel approach to hardware and software integration that allows the development and interaction of several modules in different application levels, using a medium electric vehicle platform. The platform includes an embedded system with some embedded sensors and has been validated in the context of Cybernetic Transportation System. Actual experiments in the open environment were performed in a way that validated the operations and results of the developed platform. As support for teleoperation, an application for Android system (mobile device) has been developed, allowing to control the steering and speed of the vehicle from a distance. Future works will be directed to the development of a CAN network to make the whole system more modular, which will facilitate the addition of the new devices, and the implementation of algorithms to make the vehicle autonomous using sensor fusion, machine learning, and other state-of-the-art techniques.

References

1. Rocha, R., Cunha, A., Varandas, J., Dias, J.: Towards a new mobility concept for cities: architecture and programming of semiautonomous electric vehicles. Ind. Rob.: Int. J. **34**(2), 142149 (2007). https://doi.org/10.1108/01439910710727496
2. Lucet, E., Canou, J., Dupourque, V.: Safety aspects in Cybercars design and application. Technical Report, Robosoft (2018)
3. Wang, F., Yang, M., Yang, R.: The Intelligent Vehicle Coordination of the Cybernetic Transportation System. Int. J. Adv. Rob. Syst. vol. 6. Sage (2009). https://doi.org/10.5772/6771
4. Xia, T., Yang, M., Yang, R., Wang, C.: CyberC3: a prototype cybernetic transportation system for urban applications. In: IEEE Transactions on Intelligent Transportation Systems, vol. 11, pp. 142–152. IEEE. (2010)
5. Naranjo, J.E., Bouraoui, L., Garcia, R., Parent, M., Sotelo, M.A.: Interoperable control architecture for cybercars and dual-mode cars. IEEE Trans. Intell. Transp. Syst. **10**(1), 146154 (2009)
6. Shaheen, S.A., Chan, N.D.: Evolution of EMobility in Carsharing Business Models, pp. 169–178. Springer International Publishing, Cham (2015). https://doi.org/10.1007/978-3-319-12244-1 10
7. Fernandes, L.C., Souza, J.R., Pessin, G., Shinzato, P.Y., Sales, D., Mendes, C., Prado, M., Klaser, R., Magalha˜es, A.C., Hata, A., Pigatto, D., Branco, K.C., Grassi, V., Osorio, F.S., Wolf, D.F.: Carina intelligent robotic car: architectural design and applications. J. Syst. Architec. **60**(4), 372–392 (2014). http://www.sciencedirect.com/science/article/pii/S1383762113002841

Pulse Shaping Filter Design for Filtered OFDM Transceivers

Jaime J. Luque Quispe and Luís G. Pedroso Meloni

Abstract CP-OFDM waveforms have been used successfully in the physical layer of 4G systems to overcome the channel selectivity and to provide good spectrum efficiency and high transmission data rates. 5G technology aims to support more communication services than its predecessor and considers many other scenarios such as the use of fragmented spectrum. However, this diversity of services cannot be supported by CP-OFDM since it applies a single set of parameters to the entire band to meet a particular service and it has large out-of-band spectral emission, which can affect transmissions in adjacent bands. Thus, in order to improve the spectrum containment, new waveforms more flexible than CP-OFDM have been proposed which make use of pulse shape filters. UFMC is a waveform that splits the bandwidth into several subbands which can be allocated to services with a particular set of parameters each. However, since the maximum length of UFMC filters is limited to size of CP, their stopband attenuation level can be not enough to meet spectrum emission mask requirements. This work presents a filter called semi-equiripple which has variable stopband decay rate, better frequency response than equiripple, and windowed *sinc* filters and good impulse response characteristics for reduction of ISI. The performance of the proposed filter is assessed in terms of the power spectrum density, spectrum mask compliance, MER, and operation in a multiservice asynchronous scenario.

Keywords OFDM · Filter design · Pulse shaping · 5G

J. J. Luque Quispe (✉) · L. G. Pedroso Meloni
School of Electrical and Computer Engineering,
University of Campinas, Campinas, Brazil
e-mail: luque@decom.fee.unicamp.br

L. G. Pedroso Meloni
e-mail: meloni@decom.fee.unicamp.br

Y. Iano et al. (eds.), *Proceedings of the 3rd Brazilian Technology Symposium*,
https://doi.org/10.1007/978-3-319-93112-8_15

1 Introduction

Orthogonal frequency division multiplexing (OFDM) has been successfully used in 4G long-term evolution (LTE) to overcome the selectivity of the channel and to boost the spectrum efficiency. However, its high out-of-band emission (OOBE) and constraints of orthogonality and synchronism make it unsuitable for 5G scenarios such as ultra reliable and ultra-low latency communications, massive machine type communications, and enhanced mobile broadband with ubiquitous connectivity. In order to meet the requirements of future wireless applications, the 5G communication system requires an innovative air interface with a waveform more spectrally agile than OFDM and more flexible to support multiple kinds of services [1, 2].

The OOBE results from the modulation process as the subcarriers pass through rectangular filters thus originating abrupt discontinuities between the OFDM symbols, i.e., the symbols are windowed in the time domain by rectangular functions which generate large *sinc* spectral sidelobes outside the allocated bandwidth [3], thus a potential source of interference to adjacent channels. This problem has been addressed in several current systems by employing guard subcarriers, but this approach decreases the spectrum efficiency (e.g., in LTE the guard band is 10% of the total bandwidth). Several techniques to reduce such emission were summarized in [4], but they assume orthogonality and synchronism across the entire bandwidth to avoid intercarrier interference (ICI). The synchronization processes require significant control signaling thus creating network congestion. On the other hand, new waveforms with better flexibility and performance than the conventional cyclic-prefixed OFDM (CP-OFDM) have been proposed in [5–9]. They are based in splitting and filtering the bandwidth on a per subcarrier or on a per subband basis, where the filters play a key role in mitigating the inter-subband interference (ISBI) caused by the abandonment of the full-band synchronicity and orthogonality. Based on the idea of subband splitting of universal filtered multicarrier (UFMC) transceivers [8], a waveform able to multiplex different transmissions can be realized by configuring the subbands with suitable sets of parameters, e.g., subcarrier spacing, CP size, transmission time interval (TTI), etc., as is illustrated in Fig. 1.

This work presents the design of filters for low OOBE and shows that the length of the filter impulse response can be extended beyond the size of cyclic prefix (CP) to improve the spectral containment with little bit error rate (BER) penalty related to the intersymbol interference (ISI). They have been called semi-equiripple here because of its equiripple passband and non-equiripple stopband characteristics. Three sources

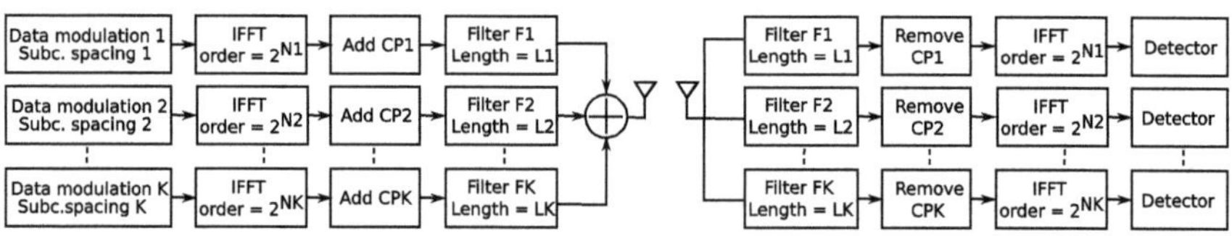

Fig. 1 Block diagram of a subband-based filtering OFDM transceiver

of ISI are analyzed: the symbol dispersion due to the filter group delay, the symbol echoes due to the passband ripples, and the fading of ISI given by the tails of the impulse response. It is found that the semi-equiripple filters provide better rejection of ISI than the equiripples and can exhibit better stopband mask than windowed truncated *sinc* filters. The remainder of this work is structured as follows: Sect. 2 addresses the filter design method. Section 3 focuses on the modeling of ISI. Section 4 is devoted to performance analysis using numerical simulations. Section 5 presents our conclusions.

2 Filter Design

The type of filter chosen is a finite impulse response (FIR) filter because of its linear-phase response which introduces the same delay to all OFDM subcarriers, so it does not cause ICI. With regard to the filter design method, windowing is not able to support arbitrary trade-offs of the parameters transition bandwidth, ripple amplitudes and order since the frequency response is defined only by the type of window employed. On the other hand, the Parks–McClellan method [10] enables to perform such trade-offs and is briefly described below.

2.1 *Optimal Design Method*

This method computes the set of coefficients of a FIR filter using the Chebyshev approximation and the Remez algorithm. It approximates the actual response $H(e^{j\omega})$ of a filter as close as possible to a desired ideal response $T(e^{j\omega})$. A frequency-domain weighted error function that measures the deviation between those responses is defined as

$$E(e^{j\omega}) = W(e^{j\omega})\left[H(e^{j\omega}) - T(e^{j\omega})\right], \tag{1}$$

where $W(e^{j\omega})$ is a weighting function. Thus, the set of coefficients is one that minimizes the maximum error given by the Chebyshev norm.

$$\left\|E(e^{j\omega})\right\|_{\infty} = \max_{\omega\in[0,\pi]} \left|W(e^{j\omega})\left[H(e^{j\omega}) - T(e^{j\omega})\right]\right|. \tag{2}$$

For a low-pass FIR, $T(e^{j\omega})$ and $W(e^{j\omega})$ can be defined as

$$T(e^{j\omega}) = \begin{cases} 1 & 0 < \omega < \omega_c \\ 0 & \omega_c < \omega < \pi \end{cases}, \tag{3}$$

$$W(e^{j\omega}) = \begin{cases} k_p & 0 \leq \omega \leq \omega_p \\ 0 & \omega_p < \omega < \omega_s \\ k_s & \omega_s \leq \omega \leq \pi \end{cases}, \tag{4}$$

where ω_c is the cutoff frequency, ω_p is the passband edge frequency and ω_s is the stopband edge frequency. In order to obtain the best solution for a type I FIR filter of order $2M$ and from the alternation theorem [10], the weighted error $E(e^{j\omega})$ alternates sign and has maximum absolute values on at least $M+2$ points ω_k with $0 \leq \omega_1 < \cdots < \omega_{M+2} \leq \pi$, therefore

$$E(e^{j\omega_k}) = (-1)^k \left\| E(e^{j\omega}) \right\|_\infty, \tag{5}$$

where $k \in \{1, 2, \ldots, (M+2)\}$. By replacing (1) in (5) and letting $\delta = \left\| E(e^{j\omega}) \right\|_\infty$, (5) can be expressed as the system of $M+2$ linear equations

$$H(e^{j\omega_k}) - \frac{(-1)^k \delta}{W(e^{j\omega_k})} = T(e^{j\omega_k}), \tag{6}$$

where $H(e^{j\omega})$ is defined by only $M+1$ coefficients of the filter. To obtain such coefficients in (6), first, the best set of frequencies ω_k has to be determined. Eq. (6) is solved for an initial set of $M+2$ points uniformly spaced between 0 and π. Then, this set is updated such that the error alternates sign in the new reference set and that the local maximum values of the error obtained from (2) exceeds the current value of $|\delta|$. These steps are performed until the following inequality is satisfied

$$\frac{\left\| E_i(e^{j\omega}) \right\|_\infty - |\delta_i|}{\left\| E_i(e^{j\omega}) \right\|_\infty} < \varepsilon \approx 10^{-6}, \tag{7}$$

thus, indicating that the Chebyshev error has been minimized.

2.2 Non-equiripple Stopband Attenuation

Wireless standards commonly define spectrum emission masks (SEMs) to limit adjacent channel interference (ACI) [11, 12]. Similar specifications could be also beneficial to reduce the ISBI in subband-based filtering waveforms operating with asynchronous multiple access schemes. In some systems, such as LTE, pulse shaping filters are not defined, opening the possibility to make trade-offs between the filter parameters (e.g., order, transition bandwidth, and size of the passband ripples) in order to meet a required performance (e.g., spectral efficiency, SEM, and ACI). Taking advantage of this fact, semi-equiripple FIR filters can be designed by modifying the constant weighting function $W(e^{j\omega})$ in (4), which produces a stopband decay of 0 dB/octave, to be non-decreasing and expressed as

Fig. 2 Weighting function in the frequency domain

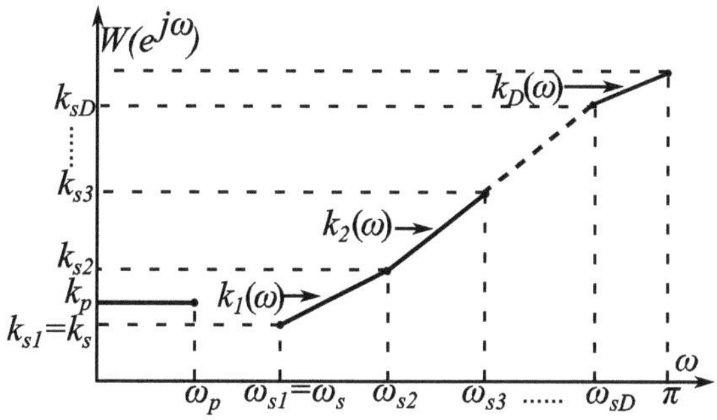

$$W(e^{j\omega}) = \begin{cases} k_p & 0 \leq \omega \leq \omega_p \\ 0 & \omega_p < \omega < \omega_s \\ k_s & \omega = \omega_s \\ k(\omega) & \omega_s < \omega \leq \pi \end{cases}, \tag{8}$$

where k_s is the weight to obtain the desired attenuation at the stopband edge and $k(\omega)$ is an increasing function to shape the stopband and can be defined as a piece-wise function (Fig. 2) to obtain different attenuation levels. In this way, (6) converges with small error sizes at frequency points with higher weights.

In Fig. 2, the stopband is divided into D sub-bands, each with an arbitrary weighting function $k_i(\omega)$, where $i = 1, 2, 3, ..., D$. Then, the reference stopband frequencies ω_{si} and weights k_{si} can be obtained to meet any SEM. Each function $k_i(\omega)$ can be obtained by interpolation of the weights at ω_{si} and ω_{si+1}. Using linear interpolation, $k_i(\omega)$ can be expressed as

$$k_i(\omega) = k_{si} + m_i \cdot (\omega - \omega_{si}), \tag{9}$$

where $m_i = (k_{si+1} - k_{si}) / (\omega_{si+1} - \omega_{si})$. The stopband can also be shaped with exponential decay ($1/f^{\gamma}$ or 6γ/octave) by defining $k(\omega)$ as

$$k(\omega) = k_s \left(\frac{\omega}{\omega_s} \right)^{\gamma}. \tag{10}$$

The performance of this filter is studied later in Sect. 4.

3 Analysis of ISI

Pulse shaping filters with narrow transition bands are useful for spectral containment. However, for the case of an equiripple filter, if the order and transition band are fixed, a specification of a higher stopband attenuation is exchanged by a higher level of passband ripple (in-band distortion). However, it was observed by simulation that

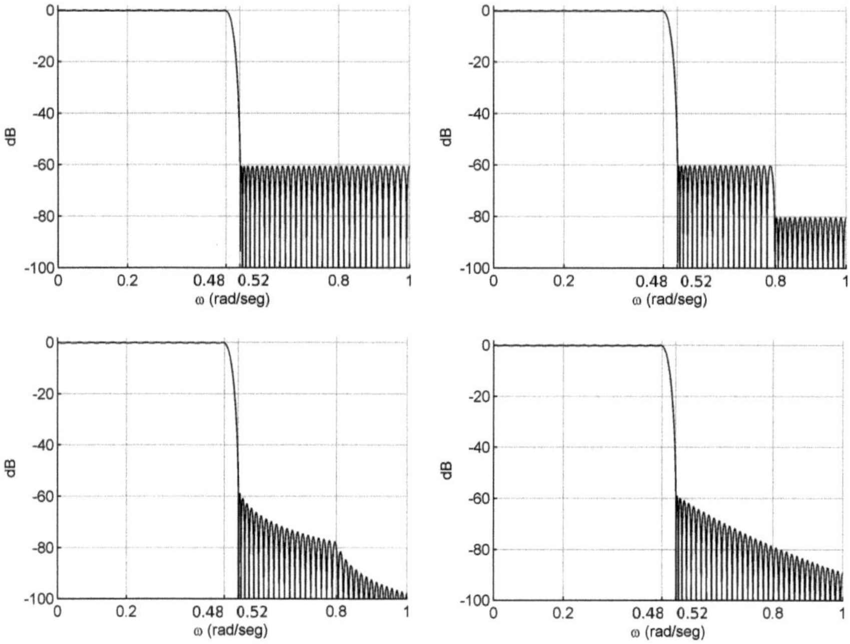

Fig. 3 Filters with different weighting functions $k(\omega)$: constant for equiripple response (top left), staircase (top right), linear (bottom left) and exponential for stopband decay $1/f^{\gamma}$ (bottom right). For all the filters: order = 130, passband edge $\omega_p = 0.48$, stopband edge $\omega_{s1} = 0.52$, $k_p = 1$, $\gamma = 5.35$

Table 1 Passband ripples for different weighting functions used in Fig. 3

Filter	# of iterations	δ	$\left\|E(e^{j\omega})\right\|_{\infty}$
1	11	0.00844060401	0.00844060401
2	13	0.0086394292	0.0086394292
3	11	0.0117381235	0.0117381235
4	10	0.0110966284	0.0110966284

the above-mentioned weighting functions for semi-equiripple filters do not lead to significant increasing of such level. For the filters shown in Fig. 3, which are of the same order and transition band but designed with different weighting functions, the actual passband ripples after the convergence of (6), are very similar, as is shown in Table 1.

The effect of these ripples is discussed below in terms of ISI. Let $H(e^{j\omega})$ be expressed as

$$H(e^{i\omega}) = \left|H(e^{i\omega})\right| e^{i \arg\{H(e^{i\omega})\}} = \left|H(e^{i\omega})\right| e^{-j\omega \cdot n_d}, \tag{11}$$

where n_d is the filter group delay, $\left|H(e^{i\omega})\right|$ is the magnitude response and $\arg\left\{H(e^{i\omega})\right\}$ is the phase response. After the convergence of (6), the ripple peaks positions are approximately uniformly spaced. Therefore, the ripples can be modeled as a sine function

$$H(e^{i\omega}) = [1 + |\delta| \cos(\varphi\omega)]e^{-j\omega n_d}, \quad \text{for}\, 0 \leq \omega \leq \omega_p, \tag{12}$$

where $|\delta|$ is the maximum passband ripple magnitude and φ is the frequency of such ripples given by $\varphi = 2\pi \cdot n_p / \omega_p$, being n_p the number of cycles of ripple in half of the filter bandwidth. Letting $x[n]$ be the input signal and $X(e^{j\omega})$ its Fourier transform, the output $Y(e^{j\omega})$ can be obtained as

$$Y\left(e^{i\omega}\right) = H(e^{i\omega})X(e^{i\omega}) = \frac{\sigma}{2}X(e^{j\omega})e^{-j\omega(n_d-\varphi)} + \\ X(e^{j\omega})e^{-j\omega n_d} + \frac{\sigma}{2}X(e^{j\omega})e^{-j\omega(n_d+\varphi)},$$

that in time domain can be written as

$$y[n] = \underbrace{\frac{\sigma}{2}x[n-(n_d-\varphi)]}_{pre\ echo} + \underbrace{x[n-nd]}_{delayed\ input} + \underbrace{\frac{\sigma}{2}x[n-(n_d+\varphi)]}_{post\ echo}, \tag{13}$$

where $y[n]$ is composed of a main component and two echoes shifted by $\pm\varphi$ at both sides which generate interference back and forth between the symbols, i.e., ISI. In order to analyze the ISI, only one path of the transceiver in Fig. 1 is taken into account. The received lth OFDM symbol can be expressed as

$$\mathbf{Y}^l = \mathbf{H}^{(0)}\mathbf{X}^l + \mathbf{H}^{(1)}\mathbf{X}^{l-1}, \tag{14}$$

where $\mathbf{X}^l$ and $\mathbf{Y}^l$ are vectors representing the filter input and output, respectively, and $\mathbf{H}^{(0)}$ and $\mathbf{H}^{(1)}$ are the Toeplitz matrices whose elements are the coefficients of the equivalent between the transmitter and receiver filters (total length $L = 2L_1 - 1$). After the CP removal, the symbol is

$$\tilde{\mathbf{Y}}^l = \left[\mathbf{0}_{N\times N_{CP}}\mathbf{I}_{N\times N}\right]\mathbf{H}^0\mathbf{X}^l + \left[\mathbf{0}_{N\times N_{CP}}\mathbf{I}_{N\times N}\right]\mathbf{H}^1\mathbf{X}^{l-1}, \tag{15}$$

where the first term contains the three components in (13) and the second term is the ISI from the previous $(l-1)$th symbol. To eliminate the second term, the CP length needs to be greater than $L-1$, otherwise ISI occurs. Furthermore, from (12), only small distortions are introduced if the filter has low order, low frequency φ and small ripple amplitude $|\delta|$.

4 Simulation Results

This section is composed of four parts in which the performance of the filtered OFDM transceiver is analyzed in terms of the power spectrum density (PSD) and compliance with arbitrary SEM, favorable characteristics of the impulse response for ISI rejection, modulation error rate (MER) in a layered OFDM waveform and bit error rate (BER) performance in a multiservice asynchronous scenario in which different filters for ISBI containment are compared. It aims to show the attenuation of the OOBE, ISI, and ISBI provided by filtering. Signal models were developed for 6-MHz ISDB-T and 5-MHz LTE systems which are used through the analysis below.

4.1 Spectral Emission Mask Compliance

The OFDM waveform considered in this part (and also in Sects. 4.2 and 4.3) follows the ISDB-T standard [11] whose layers (subbands) A, B, and C have been set with modulations QPSK, 64-QAM, and 16-QAM, respectively. ISDB-T also defines three mask types (non-critical, sub-critical, and critical) to limit the interference between adjacent 6-MHz channels. A semi-equiripple filter was designed to meet the critical SEM as is shown in Fig. 4. The stopband is shaped with an exponential function ($\gamma = 29.8$). Any other SEM could be obtained by just changing the weighting function in Fig. 2.

4.2 Impulse Response Characteristics

The attenuation rate of the Fourier transform of a function depends on how many continuous derivatives have its time-domain representation. Letting this number be $(p-1)$, the attenuation rate is proportional to $f^{-(p+1)}$.

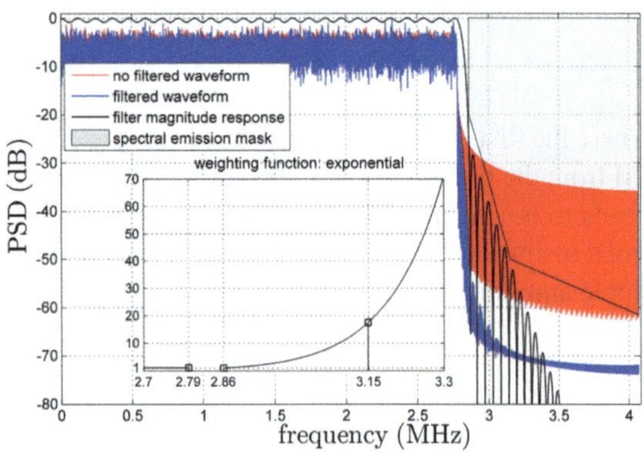

Fig. 4 Spectrum of a ISDB-T waveform, simulation with 100 OFDM symbols

Since an equiripple filter exhibit an attenuation rate of 0 dB, then $p = -1$, which means that its function has a continuous double integral and, therefore, unit impulses, which leads to the generation of outlier samples in the tails of its impulse response. On the other hand, semi-equiripple filters can have faster attenuation, and thus smoother impulse responses and smaller filter tails. The size of the outlier samples is particularly important because they might serve to attenuate the ISI echoes in (13) introduced by the filtering of a symbol, i.e., since the ISI in the symbol l is result of the convolution of the last samples of the previous symbol $l-1$ with the filter tail samples, the ISI can be attenuated by small filter tails which can increase the SIR of the filtered received signal.

An illustration of the impulse responses of semi-equiripple, equiripple, and windowed truncated *sinc* filters with the same number of coefficients is provided in Fig. 5. The filters were designed to meet the critical SEM, except the equiripple filter, which has constant stopband sidelobe level. It is observed that the outlier samples in the semi-equiripple filter are smaller than in equiripple and since the Kaiser filter exhibit more attenuation decay in frequencies near to the main lobe, see Fig. 6, their outlier samples are the smallest of the three filters. However, the semi-equiripple filter continues having much better stopband characteristics.

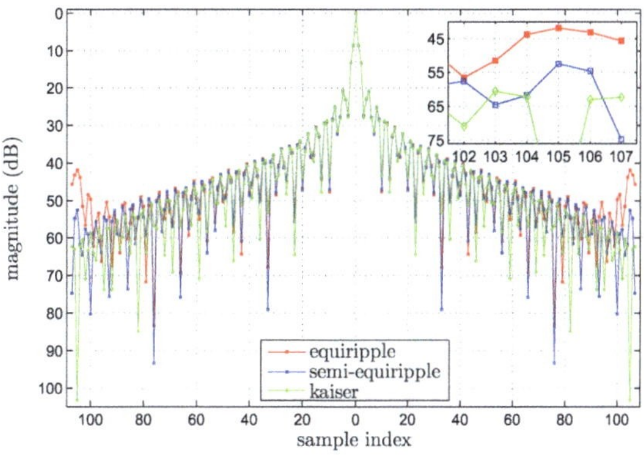

Fig. 5 Impulse responses of the equiripple, semi-equiripple, and windowed *sinc* filters. Filter order = 214 and $\delta = 0.5\%$ for the three filters. Semi-equiripple filter with exponential decay $\gamma = 29.8$. Kaiser window factor = 3.1

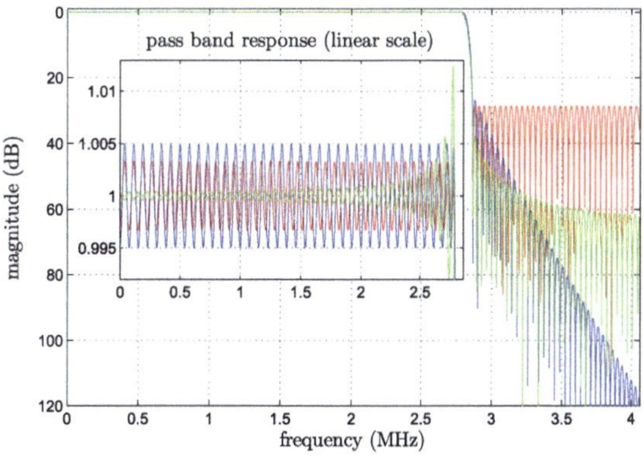

Fig. 6 Magnitude frequency responses of the filters in Fig. 5

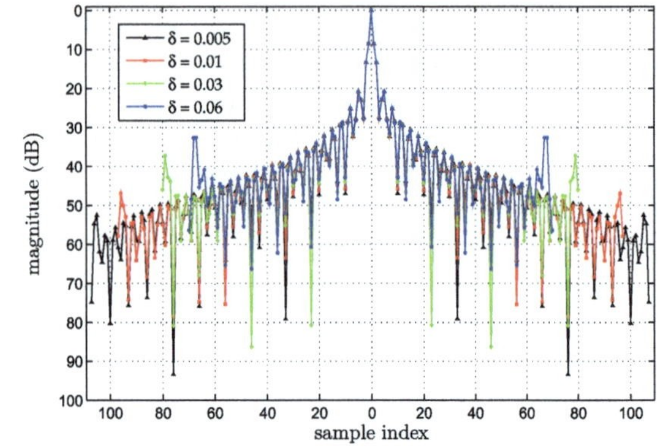

Fig. 7 Influence of passband ripples on the size of tail samples in semi-equiripple filters. Filters with exponential attenuation decay $\gamma = 29.8$

Moreover, as well as the stop-band attenuation rate, the variation rate of the in-band spectral content is also related to the smoothness of the impulse response. For instance, Fig. 7 depicts the influence of the size of the passband ripple δ on the size of the outlier samples at tails of four semi-equiripple filters with the same attenuation decay rate. It is observed that a small δ leads to smaller outlier samples, thus rejecting better the ISI.

4.3 Effect of Filtering on Modulation

The modulation accuracy is evaluated using MER measurements for several semi-equiripple filters of the same decay rate (exponential) but of different orders. The filter order is represented by a parameter called filter index such that a smaller value of it indicates that the filter is longer (or has smaller passband ripple). Each filter is applied to the entire bandwidth in the transmitter and receiver. Taking into account only the effect of the filters, the receiver sees the equivalent filter response as if it were the channel response. The MER in the transmitter (TX) considers only the TX filters and the MER in the receiver (RX) considers both TX and RX filters, which are of the same type. The MER calculated in different layers over a sequence of symbols is shown in Fig. 8. It is observed that the MER is more influenced by δ than by the filter order despite of more ISI samples are introduced, so it is possible to use long filters for better OOB emission with no significant degradation of the modulation performance. The equivalent filter will exhibit approximately twice the size of the passband ripple so the sizes of the outlier samples are expected to increase as well, thus amplifying the ISI as is shown in Fig. 8. It is also observed that the interference caused by filtering is similar for the three layers.

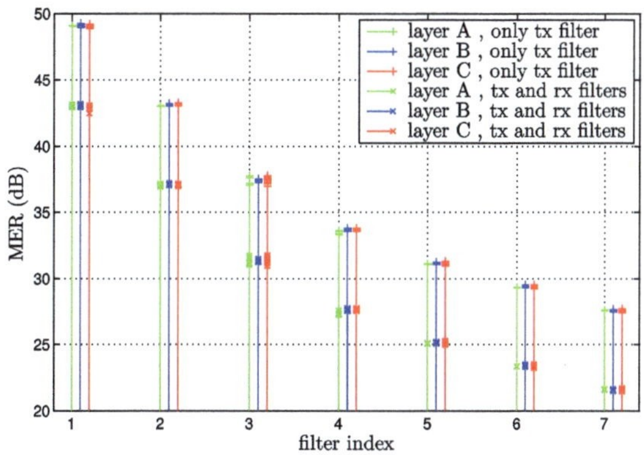

Fig. 8 MER by OFDM symbol averaged over all layer subcarriers for different shaping filters. Filters with different passband ripples but with the same attenuation rate $\gamma = 29.8$. Transmission of a burst sequence of 10 symbols in mode 1 with CP = 1/8, layer a (QPSK), layer b (64-QAM), and layer c (16-QAM)

4.4 BER Performance of Subband-Based Filtering OFDM

In order to analyze the effect of filtering on the reduction of the ISBI in a subband-based filtering transceiver, it is considered a single waveform composed of two parallel streams in adjacent subbands. Each subband occupies a fraction of a 5-MHz bandwidth and has independent parameters such as bandwidth, subband filter, subcarrier spacing, CP length, and transmission time interval (TTI). The values of the main parameters are listed in Table 2 and other unlisted values are set to follow the LTE standard.

The simulation has been performed considering that the streams corresponding to independent services (service 1 and 2) are separated in frequency by 60 kHz (the

Table 2 Waveform parameters

Parameter	Subband 1	Subband 2
Duplex mode	FDD	
Sampling rate	7.68 MHz	
Bandwidth (kHz)	750	780
FFT size	512	128
Subcarrier spacing (kHz)	15	60
Number of subcarriers	48	12
IFFT subcarrier index	$[-24, -23, \ldots, 24]$ (except 0)	$[8, 9, \ldots, 19]$
Symbol duration (μs)	66.67	16.67
TTI (ms)	1	0.1
Symbol per TTI	14	5
CP length	5.2 μs (sym. 1) 4.7 μs (sym. 2~7)	3.7 μs (sym. 1) 3.3 μs (sym. 2~5)
Modulation (QAM)	64	64

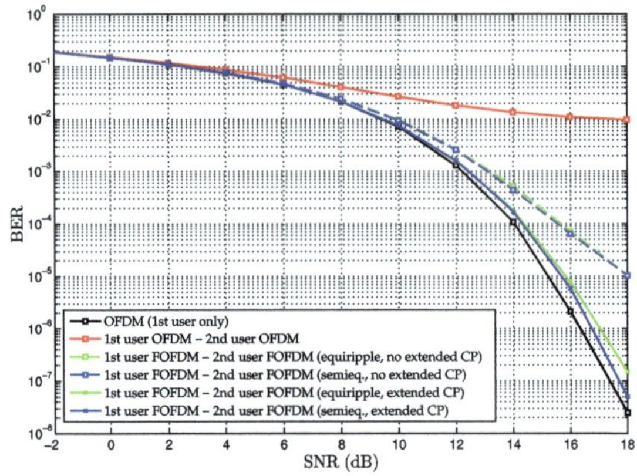

Fig. 9 BER of a subband filtered-OFDM receiver affected by ACI of other active filtered subband. Δpower = 0 dB, time offset = 150 samples, extended CP = 100. Equiripple filters with stopband attenuation of –20 dB, order = 210. Semi-equiripple with exponential attenuation that starts at –20 dB and has a decay factor $\gamma = 4.3$, order = 236. Both filters have a transition bandwidth of 45 KHz and $\delta = 1.25\%$

subcarrier spacing of service 2), subcarrier in IFFT index 0 is null and the channel is AWGN. A time delay is inserted between the streams to simulate asynchronous transmission. For a given service, the same filter is used at both transmitter and receiver sides. However, the filters of different services are different according with the bandwidth of each service. Figure 9 presents the BER results for the service 1 when equiripple and semi-equiripple filters are used in each service bandwidth. The cases when no suppression of OOBE is performed (conventional OFDM, worst case) and ACI from service 2 is not present (best case) are also included for comparison. The results show that subband filtering improves the receiver BER because it attenuates the ISBI. It is observed a BER improvement for the semi-equiripple filter in relation to the equiripple case despite the filter order is higher for semi-equiripple filters. The reason for that is that the ISI is better rejected by filters with smoother impulse response. Semi-equiripple filters do not exhibit large impulse response tails which serve to attenuate the ISI. It can be also observed in Fig. 9 an improvement in performance when the CP is extended and shifted to contain the filter tails causing ISI back and forth between adjacent OFDM symbols [7].

5 Conclusions

This work presented the design and analysis of semi-equiripple filters for spectral containment in subband-based filtering waveforms. The design uses the Parks–McClellan method with a configurable weighting function to obtain arbitrary stopband decay rates, as for instance, linearly and exponentially decaying stopbands. It is found that impulse responses longer than the CP size can be used for a lower OOBE with slight BER penalty. Semi-equiripple filters provide better reduction of OOBE than the equiripple and windowed *sinc* filters, and because of their faster stopband attenuation, they have smoother impulse responses, thus smaller outlier samples, useful for ISI rejection. It was also observed that the direct relation between the size of the passband ripples and the filter tail samples. Since the filter frequency response can be tailored to meet tight spectrum masks, different services can be accommodated within a single waveform without significant wastage of spectrum for guard bands. BER simulations for a scenario with asynchronous and non-orthogonal transmission showed that the ISBI can be contained by the subband filters, thus facilitating asynchronous multiplexing of users and services which is of interest in current 5G physical layer research.

Acknowledgements The authors gratefully acknowledge the financial support of CAPES, the research funding agency of the Brazilian government.

References

1. Talwar, S., Choudhury, D., Dimou, K., Aryafar, E., Bangerter, B., Stewart, K.: Enabling technologies and architectures for 5G wireless. In: IEEE MTT-S International Microwave Symposium (2014)
2. Wunder, G., Jung, P., Kasparick, M., Wild, T., Schaich, F., Chen, Y., Brink, S.T., Gaspar, I., Michailow, N., Festag, A., Mendes, L., Cassiau, N., Ktenas, D., Dryjanski, M., Pietrzyk, S., Eged, B., Vago, P., Wiedmann, F.: 5GNOW: non-orthogonal, asynchronous waveforms for future mobile applications. In: IEEE Communications Magazine (2014)
3. Luque, J., Meloni, L.: OFDM pulse shaping for use in 5G systems. In: XXIII Brazilian Symposium of Telecommunications (2015)
4. Loulou, A.E., Gorgani, S.A., Renfors, M.: Enhanced OFDM techniques for fragmented spectrum use. In: Future Network and Mobile Summit, pp. 1,10 (2013)
5. Fettweis, G., Krondorf, M., Bittner, S.: GFDM—generalized frequency division multiplexing. In: IEEE 69th Vehicular Technology Conference, pp. 1–4 (2009)
6. Farhang-Boroujeny, B.: OFDM versus filter bank multicarrier. IEEE Sig. Process. Mag. **28**(3), 92–112 (2011)
7. Abdoli, J., Jia, M., Ma, J.: Filtered OFDM: a new waveform for future wireless systems. In: 16th International Workshop on Signal Processing Advances in Wireless Communications, pp. 66–70 (2015)
8. Vakilian, V., Wild, T., Schaich, F., ten Brink, S., Frigon, J.F.: Universal-filtered multi-carrier technique for wireless systems beyond LTE. In: IEEE Globecom Workshops, pp. 223–228 (2013)
9. Zhang, X., Jia, M., Chen, L., Ma, J., Qiu, J.: Filtered-OFDM—enabler for flexible waveform in the 5th generation cellular networks. In: IEEE Global Communications Conference (GLOBECOM), pp. 1–6 (2015)
10. Oppenheim, A.V.: Discrete-Time Signal Processing. Prentice Hall (2010)
11. ABNT NBR 15601—Digital terrestrial television. Transmission system
12. 3GPP TS 36.104 Evolved Universal Terrestrial Radio Access

Digital TV (ISDB-Tb) Broadcasting over LTE Broadcast: A Feasibility Study

Cibele A. Makluf, Julio León and Yuzo Iano

Abstract The Brazilian Digital Television System, based on ISDB-Tb, is currently using the 700 MHz band. This frequency band is ideal for LTE due to its large area coverage and lower implementation costs for service providers. This chapter aims to demonstrate the technical feasibility of broadcasting Digital TV multimedia content over LTE Broadcast in a similar way ISDB-Tb broadcasts using the Brazil Channels and GSM Channel broadcast profiles by comparing the bit error rate over a set of different SNR scenarios.

Keywords LTE · MBMS · ISDB · Digital Television

1 Introduction

The Brazilian Digital Television System (SBTVD) has been in operation since 2007. It was based on the Integrated Services Digital Broadcasting-Terrestrial (ISDB-T) standard with improvements developed by several Brazilian Universities, therefore, called ISDB-TB (Integrated Services Digital Broadcasting-Terrestrial version B). This system was also adopted by several countries in Latin America, enabling high definition content broadcasting, multiprogramming and interactivity [1].

LTE Broadcast (Long-Term Evolution), also known as Evolved Multimedia Broadcast/Multicast Service (eMBMS) or Multicast-Broadcast Single-Frequency Network (MBSFN), is a mobile technology developed to provide transport resources for broadcasting to many users in a single cell (multicast) also using a subset of radio resources to support unicast services at the same time.

C. A. Makluf (✉) · J. León · Y. Iano
School of Electrical and Computer Engineering,
University of Campinas, Campinas, SP, Brazil
e-mail: cibeleam@decom.fee.unicamp.br

J. León
e-mail: j.leon@decom.fee.unicamp.br

Y. Iano
e-mail: yuzo@decom.fee.unicamp.br

Y. Iano et al. (eds.), *Proceedings of the 3rd Brazilian Technology Symposium*,
https://doi.org/10.1007/978-3-319-93112-8_16

Literature shows several works analysing MBSFN's performance [2–4] compared to other transmission methods using point-to-point and point-to-multipoint individually on each cell. These studies concluded that MBSFN is the most efficient mechanism for multicast communication leading to its standardisation by the 3GPP. In 2012, [5] studied further the performance of MBSFNs, linking the spectral efficiency directly with the chosen Modulation & Coding Scheme (MCS). Moreover, [6] performed analytic studies based on cost analysis to determine the ideal number of cells to optimise global performance of MBSFN transmissions.

The adoption of ISDB-Tb in Brazil brought up a problem: the 700 MHz band is currently being used by some government TV channels, and is in direct competition with LTE for it. According to the studies done by the Brazilian Television Engineering Society (SET) in partnership with the Presbyterian Mackenzie University [7, 8], the 700 MHz band assigned to LTE would render LTE technology much cheaper for service providers (lowering area coverage expansion); and end consumers, as LTE devices would be globally compatible and would benefit even further from economies of scale. This is due to the fact that current fourth-generation services (HSDPA+) in Brazil operate on the 2.5 GHz band, which has four times less reach than the 700 MHz band.

The National Telecommunications Agency in Brazil (ANATEL) decided to auction the 700 MHz band to the mobile operators in order to improve the 4G service nationwide. In order to achieve this, in 2014, ANATEL started to gradually shutdown the analogue TV signals and reallocating the digital TV channels in the 700 MHz band to other frequency bands. Complete migration is expected to finish by the end of 2018.

Nevertheless, having Digital TV broadcasted over LTE would mean the best of both worlds, with a much efficient spectrum usage and many more possibilities. This paper aims to demonstrate the technical feasibility of broadcasting Digital TV multimedia content over LTE Broadcast in a similar way ISDB-Tb broadcasts using the Brazil Channels and GSM Channel broadcast profiles.

The rest of this chapter is divided as follows: Section 3 presents the Physical Layer (PHY) characteristics of ISDB-Tb and LTE. Section 4 presents the simulations used for the comparison, with their corresponding parameters. Section 5 presents the results of said simulations, and Section 5 contains the conclusions of this work.

2 Physical Layer Characteristics

ISDB-Tb ISDB-Tb uses the Band Segment Transmission-Orthogonal Frequency-Division Multiplexing (BST-OFDM) to allow operation flexibility and robustness against frequency-selective fading. Flexibility comes from band segmentation, which allows the system to provide different services on the same channel simultaneously—this is called hierarchical transmission. Robustness comes from the use of multiple carriers and the guard interval between them [9]. Nevertheless, there are some problems using BST-OFDM, basically due to the transmission channel affecting each of the subcarriers differently [10].

The Brazilian standard specifies that BST-OFDM must divide the 6 MHz channel into 13 segments of 6/14 MHz $\approx$ 428.571 kHz, resulting in a total channel bandwidth of $6/14 \times 13 \approx 5.57$ MHz.

It is possible to operate in three transmission modes with different number of carriers using the same frequency band. The usable transmission rate does not depend on operation mode (i.e. number of carriers), but rather on the channel coding or guard interval parameters. This usable transmission rate can vary from 280.85 kbps (with OFDM's maximum redundancy) up to 1787.28 kbps (with mininum redundancy on the OFDM segment) [9].

Table 1 shows the summary of the main PHY parameters for ISDB-Tb.

LTE Broadcast LTE's aerial interface uses two different techniques for multiple access based on OFDM: for downlink, Orthogonal Frequency-Division Multiple Access (OFDMA), and for uplink, Single-Carrier Frequency-Division Multiple Access (SC-FDMA). This work is entirely focused on the downlink part (for Digital TV broadcasting).

OFDMA is an extension of OFDM specifically developed for multiple user access, which is the main requirement for mobile communication systems [11]. The LTE standard supports the QPSK, 16QAM and 64QAM modulation schemes, whose choice depends on the actual carriers allocation (i.e. modulation depends on which carrier is assigned to each user). This ensures system robustness while also enabling higher transmission rates since the carriers with the best channel conditions are used. One of LTE's features is spectral flexibility—LTE's bandwidth can vary from 1.4 to 20 MHz.

Table 1 PHY layer parameters for ISDB-Tb [9]

Parameter	Value
Layers	Up to 3
OFDM segments	13
OFDM segment bandwidth	428.571 kHz
OFDM frame length	204 symbols
IFFT sampling frequency	8.13 MHz
IFFT samples	2048 (mode 1)
	4096 (mode 2)
	8192 (mode 3)
Carriers per OFDM segment	108 (mode 1)
	216 (mode 2)
	432 (mode 3)
Guard interval	1/4, 1/8, 1/16, 1/32
Internal code rate	1/2, 2/3, 3/4, 5/6, 7/8
Modulation schemes	DQPSK, QPSK, 16QAM, 64QAM

Table 2 Typical LTE downlink parameters [12]

Parameter	Bandwidth (MHz)					
	1.4	3	5	10	15	20
Used bandwidth (MHz)	1.08	2.7	4.5	9	13.5	18
Guard interval (MHz)	0.32	0.3	0.5	1	1.5	2
Guard interval (%)	23	10	10	10	10	10
Sampling frequency (MHz)	1.92	3.84	7.68	15.36	23.04	30.72
FFT samples	128	256	512	1024	1536	2048
Used subcarriers	72	180	300	600	900	1200

Besides that, LTE also uses the Multiple-Input Multiple-Output (MIMO) technology, which takes advantage of multiple antenna arrays. The concept of multiple antennas has become very popular for mobile communications; specially with the great increase in data rates, larger coverage areas and intersymbolic interference mitigation. LTE can use the following antenna configurations: SISO (Single Input Single Output), SIMO (Single Input Multiple Output), MISO (Multiple Input Single Output) and MIMO.

Moreover, three cyclic prefixes are specified in the LTE standard [12]: normal (4.7 μs), extended (16.67 μs) with subcarrier spacing of 15 kHz and extended (33 μs) with subcarrier spacing of 7.5 kHz. The 7.5 kHz spacing can only be used on a multicast/broadcast context.

The normal cyclic prefix is appropriate for most urban and suburban transmissions, with typical propagation delays in these environments. Since the time used by each modulated OFDM symbol is around 66.7 μs, the normal cyclic prefix has an overhead of circa 7%. The extended (16.67 μs) cyclic prefix overhead is around 25%, which is considered very large and is necessary for rural environments with higher propagation delays and broadcast services.

Table 2 shows a summary of the general parameters of LTE Broadcast PHY Layer.

3 Simulations

We ran several simulations on two different types of models (one for LTE and one for ISDB-Tb), with the objective of testing the robustness and maximum transmission rates against multipath fading and AWGN channels.

3.1 LTE

We developed a model for LTE's PHY Layer on MATLAB R2015b, shown as a block diagram on Fig. 1.

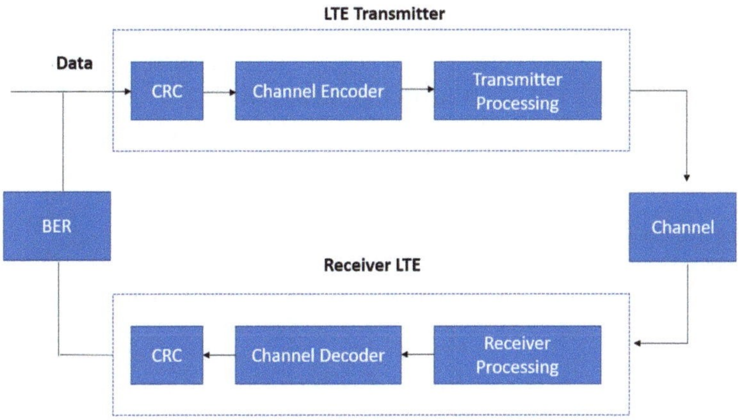

Fig. 1 LTE downlink MATLAB model block diagram

The LTE PHY Layer model is based on three main parts: the transmitter, the channel and the receiver. The BER block calculates the bit error rate by comparing the received bits with the transmitted ones. The Channel block was modelled as an AWGN channel with multipath using typical parameter values for broadcasting systems, using Brazil's Channels A, B, C, D and E for fixed reception, and for mobile reception the GSM Typical Urban Channel.

3.2 *ISDB-Tb*

We used the ISDB-Tb simulator developed by [13], implemented in C/C++. The simulator is closed source, and it simulates the PHY Layer of the modulator/demodulator in ISDB-Tb. It also allows to monitor the inputs and outputs of every block and export them to MATLAB, C and FPGA. Figure 2 shows the block diagram of the model used.

3.3 *Simulation Parameters*

LTE We ran our simulations for different SNR values measuring the BER in order to test LTE's performance on different degrading channels. For these tests, we used

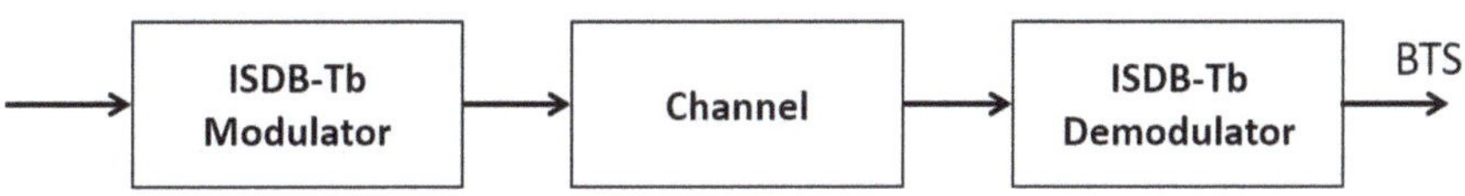

Fig. 2 ISDB-Tb high level model block diagram

Table 3 Comparison between the main parameters of LTE and ISDB-Tb used in the simulations

Parameter	LTE	ISDB-Tb
Antenna configurations	SISO, SIMO 1×2, SIMO 1×4	1×1
Modulation	64QAM	64QAM
Bandwidth	5 MHz	6 MHz
Code rate	1/3, 1/2, 3/4	1/2, 2/3, 3/4, 5/6, 7/8
Sampling frequency	7.68 MHz	8.13 MHz
Guard interval	4.7 μs	1/16
Channel model	Brazil A, B, C, D, E; GSM	Brazil A, B, C, D, E; GSM

the most robust modulation—64QAM. We also tested the performance of different antenna arrangements in SIMO and SISO: 1×2 and 1×4. MISO and MIMO were not in the scope of this work, as it would not be fair to compare it to ISDB-Tb which only uses SISO 1×1. Likewise, we used 5 MHz bandwidth only, which is the closest to the 6 MHz channel bandwidth used in ISDB-Tb.

According to 3GPP TS 36.213, LTE can have the following code rates: 1/3, 1/2, and 3/4 [14]. We used a normal cyclic prefix (4.7 μs) using 14 symbols per subframe, 15 kHz subcarrier spacing, and a sampling frequency of 7.68 MHz. The channel was modeled as an AWGN multipath channel using Brazil Channel and GSM Typical Urban Channel parameters.

Finally, we used MMSE for the MIMO receptor equalization mode due to its high performance. Channel estimation was done using interpolation techniques.

ISDB-TB We used the closed-source simulator to measure BER for different SNR values as well, using 64QAM for ISDB-Tb. Since ISDB-Tb does not support MIMO, we used a SISO configuration (1×1) and a Reed Solomon encoder with the following code rates: 1/2, 2/3, 3/4, 5/6, and 7/8. We used Mode 3 as the operation mode, with 8192 carriers and 13 OFDM segments. Finally, the guard interval was 1/16, with 6 MHz channel bandwidth and 8.13 MHz as the sampling frequency.

Parameter Comparison Table 3 presents a comparison between the main parameters used in the LTE and ISDB-Tb simulations.

4 Results

The output of both simulators can be interpreted as SNR versus BER curves. By varying the parameters in Table 3, we can generate curves for each possible scenario and compare the two technologies. Since there were several curves to be analysed, we have condensed them into tables presented in this section.

The SNR used for the simulations considered our target BER threshold, 3×10^{-6}, which is the receiver threshold for correct video decoding without artefacts [15].

The comparison between technologies was done using code rates that were common to both LTE and ISDB-Tb: 1/2 and 3/4. The performance gain was calculated using the difference of SNR between the curves of the same code rate at the BER interest threshold: lower required SNR to achieve target BER represents better performance.

Table 4 show the performance gains obtained by comparing LTE with ISDB-Tb for Brazil Channels and GSM Channel with 3×10^{-6} as the BER threshold.

4.1 Brazil Channel A

Table 4 shows performance gains between LTE and ISDB-Tb for Brazil Channel A. LTE was unable to reach the BER threshold on 1/2 and 3/4 for SISO and SIMO 1×2 while ISDB-Tb was able to do so, showing that for those cases ISDB-Tb is more robust. However, with SIMO 1×4, LTE showed a 3.5 dB SNR gain over ISDB-Tb.

4.2 Brazil Channel B

Table 5 shows performance gains between LTE and ISDB-Tb for Brazil Channel B. LTE was still unable to reach the BER threshold on 1/2 and 3/4 for SISO and SIMO 1×2. ISDB-Tb was successful on those as well. However, with SIMO 1×4, LTE showed a 4 dB SNR loss compared to ISDB-Tb. ISDB-Tb was better than LTE on every case.

4.3 Brazil Channel C

Table 6 shows performance gains between LTE and ISDB-Tb for Brazil Channel C. LTE was still unable to reach the BER threshold on 1/2 and 3/4 for SISO. LTE

Table 4 Brazil Channel A

Modulation	Antenna	Code rate	SNR (dB)			Best
			LTE	ISDB-Tb	Δ	
64QAM	SISO	1/2	–	13.5	−13.5	ISDB-Tb
		3/4	–	17	−17	ISDB-Tb
64QAM	SIMO 1×2	1/2	–	13.5	−13.5	ISDB-Tb
		3/4	–	17	−17	ISDB-Tb
64QAM	SIMO 1×4	1/2	9	13	4	LTE
		3/4	14	17	3	LTE

Table 5 Brazil Channel B

Modulation	Antenna	Code rate	SNR (dB)			Best
			LTE	ISDB-Tb	Δ	
64QAM	SISO	1/2	–	13.5	−13.5	ISDB-Tb
		3/4	–	17	−17	ISDB-Tb
64QAM	SIMO 1×2	1/2	–	13.5	−13.5	ISDB-Tb
		3/4	–	17	−17	ISDB-Tb
64QAM	SIMO 1×4	1/2	15	14.5	−0.5	ISDB-Tb
		3/4	25	17	−8	ISDB-Tb

Table 6 Brazil Channel C

Modulation	Antenna	Code rate	SNR (dB)			Best
			LTE	ISDB-Tb	Delta	
64QAM	SISO	1/2	–	13.5	−13.5	ISDB-Tb
		3/4	–	17	−17	ISDB-Tb
64QAM	SIMO 1×2	1/2	16	13	−3	ISDB-Tb
		3/4	–	17	−17	ISDB-Tb
64QAM	SIMO 1×4	1/2	9	14	5	LTE
		3/4	13	17	4	LTE

managed to reach the BER threshold for SIMO 1×2, but only with 1/2 as code rate, where it showed a SNR gain of 2 dB. For SIMO 1×4, LTE showed a 4.5 dB SNR gain over ISDB-Tb.

4.4 Brazil Channel D

Table 7 shows performance gains between LTE and ISDB-Tb for Brazil Channel D. LTE was unable to reach the BER threshold on 1/2 and 3/4 for SISO and SIMO 1×2. For SIMO 1×4, LTE showed an average 4 dB SNR loss compared to ISDB-Tb. ISDB-Tb was better than LTE on every case.

4.5 Brazil Channel E

Table 8 shows performance gains between LTE and ISDB-Tb for Brazil Channel E. LTE managed to reach the BER threshold for every case. For SISO, ISDB-Tb had an

Table 7 Brazil Channel D

Modulation	Modulation	Modulation	SNR (dB)			Best
			LTE	ISDB-Tb	Δ	
64QAM	SISO	1/2	–	13.5	−13.5	ISDB-Tb
		3/4	–	17.5	−17.5	ISDB-Tb
64QAM	SIMO 1×2	1/2	–	13.5	−13.5	ISDB-Tb
		3/4	–	17.5	−17.5	ISDB-Tb
64QAM	SIMO 1×4	1/2	16	13	−3	ISDB-Tb
		3/4	26	18	−8	ISDB-Tb

Table 8 Brazil Channel E

Modulation	Antenna	Code rate	SNR (dB)			Best
			LTE	ISDB-Tb	Δ	
64QAM	SISO	1/2	20	13	−7	ISDB-Tb
		3/4	25	17.5	−7.5	ISDB-Tb
64QAM	SIMO 1×2	1/2	11	13	2	LTE
		3/4	17.7	17.5	−0.2	ISDB-Tb
64QAM	SIMO 1×4	1/2	9	13	4	LTE
		3/4	15	17.5	2.5	LTE

average SNR gain of 7 dB. SIMO 1×2 shows an SNR gain of 2 dB for LTE on 1/2, while both technologies are tied on 3/4. For SIMO 1×4, LTE showed a 3 dB average SNR gain over ISDB-Tb.

4.6 GSM Channel

Table 9 shows performance gains between LTE and ISDB-Tb for GSM Channel. LTE was unable to reach the BER threshold on for SISO. However, LTE managed to reach the BER threshold for both SIMO configurations. Nevertheless, ISDB-Tb proved to be better for SIMO 1×2, while LTE still excelled in SIMO 1×4, with an average SNR gain of 3 dB.

5 Conclusions

We were able to do a fair comparison between both LTE and ISDB-Tb physical layers by using similar antenna configurations such as SISO 1 × 1 and current broadcast profiles used in Brazil. ISDB-Tb proved better in every test; furthermore, LTE was

Table 9 GSM Channel

Modulation	Antenna	Code rate	SNR (dB)			Best
			LTE	ISDB-Tb	Δ	
64QAM	SISO	1/2	–	13.5	−13.5	ISDB-Tb
		3/4	–	17	−17	ISDB-Tb
64QAM	SIMO 1×2	1/2	15	14.5	−0.5	ISDB-Tb
		3/4	27	17	−10	ISDB-Tb
64QAM	SIMO 1×4	1/2	9	13	4	LTE
		3/4	16	18	2	LTE

not able to reach target BER threshold in some of the tests. This may be due to the multipath echoes being larger than the LTE guard intervals. However, LTE showed SNR gains over ISDB-Tb when taking advantage of multiple antennas at the reception. This is even more interesting considering that we were trying to compare LTE to ISDB-Tb by using only 1 antenna for transmission. We expect that tests using more antennas (MIMO), or greater bandwidth (thus removing our self-imposed restriction) would show greater gains over ISDB-Tb.

References

1. Digital Broadcasting Expert Group (DIBEG). ISDB-T (2017). http://www.dibeg.org/
2. 3GPP R1-071049, Spectral Efficiency comparison of possible MBMS transmission schemes: additional results. Tech. Rep. (2007)
3. 3GPP R1-071433, Additional Results on eMBMS Transmission Configurations. Tech. Rep. (2007)
4. 3GPP- R1-070051, Performance of MBMS Transmission Configurations. Tech. Rep. (2007)
5. Alexiou, A., Bouras, C., Kokkinos, V., Tsichritzis, G.: Performance evaluation of LTE for MBSFN transmissions. Wirel. Networks **18**(3), 227 (2012). https://doi.org/10.1007/s11276-011-0341-z
6. Alexiou, A., Bouras, C., Kokkinos, V., Papazois, A., Tsichritzis, G.: Modulation and coding scheme selection in multimedia broadcast over a single frequency network-enabled long-term evolution networks. Int. J. Commun. Syst. **25**(12), 1603 (2012). https://doi.org/10.1002/dac.1326
7. ANATEL, Relatório do Teste em Campo sobre a Convivência do LTE na faixa de 700 MHz com ISDB-T. Tech. Rep. (2014)
8. Sociedade Brasileira de Engenharia de Televisão (SET), Relatório dos Testes de Interferência do Sinal LTE na TV digital na faixa de UHF. Tech. Rep., São Paulo (2013)
9. Paiva, M.: Uma implementação em software do subsistema de transmissão do padrão ISDB-TB. Msc, Instituto Nacional de Telecomunicações (2010)
10. Takada, M., Saito, M.: Transmission system for ISDB-T. Proc. IEEE **94**(1), 251 (2006). https://doi.org/10.1109/JPROC.2005.859692
11. M. Mendes, Utilização da estimativa de canal (sounding) na alocação de recursos de radio no enlace reverso (Uplink) de redes Long Term Evolution—LTE. M.Sc. Dissertation (2009)
12. Ghosh, A., Zhang, J., Andrews, J.G., Muhamed, R.: Fundamentals of LTE. Pearson Education (2010)

13. Akamine, C.: Contribuições para Distribuição. Universidade Estadual de Campinas, Modulação e Demodulação do Sistema Brasileira de TV Digital ISDB-Tb. Ph.D. Thesis (2011)
14. 3GPP, LTE; Evolved Universal Terrestrial Radio Access (E-UTRA); Physical layer procedures (3GPP TS 36.213 version 8.8.0 Release 8). Tech. Rep. Release 8 (2009)
15. Yamada, F., Sukys, F., Bedicks, Jr, G., Akamine, C.: Parte I - Sistema de TV Digital. Revista Mackenzie de Engenharia e Computação **283**, 99 (2010)

A Packet Scheduling Algorithm with Traffic Policing in LTE Networks

Jeanette Quiñones Ccorimanya and Lee Luan Ling

Abstract The packet scheduling mechanism is a very important technical issue for long-term evolution (LTE) systems. In this paper, we improve the performance of traditional packet scheduling algorithms in LTE systems for multimedia service applications using traffic policing strategies known as leaky bucket regulation. We seek and achieve fairness between classes of services, controlling the arrival rates of packets to transmission queues of the scheduler. Simulation results confirm that schedulers with policed input traffic provide better performance for real-time services, especially those involving video traffic. The performance evaluation was carried out in terms of quality of service parameters, such as packet delay, packet loss ratio, and system throughput for video and VoIP traffic.

Keywords Long-term evolution (LTE) · Traffic policing · Packet scheduler Quality of service (QoS)

1 Introduction

The growing demand for mobile devices offering Internet access has provoked an increase in the volume of real-time data traffic in mobile networks. In this context, the wireless telecommunications industry has defined a new air interface of mobile communications known as long-term evolution (LTE) promoted by the standard of communication 3ra Generate Partnership Project (3GPP) that provides high data rates, low latency, and packet-optimized radio access. The LTE system employs orthogonal frequency division multiplexing access (OFDMA) technology in downlink transmission, where the bandwidth is divided into multiple orthogonal subcarriers, allowing parallel information transmission and a flexible management of radio resources [1].

J. Q. Ccorimanya (✉) · L. L. Ling
Department of Communications, State University of Campinas - UNICAMP,
Campinas, SP 13083-852, Brazil
e-mail: jeanetqc@decom.fee.unicamp.br

L. L. Ling
e-mail: lee@decom.fee.unicamp.br

Y. Iano et al. (eds.), *Proceedings of the 3rd Brazilian Technology Symposium*,
https://doi.org/10.1007/978-3-319-93112-8_17

The radio access network (RAN) for LTE is known as evolved universal terrestrial access network (E-UTRAN) consisting mainly of evolved node B (eNodeB) so-called base stations [2] where all functions of radio resource management (RRM) are executed. One of the most important functions of RRM is the dynamic allocation of radio resources for users in the downlink direction performed by the packet scheduler. The packet scheduler is responsible for careful selection of users (UE) so that the air interface resources can be effectively used. In this paper, we proposed a strategy to improve the performance of two scheduler algorithms, M-LWDF (Modified Largest Weighted Delay First) and EXP/PF (Exponential Proportional Fair), for multimedia service applications using the traffic policing algorithms, with aim of minimizing packet delay of the LTE system while maintaining a desirable throughput. We compared the performance of these scheduling algorithms based on the quality of service parameters for real-time traffic applications, such as video and VoIP, in terms of packet delay, throughput and packet loss rate. The paper is organized as follows. In Sect. 2, we present the traffic policing algorithms. In Sect. 3, we describe the packet schedulers. In Sect. 4, we propose a new scheduling algorithm for resource allocation for LTE networks. Sect. 5 is dedicated to the experimental research and performance analysis of the scheduling algorithms. Finally, we conclude in Sect. 6.

2 Traffic Policing Algorithm

The traffic policing algorithm is an important tool in network architecture. It controls the entrance of data packets into the network, establishing the connection for those well-behaved ones while discarding or attending with low priority those badly behaved ones [3]. In this section, we describe the some traffic policing algorithms widely reported in the literature.

2.1 Leaky Bucket—LB

The control mechanism of the leaky bucket algorithm (LB) can be interpreted as a sequential test to analyze the behavior of an input traffic flow in order to establish and adjust data transmission conformities [3, 4]. This LB algorithm is illustrated in Fig. 1 showing the conceptual analogy between the leaky bucket water storage system and the data stream transmission scheme. The leaky bucket modifies the input flow process in terms of the bucket storage condition and the output flow rate determined by the hole size [5]. When the bucket is full, the overflow of the bucket (or buffer) will cause the loss of water (data). If we assume that the bucket is initially empty ($t = 0$), the maximum amount of work processed by the leaky bucket at time t is given by

$$LB(t) = \bar{a}t + S \tag{1}$$

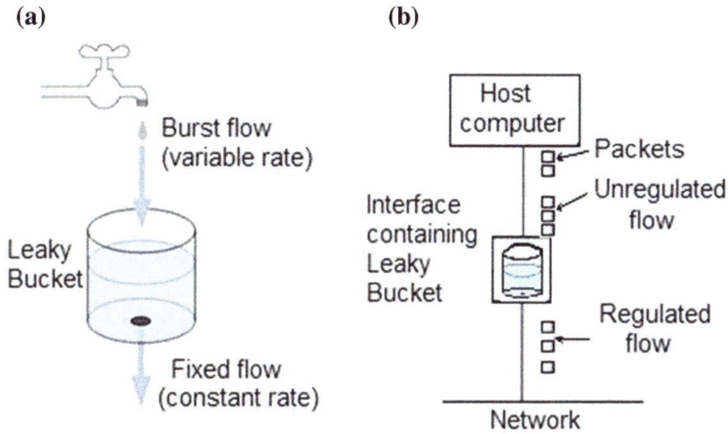

Fig. 1 **a** Leaky bucket water control **b** Leaky bucket data traffic control

where $\bar{a}$ is the average rate of the incoming traffic and S is the size of the bucket.

2.2 *Fractal Leaky Bucket—FLB*

The fractal leaky bucket (FLB) is a traffic policing mechanism introduced in [6] based on the fBm (fractional Brownian motion) envelope process concept. The fBm model is considered accurate in describing traffic flow with monofractal characteristics [4]. The maximum amount of work accepted by the fractal leaky bucket is given by the equation

$$FLB(t) = \bar{a}t + k\sigma t^H + S \tag{2}$$

where H is the Hurst parameter [7], σ represents the standard deviation of the traffic rate, and k is a constant related to the probability of violation ε of the envelope process by Eq. (3) [8].

$$k = \sqrt{-2ln\varepsilon} \tag{3}$$

The FLB algorithm can also be seen as a sequential test described by the following set of equations:

$$E_n = max\,\{0, E_{n-1} + X_n - \bar{a}\Delta\} \tag{4}$$

$$K_n = \begin{cases} S, & \text{if } E_n = 0; \\ k\sigma\Delta^H\left[n^H - (n-1)^H\right] + K_{n-1}, & \text{otherwise} \end{cases} \tag{5}$$

$$J_n = \begin{cases} 0, & \text{if } E_n \le K_n; \\ X_n - \bar{a}\Delta - k\sigma\Delta^H \left[n^H - (n-1)^H\right], & \text{otherwise} \end{cases} \tag{6}$$

Note that E_n is a sequence testing term, K_n is the decision threshold to E_n and Δ is the updating time interval. The sequence J_n denotes the number of packets marked with lower priority, or that will be discarded, J_n is the control rule applied to the input traffic X_n sequence.

2.3 Multifractal Leaky Bucket—MLB

The Multifractal Arrival Policing Mechanism (MAPM) is introduced in [3] using an envelope process known as the Multifractal Bounded Arrival Process (MFBAP) [7] represented by the time process of Eq. (7).

$$MLB(t) = \bar{a}t + k\sigma t^{H(t)} + S \tag{7}$$

where $H(t)$ is the Hölder exponent [9]. The other parameters are similar to those of the previous algorithm.

$$E_n = max\left\{0, E_{n-1} + X_n - \bar{a}\Delta\right\} \tag{8}$$

$$K_n = \begin{cases} S, & \text{if } E_n = 0; \\ k\sigma\Delta^{H(t)} \left[n^{H(t)} - (n-1)^{H(t)}\right] + K_{n-1}, & \text{otherwise} \end{cases} \tag{9}$$

$$J_n = \begin{cases} 0, & \text{if } E_n \le K_n; \\ X_n - \bar{a}\Delta - k\sigma\Delta^{H(t)} \left[n^{H(t)} - (n-1)^{H(t)}\right], & \text{otherwise} \end{cases} \tag{10}$$

3 Scheduling Algorithms

3.1 Modified Largest Weighted Delay First (M-LWDF)

The M-LWDF scheduling algorithm was developed to support multiple user data with different quality of service (QoS) requirements [10]. It prioritizes the real-time flows with the longest delay and consider the best channel condition. The assigned metric is found by the following expression:

$$w_{i,j} = -\frac{log\delta_i}{\tau_i}.D_{HOL,i}\frac{r_{i,j}}{\bar{R}_i} \tag{11}$$

where τ_i is the packet delay threshold value supported by the user application i, δ_i is defined as the maximum probability that the delay $D_{HOL,i}$ exceeds the threshold time, $r_{i,j}$ denotes as the current data rate that could be used by the ith flow in the jth sub-channel at time t. $\bar{R_i}$ is the estimated average transmission data rate. $D_{HOL,i}$ calculates the delay of the first packet to be transmitted in the queue.

3.2 *Exponential Proportional Fair (EXP-PF)*

The EXP/PF scheduling algorithm was designed to increase the priority of real-time flows with respect to non-real-time flows. The assigned metric is estimated by the following expression [10]:

$$w_{i,j} = exp\left(\frac{\alpha_i D_{HOL,i} - \lambda}{1 + \sqrt{\lambda}}\right) \frac{r_{i,j}}{\bar{R_i}} \tag{12}$$

$$\alpha_i = -\frac{log\delta_i}{\tau_i} \tag{13}$$

$$\lambda = \frac{1}{N_{rt}} \sum_{i=1}^{N_n} \alpha_i D_{HOL,i} \tag{14}$$

where N_{rt} represents the number of active real-time flows in the downlink transmission direction.

4 The Proposed Scheduling Algorithm

In this section, we present the proposed scheduling method, focusing on the arrival rate of packets in the transmission queue in the eNodeB of the LTE downlink system. Figure 2 shows the transmission queue model of eNodeB based on the leaky bucket method. We associate with each data transmission queue in the eNodeB, a traffic policing mechanism known as a leaky bucket regulator. The leaky bucket algorithm regulates and controls input traffic where tokens or permits are generated at a constant rate ρ and are queued into the S-sized bucket, the output traffic is released into the network only when the corresponding token number is removed from the leaky bucket [4]. Notice that the leaky bucket algorithm does not work well when the input process receives packets in bursts. These events occur frequently when inputs traffic flows are monofractal or multifractal [11].

We define $P_{HOL,i}$ to be the delay of the head of line policed in the ith flow of leaky bucket. The delay $P_{HOL,i}$ of the leaky bucket can be determined using Eq. (15) where $Q_{P,i}(t)$ denotes the number of tokens accumulated in the leaky bucket of the

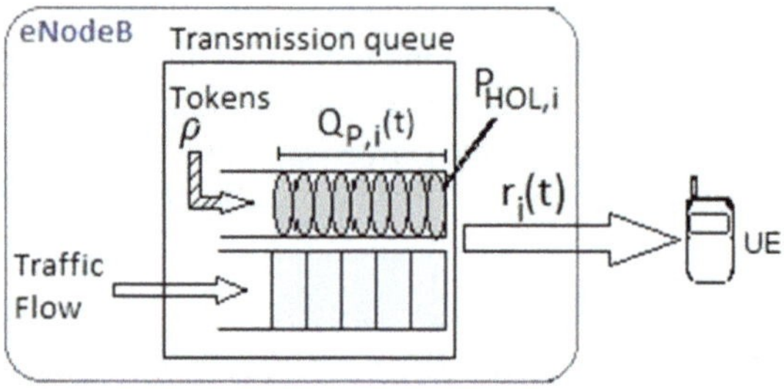

Fig. 2 The transmission queue model of eNodeB based of the leaky bucket method

i^{th} flow at the time t and $r_i(t)$ is the maximum data rate that can be supported.

$$P_{HOL,i} = \frac{Q_{P,i}(t)}{r_i(t)} \tag{15}$$

On the other hand, the M-LWDF and EXP/PF scheduling algorithms take decisions of scheduling based on the queue status information such as current packet delay, queue length, and channel conditions. We replace the $D_{HOL,i}$ parameter of M-LWDF and EXP/PF schedulers by $P_{HOL,i}$ to get new metric expression (Eqs. 16 and 17), respectively.

$$w_{i,j}^{(p)} = -\frac{log\delta_i}{\tau_i} P_{HOL,i} \frac{r_{i,j}}{\bar{R}_i} \tag{16}$$

$$w_{i,j}^{(p)} = exp\left(\frac{\alpha_i P_{HOL,i} - \lambda}{1 + \sqrt{\lambda}}\right) \frac{r_{i,j}}{\bar{R}_i} \tag{17}$$

$$\lambda = \frac{1}{N_{rt}} \sum_{i=1}^{N_n} \alpha_i P_{HOL,i} \tag{18}$$

More precisely, $w_{i,j}^{(p)}$ denotes the new policed metric assigned to ith flow in the jth sub-channel.

5 Experimental Analysis and Results

In this section, we analyze the performance of the M-LWDF and EXP/PF packet schedulers in LTE downlink transmission by simulation. We consider a scenario of a single cell of 1km of radius, with cellular interference and number of users ranging from 10 to 40 moving randomly at a speed of 3km/h inside the cell region. Data flows are distributed as follows: 40% of users use video flows, 40% of users use VoIP flows, and the remaining 20% use best effort flow services. Table 1 lists the main parameters of simulation used in the simulator LTE-Sim [12]. For the video

Table 1 LTE downlink simulation parameters

Parameters	Values
Simulation duration	150 s
Flow duration	120 s
Slot duration	0.5 ms
Bandwidth	10 MHz
Scheduling time (TTI)	1 ms
Frame structure	FDD
Number of RBs	50
Radius	1 km
Max delay	0.1 s
Number of users	10, 15, 20, 25, 30, 35, 40
UE velocity	3 km/h
Video bit-rate	242 kbps
VoIP bit-rate	8.4 kbps

flow, an application was used generating realistic video traces with a rate of 242 kbps [13]. For VoIP flow, a G.729 voice flow with a rate of 8.4 kbps was considered. The voice flow is generated with an ON/OFF Markov chain [12].

5.1 Numerical Results

We evaluate the performance of the proposed LTE downlink system using two traffic types: video traffic and VoIP traffic. The employed performance measures are packet delay, packet loss rate, and system throughput. Figure 3 shows the average delay of packets for video flows. Note that the delay grows with the increase in the number of users for all considered scheduling algorithms. Clearly all EXP/PF schedulers outperform the most M-LWDF schedulers.

Fig. 3 Packet delay for video flow

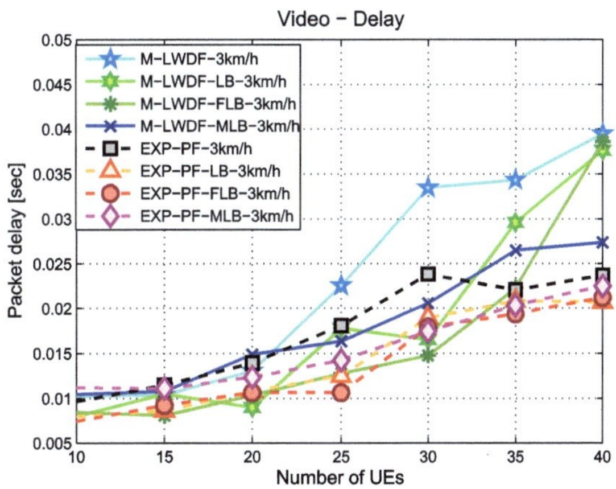

The performance of the average throughput for video flow is shown in Fig. 4. As the number of users increases, the average throughput of video flow decreases for all scheduler algorithms. The EXP/PF and EXP/PF-MLB schedulers almost the same performance with practically the same slow degradation tendency compared to the other schedulers.

Figure 5 shows the packet loss rate for video flows remains stable below 10% for all schedulers when the number of users is less than 20. Then, the loss rate grows rapidly when the number of users exceeds 20. The M-LWDF and EXPPF-MLB schedulers maintain a relatively slower loss rate compared to other schedulers. We observe that for video flows with 40 users, the M-LWDF-MLB scheduler presents lower delay and a lower loss rate. On the other hand, the EXP/PF-MLB scheduler presents better performance in the simulation with lower delay, higher throughput, and lower packet loss.

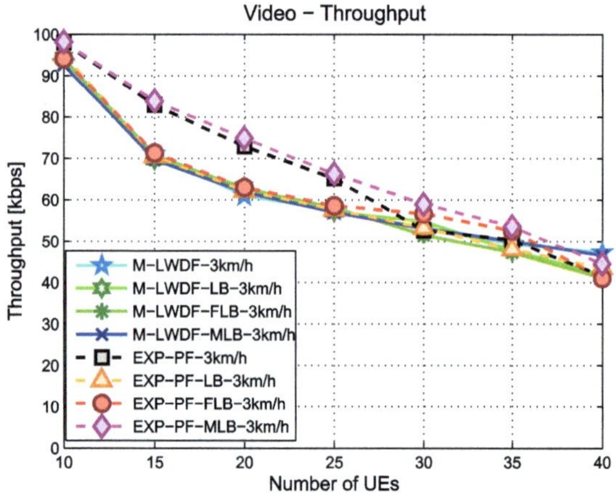

Fig. 4 Throughput for video flow

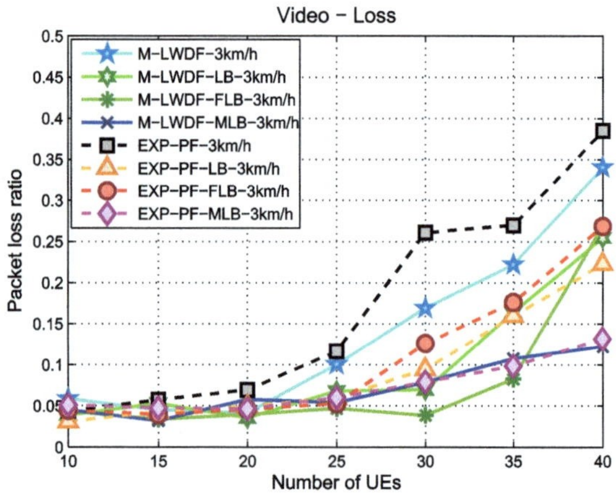

Fig. 5 Packet loss rate for video flow

The average delay for the VoIP flow is shown in Fig. 6. We note that the average delay is immutable around 2.3 ms for all schedulers when the number of users is less than 25. However, the average delay becomes larger under all M-LWDF scheduling algorithms. However, under the M-LWDF approached, the M-LWDF-MLB scheduler achieves the lowest delay when the user number exceeds 25.

Figure 7 shows the average throughput for the VoIP flow. It should be noted that the flow rate is almost constant from 15 users for all schedulers. A possible explanation for this non-variation could be the lack of the fractal nature under the traffic model and the ON/OFF periods used in the simulation.

Figure 8 shows the packet loss rate for VoIP flows. We note that there is no significant difference in packet loss for all schedulers.

Fig. 6 Packet delay for VoIP flow

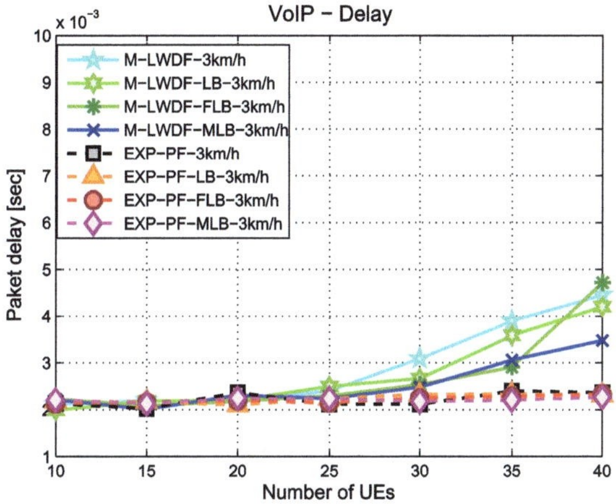

Fig. 7 Throughput for VoIP flow

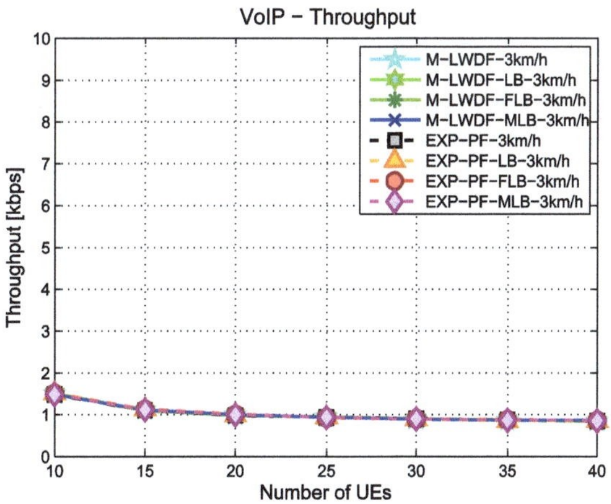

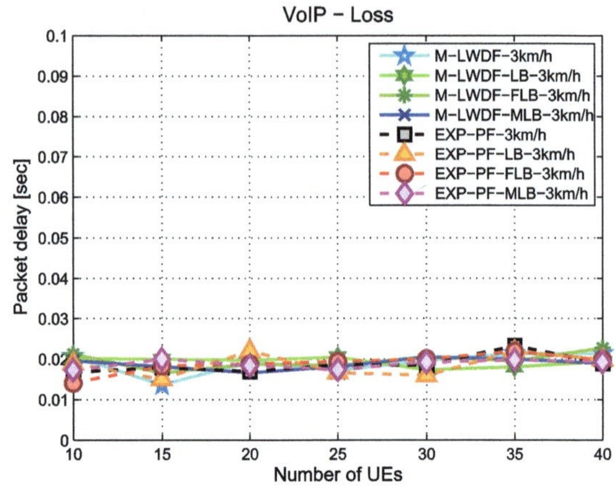

Fig. 8 Packet loss rate for VoIP flow

6 Conclusion

In this paper, we focused mainly on the performance of the scheduling algorithms of real-time service packets in the LTE downlink system, using traffic policing algorithms, known as leaky bucket regulation. We investigated the following scheduling approaches: M-LWDF and EXP/PF. The schedulers make decisions based on packet current delay using the $P_{HOL,i}$ delay information for each transmission queue in the eNodeB. The policing algorithms regulate incoming traffic in the transmission queue, taking into account the fractal behavior (monofractal and multifractal) of wireless traffic flows. To evaluate the performance of the proposed scheduling method, we adopted three performance metrics: loss rate, average packet delay, and system throughput. The results of the simulations showed that EXP/PF scheduler with traffic policing presented better performance in terms of packet delay for Video and VoIP flows. Specifically, the EXP/PF-MLB scheduler offers a higher throughput average in video flow compared with other schedulers. In general, resource allocation with traffic policing improves the performance of throughput and packet delay in LTE systems. As future work, we intend to develop algorithms for traffic policing in the advance LTE system, as well as other complex traffic policing algorithms will be explored.

References

1. 3GPP, Tech. Specific.: Technical Specification Group Radio Access Network; Feasibility Study for orthogonal Frequency Divition Multiplexing (OFDM) for UTRAN enhancement (3GPP TR 25.892. Release 6). http://www.qtc.jp/3GPP/Specs/25892-600.pdf
2. 3GPP, Tech. Specific. LTE; Evolved Universal Terrestrial Radio Access (E-UTRA) and Evolved Universal Terrestrial Radio Access Network (E-UTRAN); Overall description; Stage 2 (3GPP

TS 36.300 version 13.2.0 Release 13). http://www.etsi.org/deliver/etsi_ts/136300_136399/136300/13.02.00_60/ts_136300v130200p.pdf
3. Junior JAS, Vieira FHT: Estimation of backlog and delay in OFDM/TDMA systems with traffic policing through network calculus. IEEE Latin Am. Trans. 796–803 (2015). https://doi.org/10.1109/TLA.2015.7069107
4. Takada, HH, Anzaloni, A.: A sequential test approach for policing mechanisms to deal with monofractal and multifractal traffic. In: Next Generation Internet Design and Engineering, 2006. NGI'06. 2006 2nd Conference on. IEEE, p. 6 pp.-268. Valencia, Spain (2006). https://doi.org/10.1109/NGI.2006.1678250
5. Tanenbaum, A.S.: Redes de Computadores, 4th edn. Campus, São Paulo (2003)
6. Fonseca, N.L.S., Mayor, G.S., Neto, C.A.V.: On the equivalent bandwidth of self-similar sources. ACM Trans. Model. Comput. Simul. (TOMACS) **10**(2), 104–124 (2000)
7. Pereira, FdM., Fonseca, N.L.S., Arantes, D.S.: Modelagem de trfego fractal utilizando processos envelope. Technical Report Campinas, SP, Brazil, State University of Campinas. Revista da Sociedade Brasileira de Telecomunicacões 19.2 pp. 94–106 (2004)
8. Takada, H.H., Anzaloni, A.: A multifractual traffic policing mechanism. IEEE Commun. lett. **10**(2), 120–122 (2006). https://doi.org/10.1109/LCOMM.2006.02003
9. Le, G.R., Véhel, J.: Incremental moments and Hlder exponents of multifractional multistable processes. ESAIM: Probab. Statist. **17**, 135–178 (2013). https://doi.org/10.1051/ps/2011151
10. Nasralla M.M., Martini M.G.: A downlink scheduling approach for balancing qos in lte wireless networks. In: Personal Indoor and Mobile Radio Communications (PIMRC), 2013 IEEE 24th International Symposium on. IEEE, pp. 1571–157. London, UK (2013). https://doi.org/10.1109/PIMRC.2013.6666392
11. Ccorimanya, JQ., LING, L.L.: Traffic control in the transmission queue of LTE system downlink based on policing algorithms. In: Electronics, Electrical Engineering and Computing (INTERCON), 2017 IEEE XXIV International Conference on. IEEE, Cusco, Peru (2017). https://doi.org/10.1109/INTERCON.2017.8079709
12. Piro, G., Grieco, L.A., Boggia, G., Capozzi, F., Camarda, P.: Simulating LTE cellular systems: an open source framework. In: IEEE Transactions on Vehicular Technology, 60.2 pp. 498–513. IEEE (2011) https://doi.org/10.1109/TVT.2010.2091660
13. Vídeo trace library. http://trace.eas.asu.edu

Programmable Data Plane with Stateful Flow Processing for NFV Acceleration

Marcelo Nascimento, Tiago Primini, Eric Baum, Pedro Martucci, Francisco Cabelo and Leonardo Mariote

Abstract Network Function Virtualization (NFV) and Software Defined Networking (SDN) are software-based approaches that can advance each other's capabilities to build more scalable, agile, and innovative networks Kreutz et al. (Proc IEEE 103(1), 2004) [1]. This work's framework Nascimento et al. (Acceleration mechanism for high throughput and low latency in nfv environments, pp. 1–3, 2017) [2] combines both technologies together for accelerating NFV environments. Although the results obtained from the evaluation are meaningful regarding performance, the network functions started relying on idle timeouts to release unused resources, which is very inefficient. In this article, we propose a mechanism to address this issue, allowing flows to be removed based on the state of the TCP sessions. The technique can significantly optimize the flow table utilization and improve scalability. The proposal was implemented and evaluated, where an expressive reduction of 80% in resources allocation was achieved in the laboratory.

Keywords Network functions virtualization · Software-Defined networks Acceleration · Stateful

M. Nascimento (✉) · T. Primini · E. Baum · P. Martucci · F. Cabelo · L. Mariote
CPqD Foundation, Campinas, SP, Brazil
e-mail: marcelon@cpqd.com.br
URL: http://www.cpqd.com.br

T. Primini
e-mail: tiagopri@cpqd.com.br

E. Baum
e-mail: ebaum@cpqd.com.br

P. Martucci
e-mail: pedrom@cpqd.com.br

F. Cabelo
e-mail: fangelo@cpqd.com.br

L. Mariote
e-mail: lmariote@cpqd.com.br

Y. Iano et al. (eds.), *Proceedings of the 3rd Brazilian Technology Symposium*,
https://doi.org/10.1007/978-3-319-93112-8_18

1 Introduction

SDN [3] is a modern paradigm that enables rapid application deployment and service delivery through a programmable data plane and a logically centralized control plane. In parallel, NFV [4] brings agility, flexibility and scalability, delivering network services with cost efficiency to network operators [5]. Both technologies can be combined to build innovative networks.

In a previous work [2], we proposed an architecture that combines NFV and SDN with the goal of enhancing the performance of NFV platforms by offloading the service function chains while reducing latency. The idea is to have the service function chain acting as the control plane of a programmable packet forwarder, called Acceleration Mechanism (AM). So, after a flow is installed on the AM, the upcoming packets that match the flow are processed and forwarded to the destination, rather than going through every Virtual Network Functions (VNFs) within the chain. The proposal uses a reactive flow-based approach, considering the first packets of every session are processed by the network functions, as well as in a standard NFV architecture. At the end of the service function chain, the flow rule is built based on the comparison between the original packet and the one that was modified by the VNFs, thus the flow is installed in the data plane.

The reactive approach supports more dynamic applications, which is the essence of NFV. Some VNFs have very complex logic that relies on advanced analysis using internal state and L4-L7 information to make decisions. Even though this approach has a poor performance in most SDN scenarios, in the proposed architecture the control plane's role is played by VNFs, which are scalable, reliable, and designed for high-performance packet processing.

Although the performance is improved when bypassing the service function chain and the scalability is increased by offloading the VNFs, the AM can be a bottleneck. It must be very efficient when performing table lookup operation and re-writing packets in order to achieve high throughput. In the virtual CPE (vCPE) [6] use case, where a network operator can provide Internet access to tens of thousands of subscribers, the total amount of simultaneous TCP sessions can be huge. As a consequence, the flow table of the AM has to deal with all these sessions. In addition, unused allocated resources will be held, waiting for the flows to expire after the idle timeout. This is a critical scenario that has to be addressed in order to make the proposed architecture feasible in production.

In this paper, we propose the implementation of a TCP state machine in the AM to control the flow expiration. The flow table is modified to have a state attribute that will be updated locally. The main goal is to optimize the resource utilization by removing the flow as soon as the TCP session has finished, instead of waiting for an idle timeout. This technique can significantly reduce the number of entries in the flow table, increasing the scalability of the architecture. In addition, the allocated resources from VNFs are optimized considering they are released more quickly. When the flow is installed, the VNFs stop receiving packets to update their state machines. The AM is in charge of controlling the sessions and advertising their state VNFs through flow events.

The rest of this paper is organized as follows. In Sect. 2, we provide an overview of the NFV challenges and briefly introduce related works, while in Sect. 3, we present the design and implementation of our optimized flow table with stateful flow processing for the AM. Evaluation tests are described in Sect. 4, while in Sect. 5 we conclude the paper and discuss future works.

2 Background and Related Works

NFV decouples network functions from proprietary hardware devices and allows network services, such as Network Address Translation (NAT), firewalling, intrusion detection, load balancing and so on, to be hosted on virtual environments. However, ensuring that a network built upon software performs as well as a traditional network is one of the key challenges for NFV [7]. In parallel, SDN [1] decouples the forwarding- and control-functions, enabling the network to be intelligently and centrally controlled.

Aiming for increasing performance of NFV environments, we proposed an NFV architecture [2] that includes an acceleration mechanism to offload the service function chain and reduce latency. The strategy is to add a programmable packet forwarding engine to perform a fast-path forwarding for known traffic flows, as depicted in Fig. 1. It improves forwarding speed significantly. The slow-path is applied for the new session only when the packet has to go through the service function chain to be processed by the VNFs.

The packet processing starts at the AM, where the flow table performs a packet lookup. When an incoming packet does not match any table entry, it encapsulates the

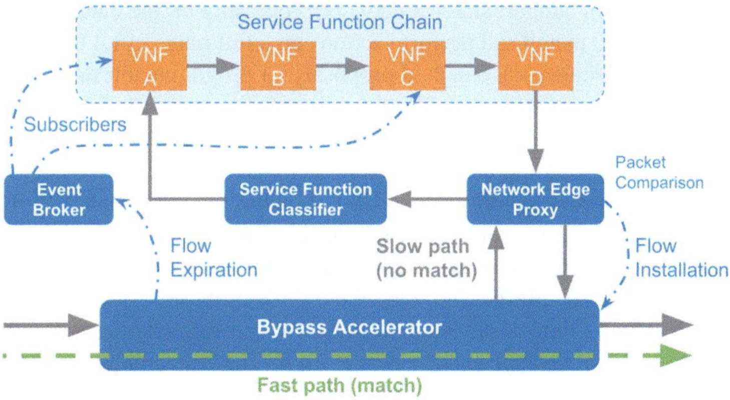

Fig. 1 System architecture

packet with a Service Function Header (SFH), where additional metadata information and the packet's original headers are stored. Right after being encapsulated, the packet is sent through the chain interface to the corresponding Network Edge Proxy (NEP) module. Each external interface can have its NEP to process the traffic for a specific technology, such as MPLS, QinQ, PPPoE, IPSec and so on. The NEP is always the first and the last network function in the service chain. At first, it is responsible for parsing the packet, de-encapsulating when applicable, identifying the customer and adding to the packet's metadata the Customer Unique ID attribute, delivering the packet to the Service Function Classifier (SFC) module, which is responsible for classifying the packet, adding a traffic profile attribute and enriching the packet's metadata with such information.

Each VNF can modify the packet by changing its headers. Then the packet is forwarded to the next VNF in the chain. After passing through all VNFs, the packet reaches the NEP. The two main NEP responsibilities at the end of the service chain are (i) to make the changes required by the adjacent destination network, such as adding a new header related to the bordering network technology; and (ii) to detect differences in the packet headers by comparing the original headers stored in the metadata with the actual header of the packet. All modifications in the packet are translated into specific actions which modify the packet headers. These actions are combined with the fields for the match rule, and an install flow message is built and sent to the AM. From then on the subsequent packets for such flow will be directly forwarded by the AM instead of going through all the VNFs of the chain. Finally, when the flow expires, the AM sends an event to the message broker, which forwards the event to all subscribers.

The process of removing flows, which recycles resources, is not efficient when it relies on idle timeout to detect the flow is no longer needed. Stateless programmable data planes, such as OpenFlow switches, provide only idle and hard timeouts as flow expiration mechanisms. In other to quickly detect termination, the flow requires a state machine attached to it, so that the termination state can be achieved based on the type of packets, e.g., TCP-FIN and TCP-RST.

OpenFlow cannot support stateful network processing on the data plane, which includes the need to match and perform actions on a full range of L4 to L7 header's field. In this sense, there are a bunch of proposals for programmable data planes that offers persistent state. Mekky et al. [8] proposes an application-aware data plane that includes stateful actions and L4-L7 information with the goal of keeping most of the packet processing in the data plane and limits the need to send packets to the controller, which incurs significant delays. Following the same path, [9] proposes FAST (Flow-level State Transitions) as a switch abstraction that allows the controller to pro-actively program state transitions, and allows the switches to run dynamic actions based on local information. OpenState [10] proposes a viable abstraction, based on eXtended Finite State Machines (XFSM), to formally describe a desired stateful processing of flows inside the device itself. The list continues with P4 [11], POF [12], Domino [13] and OVS [14].

All these stateful data planes can significantly offload the controller, reduce response time, and improve scalability. They provide mechanisms to move the application's logic from control plane to data plane by implementing state machines in the flow table. On the other hand, there is no optimization of the flow table utilization, because the flows continue to expire based on either hard or idle timeouts. So when a TCP session is finished, the flow rule starts dropping the packets instead of removing the flow entry. There should be no more packets after the session finishes so that the flow can be removed and the session resources released from stateful VNFs.

In regards to the proposed architecture for accelerating NFV platforms, the goal is to keep the programmable data plane as simple as possible by avoiding the addition of application's logic to it, while optimizing the flow table utilization. For that, it is necessary to implement a TCP state machine in the flow table for flow expiration only. This way, the solution can still maintain the high throughput while improving the scalability.

3 Design and Implementation

The system architecture follows the implementation described in [2]. This work focuses on how some resources that were allocated during packet processing in our NFV platform, for example a public IPv4 address and port provided by a Carrier Grade NAT (CGNAT), have their control taken by the AM so it can handle packets that are bypassing the service chain, and mainly how those resources are returned as quickly as possible to the related network functions.

Some network functions work in a stateful mode, where it is important to keep the session's state machine up to date with the packets that are being processed, or keeping allocated resources during its lifetime. In our NFV platform, that is equipped with the AM, the synchronization between this module and other deployed network functions, mainly the stateful ones, is extremely important, because as soon as a resource is released from the AM, it must be given back to the network function, where the resource was initially allocated.

As explained in [2], the NEP is responsible for extracting differences applied to a packet during its path over the chain of services, and to convert those differences in specific actions that will be installed in the AM. These actions can be simple and may represent a stateless network function, as an example a Remove Tunnel Header Action (QinQ, VLAN, MPLS), or they can even be more complex actions associated with stateful network functions, as an example a NAT44 Action. In the first case, just some basic information involving the action itself is enough for the install flow message, but in the second case, some additional information must be attached, for example, the source and public IP/Port and the current state of a TCP NAT44 session. TCP connection state machine follows the standard defined in [15]. The install flow message may contain the following data: instruction type: add/remove a flow, or flush table; idle timeout; customer identification number; device MAC Address; action Add/Remove Tunnel (EoMPLS, MPLS, VLAN, QinQ); action Change MAC Address; action SRC/DST NAT; action increment/decrement TTL; and action TCP state machine.

The NAT actions contain all information regarding its specific translation mode; for example, a source NAT44 action will have the following data: source private/public IPv4 address; source private/public port; destination IPv4 and port; and protocol. The TCP state machine action will only be instantiated when the protocol is TCP. This action will contain the following information: current sequence number; current acknowledgment number; expected acknowledgment number; and session state.

Once a TCP connection is established, an acknowledgment number is sent. The TCP segments carry this variable which is the sequence number of the next expected data octet of transmissions in the reverse direction. Our TCP state machine calculates the expected acknowledge number to compare it with the received acknowledgment number, so we can verify whether the session state must be changed, avoiding changing state in an undesirable moment, as in an event of a duplicated or an out-of-order packet.

The install flow message was implemented using Google Protocol Buffers [16]. The message is binary serialized and sent to the AM using a push/pull schema. Each flow entry may consume no more than 161 bytes in memory to store the parameters of the flow message as shown in Fig. 2.

The destination IPv4 and Port are important parameters to store because some NAT functions work with different types of filtering behaviors, for example, EIF—End-

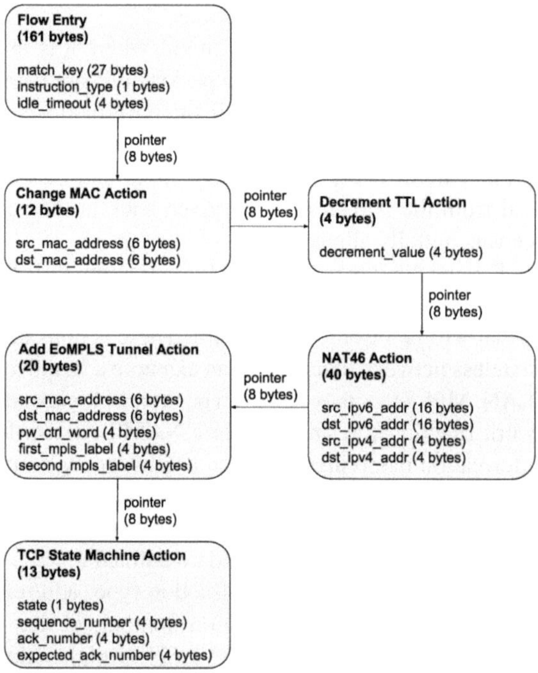

Fig. 2 AM flow entry example

point independent filtering, ADF—Address dependent filtering and APDF—Address and port dependent filtering. They are used to filter inbound packets and determine if a packet will be passed or dropped, depending on the configured behavior mechanism. In the case of APDF, the destination IP and port are important parameters to keep in memory, because they will be used to recover a NAT session. More specifically, destination IP and port are used to filter out undesirable packets, because only those packets sent as a response to the outbound packet previously sent by an internal endpoint are allowed to come back to the related internal endpoint.

When the AM receives the install flow message, it extracts all the data from the message and builds a new flow entry to be installed in its mapping table. Every installed flow entry has an idle timeout and will expire after *n* seconds of inactivity. Besides that, it is possible to configure a hard timeout as well. For TCP sessions, each flow entry is equipped with a TCP state machine, initialized in the network function that has processed the first flow packet, as an example, a CGNAT VNF. In this way, every TCP flow entry can be finished gracefully and the resources allocated for the session can be freed to be reused faster.

To guarantee that the AM is controlling the resources allocated for a session when the flow entry is successfully installed, and also to make sure that those resources have been released after a flow entry have expired or after it has been removed due to a TCP connection close, there are two event types that are exported by the AM. These events use a publish/subscribe schema to which all the network functions that are interested in receiving those events must be subscribed.

4 Evaluation

A representation of the simulation environment can be seen in Fig. 3. To evaluate the AM's performance and gains, a program developed in Python was used to simulate a customer accessing the Internet. In the test environment, each of the namespaces (from C001 up to C100) ran this program to simulate multiple customers consuming data from the Internet. Each of the customers was simulated as containing 10 different devices (from D01 up to D10) generating HTTP requests, randomly accessing websites from a list of top 500 commons websites [17] for 1 h. The simulation environment was installed on hardware with the following configuration: Intel(R) Xeon(R) CPU X5660 @ 2.80 GHz, 24 cores, 16 GBytes of RAM—DIMM DDR3 1333 MHz, Ethernet card—1000 Mb/s Full Duplex.

Two scenarios were configured to verify the amount of used and available resources on the Platform. In the first scenario, the TCP state machine was deactivated in the AM's flow table, so all the flows, independently of the protocol type, were removed only after the expiration of the idle timeout, which was set to 60 s. In the second scenario, the TCP state machine was activated, so TCP flows could be removed as soon as the session was closed, enabling the reuse of ports and IP addresses more efficiently by VNFs, as in the case of CGNAT network function. Another advantage is related to the reduction of flow entries in the AM's flow table,

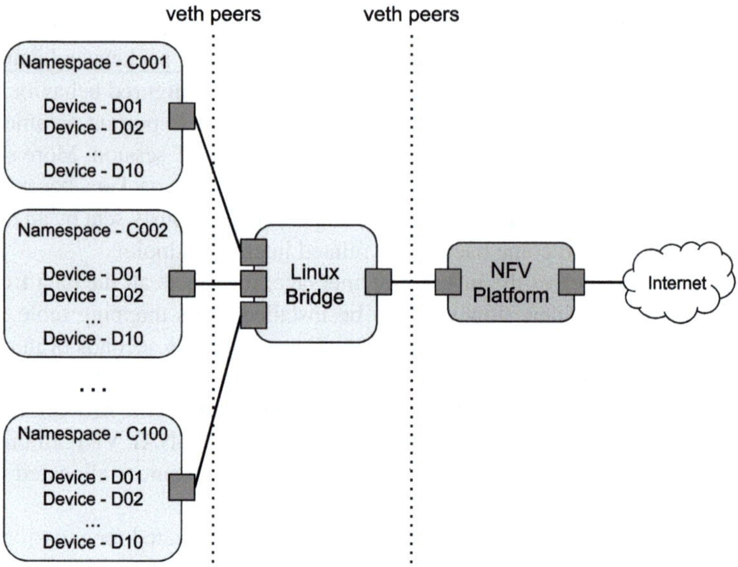

Fig. 3 Simulation environment

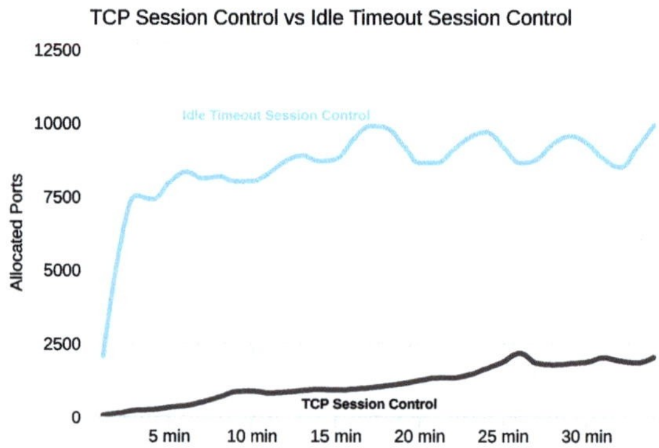

Fig. 4 Allocated resources: TCP state machine versus idle timeout

because only the active TCP sessions must be handled in memory. The comparison of the allocated resources in both scenarios is shown in Fig. 4.

It is also possible to verify that when the TCP state machine was deactivated, the consumption of resources has increased sharply at the beginning of the test, while in the scenario equipped with TCP state machine, the consumption line has increased softly. When the resources consumption is established, the gain in resources allocation that was found when using TCP's state machine was on the order of 80%. This can be checked in Fig. 5.

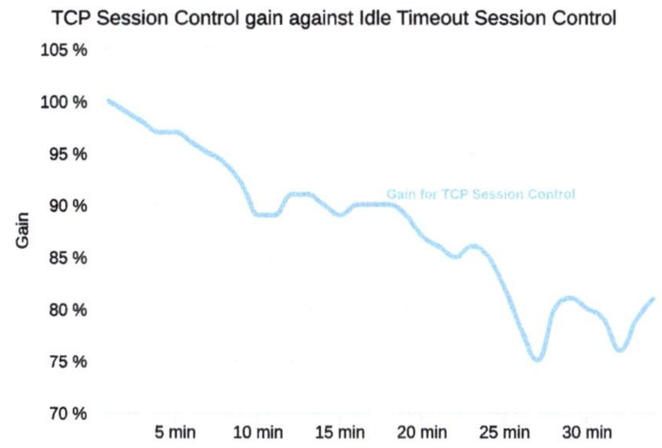

Fig. 5 Resources gain using TCP session control

5 Conclusion and Future Works

The acceleration mechanism proposed in our previous work proved to be very effective in reducing the latency and in offloading the network service functions. On the other hand, the VNFs started depending on flow removed events from the AM to release allocated resources, incurring significant delay when detecting the termination of sessions. To address this issue, we implemented a TCP state machine in the AM, so the flow is removed as soon as the TCP session is finished. The results show that the flow table utilization has been optimized in 80% compared to the use of an idle timeout of 60 s. It means the AM releases the unused resources much sooner, and so do the VNFs. Considering stateful network functions, such as a CGNAT, there would be more IP addresses and ports available for translation, improving the scalability of the network.

In the future, we intend to design a model that describes the behavior of resource utilization based on idle timeout, average TCP session duration time and flow installation rate. This model will provide us information to exploit the limitation of our current implementation and to make an analysis of deployment sizing. We also intend to integrate new features related to the state machine to the AM flow table to accelerate next-generation firewall applications.

References

1. Kreutz, D., Ramos, F.M., Verissimo, P.E., Rothenberg, C.E., Azodolmolky, S., Uhlig, S.: Software-defined networking: a comprehensive survey. Proc. IEEE **103**(1), 14–76 (2015)
2. Nascimento, M., Primini, T., Baum, E., Martucci, P., Cabelo, F., Mariote, L.: Acceleration mechanism for high throughput and low latency in nfv environments. In: High Performance Switching and Routing (HPSR), 2017 IEEE 18th International Conference on, IEEE, 1–6 (2017)
3. McKeown, N.: Software-defined Networking. INFOCOM Keynote Talk **17**(2), 30–32 (2009)

4. European Telecommunications Standards Institute (ETSI), Network Function Virtualisation. Available: http://www.etsi.org/technologiesclusters/technologies/nfv, Last Accessed 22 Mar 2017
5. Paper, N.W.: Network functions virtualisation: an introduction, benefits, enablers, challenges and call for action. Issue 1 (October 2012)
6. NFV, G.: 001: Network functions virtualisation (nfv); use cases, v 1.1. 1. ETSI, December (2013)
7. Hwang, J., Ramakrishnan, K., Wood, T.: Netvm: high performance and flexible networking using virtualization on commodity platforms. IEEE Trans. Netw. Serv. Manage. **12**(1), 34–47 (2015)
8. Mekky, H., Hao, F., Mukherjee, S., Zhang, Z.L., Lakshman, T.: Application-aware data plane processing in sdn. In: Proceedings of the Third Workshop On Hot Topics in Software Defined Networking, ACM, 13–18 (2014)
9. Moshref, M., Bhargava, A., Gupta, A., Yu, M., Govindan, R.: Flow-level state transition as a new switch primitive for sdn. In: Proceedings of the Third Workshop on Hot Topics in Software Defined Networking, ACM, 61–66 (2014)
10. Bianchi, G., Bonola, M., Capone, A., Cascone, C.: Openstate: programming platform-independent stateful openflow applications inside the switch. ACM SIGCOMM Comput. Commun. Rev. **44**(2), 44–51 (2014)
11. Bosshart, P., Daly, D., Gibb, G., Izzard, M., McKeown, N., Rexford, J., Schlesinger, C., Talayco, D., Vahdat, A., Varghese, G., et al.: P4: Programming protocolindependent packet processors. ACM SIGCOMM Comput. Commun. Rev. **44**(3), 87–95 (2014)
12. Song, H.: Protocol-oblivious forwarding: unleash the power of sdn through a future-proof forwarding plane. In: Proceedings of the second ACM SIGCOMM workshop on Hot topics in software defined networking, ACM 127–132 (2013)
13. Sivaraman, A., Cheung, A., Budiu, M., Kim, C., Alizadeh, M., Balakrishnan, H., Varghese, G., McKeown, N., Licking, S.: Packet transactions: High-level programming for line-rate switches. In: Proceedings of the 2016 conference on ACM SIGCOMM 2016 Conference, ACM 15–28 (2016)
14. Pfaff, B., Pettit, J., Koponen, T., Jackson, E.J., Zhou, A., Rajahalme, J., Gross, J., Wang, A., Stringer, J., Shelar, P., et al.: The design and implementation of open vswitch. In: NSDI. 117–130 (2015)
15. University of Southern California.: Transmission control protocol. Available: https://tools.ietf.org/rfc/rfc793.txt (September 1981)
16. Google: Google protocol buffers. Available: https://developers.google.com/protocol-buffers/. Last Accessed 12 May 2017
17. Top, T.M.: themoztop. Available: https://moz.com/top500. Last Accessed 27 Mar 2017

Design of the Voltage-Controlled Ring Oscillator Using Optimization Tools (MunEDA® WiCkeD)

Agord de Matos Pinto Jr., Raphael R. N. Souza, Leandro Tiago Manera, Jorge Enrique Vargas Solano, Cássia Maria Chagas and Saulo Finco

Abstract Integrated circuit design is a very complex and time-consuming task, despite several SPICE-like simulator tools being used to improve productivity and reduce the expended time, make use of auxiliary tools is a must when trying to improve circuit performance and yield, as to get a much more robust design. This paper presents the design flow using a worst-case distance analysis tool WiCkeD™, explaining its algorithms and showing the results of designing a Voltage-Controlled Ring Oscillator (VCRO) as a test case. Where the simulation proved that parameter variations with the SQP algorithm are more efficient and more reliable compared to other algorithms. With a considerable improvement in two essential items, consumption (4.602 mW) and phase noise (−111.8 dBc/Hz). The circuit was designed using 180 nm CMOS technology from UMC and with the DARE library from IMEC.

Keywords Voltage-controlled ring oscillator · MunEDA® WiCkeD

A. de Matos Pinto Jr. (✉) · L. T. Manera
State University of Campinas (Unicamp), University City - Campinas-SP, Brazil
e-mail: agord@dsif.fee.unicamp.br

L. T. Manera
e-mail: manera@dsif.fee.unicamp.br

R. R. N. Souza · J. E. V. Solano · C. M. Chagas · S. Finco
Center of Information Technology Renato Archer, CITAR, Campinas-SP, Brazil
e-mail: raphael@cti.gov.br

J. E. V. Solano
e-mail: jorge.solano@cti.gov.br

C. M. Chagas
e-mail: cassia.chagas@cti.gov.br

S. Finco
e-mail: saulo.finco@cti.gov.br

Y. Iano et al. (eds.), *Proceedings of the 3rd Brazilian Technology Symposium*,
https://doi.org/10.1007/978-3-319-93112-8_19

1 Introduction

Nowadays, with the popularization of day use gadgets, a need for cheaper and more efficient systems results on bigger challenges for the IC designer in terms of robustness and power efficiency, which in turn results in a more robust integrated circuit design. In this context, the development of platforms that integrates all the necessary tools for the creation of a complete electronic product development environment, allowed the designers increase their productivity in line with the growing complexity of systems. This design environment involving SPICE-like simulators and their auxiliary tools are becoming more important to provide robustness of designs, resulting in a safer design.

As we explore the limits of very deep submicron scaling, we are approaching a point where many factors become questionable, e.g., leakage current. As a result, many fundamental aspects related to circuit and system design will have to be reconsidered, especially when taking into account the radically different properties of future nanometer scale devices [1].

Considering those factors and following the analog design flow methodology, from schematic conception to signal integrity analysis, a Voltage-Controlled Ring Oscillator (VCRO) was designed. The VCRO used the UMC 180 nm technology with Design Against Radiation Effects (DARE-library); with the aid of the MunEDA® WiCkeD tool, which is a product for optimizing circuits performance and yield by the use of automation routines, based on the worst-case operation distance methodology. Through the MunEDA® WiCkeD was performed the analysis and optimization of the circuit behavior and specifications on several operation conditions, given the restrictions on power and area consumption, and emulating a few the radiation effects.

In this context, this paper describes the features and optimization process applied to VCRO project implementation along the design flow. Thus, Sect. 2 gives a brief consideration on DARE technology use, Sect. 3 presents the VCRO circuit, Sect. 4 explain the operation tool MunEDA® WiCkeD, and Sect. 5 summarizes the obtained results.

2 Technology Rad-Hardening (DARE)

Radiation Hardened libraries are meant to be used in combination with commercial foundry technology and seek for foundry-independence while providing competitive, high-performance, low-power, low mass solutions for components to be used in harsh radiation environments. The cells of the IMEC's DARE library (using UMC 180 nm technology) are continually being improved, and new elements are added to increase maturity to an appropriate level. The library has evolved from an initial offering of only purely digital standard cells to a technology platform capable of delivering a robust mixed-signal and analog signal designs.

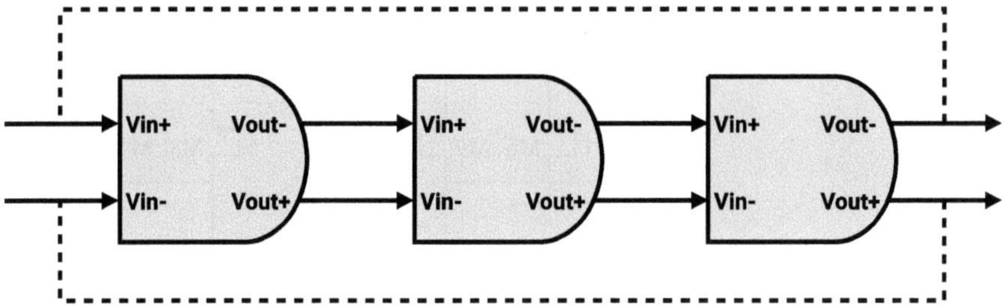

Fig. 1 Voltage-controlled ring oscillator (VCRO)

This technology has been chosen because the availability of ELT transistors cells, which have no edges and improvement in leakage. Furthermore, the availability of rad-hard digital standard cells with a maximum achievable gate density of 25 kgates/mm^2 in addition to the six metal layers [2].

3 Voltage-Controlled Ring Oscillator

Most electronic signal processing systems require frequency or time reference signals, to use the full capacity of communication channels. For the lower end of the spectrum, one can use the stable properties of quartz crystals as a resonator to build very accurate fixed frequency or time reference signals. For higher frequencies (> few hundred MHz), the quality of the crystal resonators degrades due to physical limitations and material properties [3].

Compared to LC oscillators, ring oscillators have several advantages, such as (i) they can be easily modified to enable the change in the oscillation frequency given a control voltage; (ii) do not require any resonator; (iii) have large tuning range; (iv) do not make use of inductors; and (v) have a smaller area. Because of these and some other reasons, the ring oscillator topology has been chosen to implement the VCRO.

Due to their integrated nature, ring oscillators have become an essential building block in many digital and communication systems [4]. Through frequency synthesizers [5, 6], they are used as VCROs in applications such as clock recovery circuits for serial data communications [7, 8], disk-drive read channels [9], and on-chip clock distribution [10]. The design of a ring oscillator encompasses many trade-offs regarding speed, power, area, and application domain [11].

The proposed VCRO is composed of three differential delay stages, with 1.8 V power supply as illustrated in Fig. 1.

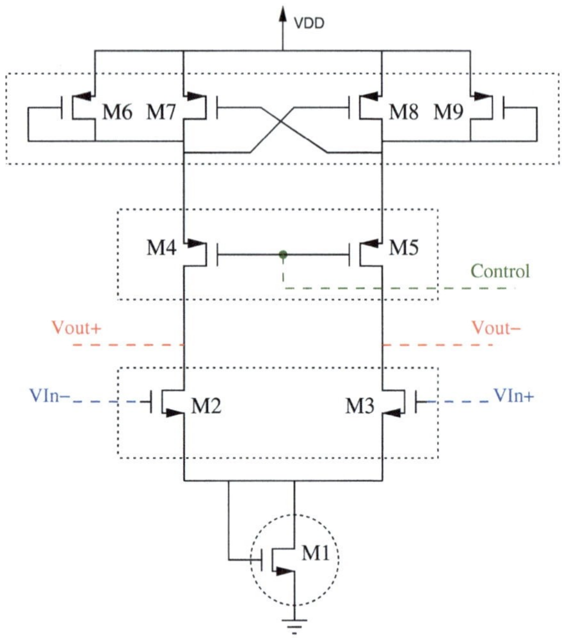

Fig. 2 Delay cell differential

3.1 Delay Cell Topology

In general, some delay cells, which are connected in a positive or regenerative feedback loop for building a basic ring oscillator, are the main basis of VCRO [12].

According to its delay cell's topology, a VCRO can be classified into two types: single-ended and differential. In contrast to its single-ended counterpart, a differential VCRO can provide both, better common-mode noise suppression performance and desired differential clocks. Therefore, this type of VCRO is widely used [13]. The differential logic can be fully differential or pseudo-differential, according to whether or not there exists a tail current source [14]. Choosing an optimum number of stages for construction of high-frequency oscillator is an essential part of designing VCRO. Two, three, and four stages are common structures in communication systems [12].

Currently, differential circuit topology is getting acceptance among designers as it has a common-mode rejection of supply and substrate noise [15]. Moreover, it could have an odd or even number stages, and it is possible to achieve both in-phase and quadrature outputs [16].

Circuit Implementation This VCRO was designed by using three delay cells as illustrated in Fig. 1, with a differential topology, as illustrated in Fig. 2.

Oscillators with differential ring topology, work when they are in a suitable bias point, so their DC gains do not vary with time, and all transistors are operating in the active region as they should be [17, 18].

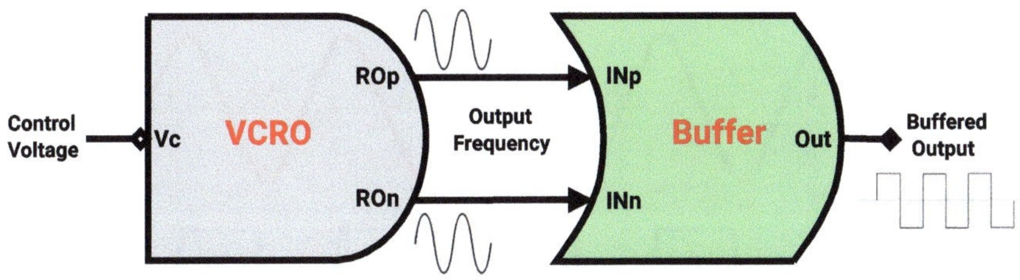

Fig. 3 VCRO interconnected to buffer

Therefore, the delay cell is composed of a differential pair of negative channel Metal-Oxide-Semiconductor (nMOS) M2 and M3, with a cross-coupling pMOS load (M8 and M7). Such connection uses symmetrical loads with replica feedback biasing and the self-biasing concept, which lessens the sensitivity of the oscillator sensitivity to supply and substrate noise.

Additionally to the cross-coupling, the diode-connected MOS devices (M6 and M9) are used to discharge the output, whose current is controlled by a pair of pMOS transistors (M4 and M5), working as varactors.

Where these control devices driven by a gate voltage, control the stage above, thereby limiting the strength of the block. When changing the control voltage (control - in green) on the gate of control devices (M4 and M5), the charge down current of the delay cell output load is changed. Therefore its delay time and thus the frequency of the whole VCRO are controlled.

By adjusting the transconductance of transistors M7 and M8, a wide tuning range can be achieved. They (M7 and M8) are connected to a supply voltage directly to maximize the voltage swing, increase the carrier power, and lower the noise power. pMOS Transistors M6 and M9 are connected directly to a supply voltage to maximize the tuning range.

3.2 *Buffer*

Output buffer circuit magnifies and regulates the output signal of the front circuit, at the same time it provides large enough current and voltage to drive follow-up circuit. Even more important, it provides the ability of better capacitive isolation, and avoids the VCRO's oscillation frequency to increase the frequency traction's impact when there are external effects of loads; moreover, it achieves relatively robust insulation of loads around 6 pF; as illustrated in Fig. 3.

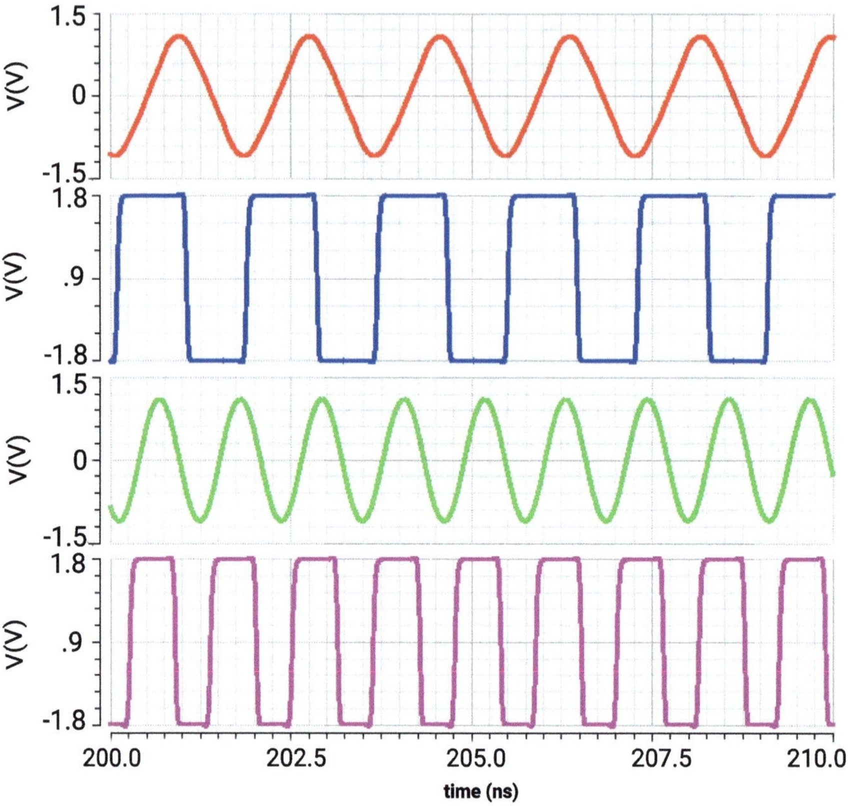

Fig. 4 Output frequency waveforms

4 Muneda WiCkeD

The WiCkeD™ tool suite provides a methodology solution for full-custom circuit analysis and optimization. In WiCkeD™, considers transistors devices geometries and passive devices as design parameters. The circuit's working conditions such as temperature, voltage supply or load are treated as operating parameters. Process variation and mismatch are modeled as statistical parameters [19].

The VCRO fulfilling the criterion of oscillation, where the gain module mesh is already less than unity, e.g., the circuit is already serving the Barkhausen criterion, in full operation in the PSpice simulator (*Cadence Analog Design Environment*—ADE), as illustrated in Fig. 4. The waves in red and green represent the outputs of the VCRO (F_{Min} and F_{Max}), and the waves in blue and pink represent the outputs of the Buffer (F_{Min} and F_{Max}), respectively.

In this sense, Fig. 5 illustrates the design flow of the WiCkeD™ tool integrated to the PSpice simulator (Cadence Analog Design Environment—ADE), for the purpose

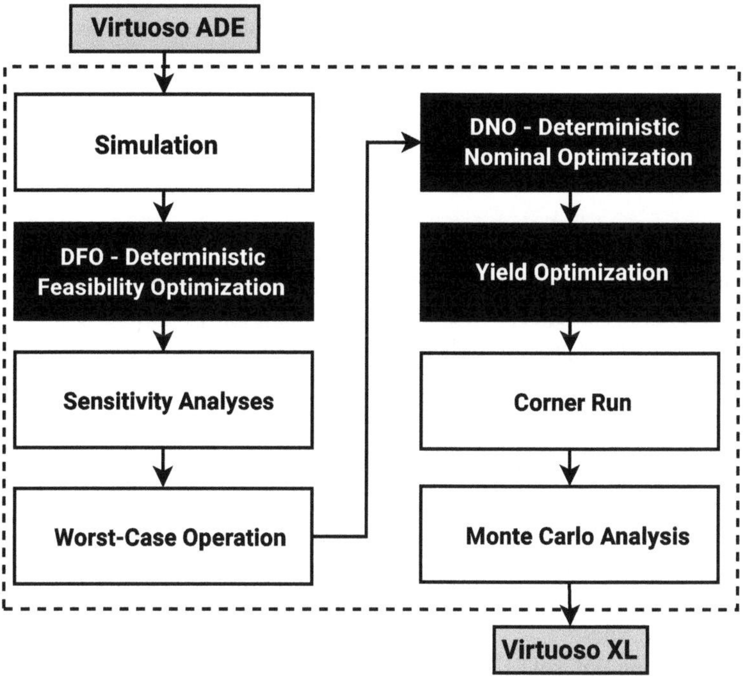

Fig. 5 Analog flow with increased WiCkeD™ tool

of to having specific design improvements and to optimize the circuit through their algorithms: SQP (Sequential Quadratic Programming), least squares, and parameter distances.

These analysis methods can be combined to design flows, which give the designer greater flexibility into the circuit or system and list all effective influences with priority.

After defining the proper performance metrics for each circuit at ADE, very first step inside WiCkeD™ tool's environment is to translate the schematic and the metrics into the Constraint Editor (CE). In this design stage, parameter setup section-based interface allows the specification of the range for the reference parameters:

- Design parameters: W/L and levels for active components, and C for the passive component;
- Operating parameters: voltage and temperature;
- Performance Parameters: parameters previously defined for measurement at Cadence ADE with the corresponding analysis specification.

The following steps in the flow are the different analysis WiCkeD™ does to verify the defined constraint and later to optimize the design; those can be seen in the diagram in Fig. 5. According to this diagram, the verification-based steps are inside white boxes, and the optimization-based steps are the black ones. It is worth to notice that inside each step several simulations are performed [20].

4.1 Simulation

After all parameters are imported from Cadence's ADE into WiCkeD™, an initial point simulation is executed, serving as starting point for the next steps.

This simulation also checks whether the constraints and performance are fulfilled, for either operation or process variations.

4.2 Deterministic Feasibility Optimization (DFO)

The restricted design space that satisfies all sizing rules is called the "feasible region". For many circuits, the feasible region is very small (perhaps 1% of the total design space), and it is sometimes difficult to find a set of design parameter values that satisfy all sizing rules.

The underlying WiCkeD's™ optimization algorithms change the values of design parameters such that all given constraints are fulfilled under the influence of operating and corner conditions. As structural constraints define the correct operating conditions for the circuit technically under construction (e.g., transistors operate in saturation region), finding a feasible design represents the first step toward a successful circuit optimization.

This is useful when a few rules are slightly violated. Otherwise, the design is otherwise considered acceptable. "Find central" overfulfills all sizing rules as far as possible. This approach is useful for quickly finding an initial sizing, thus leaving all the transistors in the correct region of operation.

4.3 Sensitivity Analysis

WiCkeD™ also has the ability to identify dependencies between specific transistors and circuit performance parameters. This capability informs designers of which transistors influence a particular performance parameter and the degree of that influence.

In this case, we observed that the transistor's size had more influence on the capacitance (load) of the block; this was important to realize it minimally, for generating a filter with smaller dimensions, leaving the PLL marketable and economically viable.

4.4 Worst-Case Operation

It consists of the worst combination of operation conditions and PVT corners according to a given performance specification or constraint. Therefore, it changes operating parameters and the corner parameters such that for a lower specification, the performance value or constraint value is minimized, and for an upper specification, the performance value is maximized, in other words, the parameter combination with worst performance within an x-sigma distance.

4.5 Deterministic Nominal Optimization (DNO)

It makes use of the initial design solution obtained by the DFO option to perform a nominal optimization of all design specifications simultaneously.

Here the tool provides three algorithms for automatic performance optimization. The designer enters performance specifications and selects parameters to be modified. Available algorithms are [21, 22]:

- Sequential Quadratic Programming (SQP): Especially suitable for nonlinear performances, it solves the optimization problem by creating quadratic sub-models of performances in every algorithm iteration, without extrapolating the settings;
- Least Squares (LS): This algorithm changes the values of the selected design parameters as little as necessary to meet the reference specifications. Performance specifications are fulfilled, but the optimizer does not try to overfulfill them;
- Parameter Distances (PD): This algorithm tries to exceed the specifications limits to achieve a better convergence. The optimizer tries to overfulfill each specification as far as possible. Weight factors can be changed by the designer.

The first two optimizers are gradient-based. They converge quickly to the nearest local maximum of their goal function. Since the design space is restricted to the small feasible region, this is usually the only maximum. In rare cases with ill-defined, noncontinuous performance functions over a large design space with few constraints, the latter two optimizers provide a global approach to optimization. For most practical circuits, gradient-based optimization outperforms global optimization regarding both algorithmic effort and quality of results.

4.6 Yield Optimization

In particular, the mismatch effect between identically sized devices needs to be modeled accurately, because it could be a serious limiting factor of the accuracy of both analog circuits [23–25]. An accurate model of device mismatch allows the designer not only to effectively predict the parametric yield of the circuit under analysis but also to improve it through appropriate use of the insight gained from the mismatch model [25–27].

With a yield optimization, a circuit is sized such that given specifications are met. However, for low-power circuits, the circuit should also have an upper limit to total power consumption. The yield thus depends on the cutoff point taken for the circuit, for instance, the yield for this design is the percentage of total circuits whose delay falls under a critical delay cutoff.

The result of a yield optimization is also feasible, i.e., all constraints are fulfilled. Process variations are taken into account as well as operating and corner parameters.

4.7 Corner Run

The Corner Run algorithm analyzes the influence of corner cases (PVT corners/ verifying the process variation) on given performances and constraints [20]. Underlining, in this regard, the tool uses the models of technology, which is used in this paper is the DARE-UMC 180nm.

4.8 Monte Carlo Analysis

Gives the designer insight on how process variation influences a given design. In each Monte Carlo run, process parameters are varied in a random manner and the design is simulated for many different values of process parameters.

The statistics of the simulation results are then analyzed. In one Monte Carlo analysis node, multiple Monte Carlo runs can be performed subsequently. Two different Monte Carlo analysis nodes in the main history do not influence each other. Monte Carlo analysis, however, reuses simulations that were already performed and are therefore kept in the simulation database. Monte Carlo analysis supports not only performance specifications but also constraints [20].

5 Results

To evaluate the performance of the designed oscillator relative to its alternatives, a Figure of Merit (FoM) was used [28–30]. FoM that takes into consideration the Power Consumption (P_{DC}), the Phase Noise (PN) (associated with a reference frequency offset Δf) and the oscillation frequency f_0, as represented in the expression 1.

$$FoM = 10log\left(\frac{P}{1\,\mathrm{mW}}\right) + PN - 20log\left(\frac{f_0}{\Delta f}\right) \tag{1}$$

Table 1 Comparison of algorithm results

Simulations	Initial value	SQP	LS	PD
FoM	−99.5	−104.6	−101.2	−101.8
PN@1 MHz (dBc/Hz)	−104.2	−111.8	−109.2	−110.1
Period Jitter (ps)	40.02	20.05	22.50	25.25
Cycle-to-cycle Jitters (ps)	69.32	34.74	38.98	43.73
P_{DC} (mW)	5.23	4.60	5.01	4.87

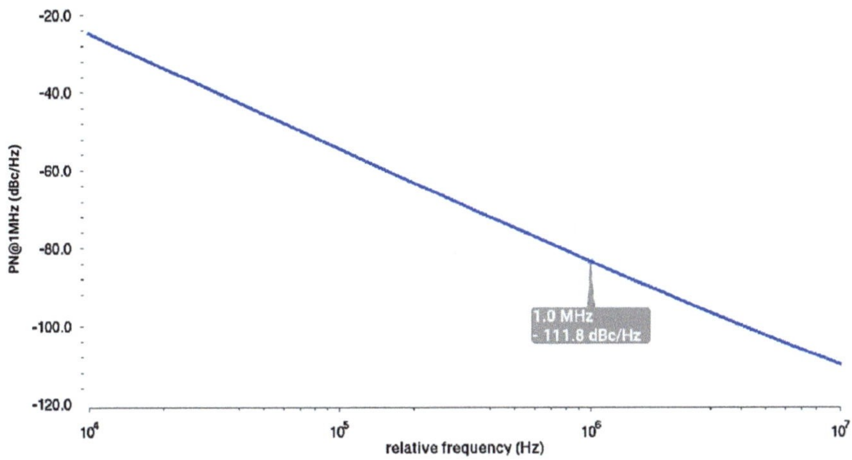

Fig. 6 Best performance achieved for PN optimization (SQP algorithm)

After all these simulations and optimization through the DNO, Table 1 provides a summary of the numerical characterization through a set of reference parameters for comparative performance evaluation, inside a common context, such as the same frequency range (550 MHz–1 GHz) and capacitive load (50 fF).

A certain improvement was observed in all cases, paying more attention to the phase noise, by the fact, that perhaps it is the most important parameter in many oscillators, reflecting frequency stability of a signal. On this matter, the best performing algorithm was the SQP, in Fig. 6 a detailed result is illustrated.

6 Conclusion

A circuit analysis and yield improvement methodology have been presented in this paper to study a VCRO with the goal of putting in place an approach to deeply analyze the circuit and find the best feasible solution with increased reliability in junction with the use of radiation tolerant devices.

Whether it was to explore different circuit analysis, WiCkeD™ module interfaces, and consequently, the separate databases created by the tool (feasibility analysis, sensitivity analysis, mismatch analysis and worst-case analysis). We have been able to gather plenty of useful information, needed to really understand the impact of each effect on performances and then concentrating our efforts on the most relevant ones. Those different analyses were successful on the task of emulating various tests (e.g., operating conditions—PVT), in order to solve the problems more accurately, and have a starting point efficiency and security.

Simulation results show that yield analysis for local parameter variations with the SQP is more efficient and more reliable compared to other algorithms for this application. Parameter variations with the SQP are more efficient and more reliable compared to other algorithms. With a considerable improvement in two essential items, consumption (4.602 mW) and phase noise (−111.8 dBc/Hz). Thus, the presented deterministic approach provides the necessary optimization including the dependency of parameter variations.

Significant results were obtained. Thus the tool has demonstrated powerful effects through their optimization algorithms. In addition, having an easy-to-use interface allied to user's circuit knowledge and design experience provides the project an optimized design cycle time.

Acknowledgements The authors would like to thank CAPES, CNPq, and FAPESP for the financial support, CITAR Program for providing the infrastructure for the accomplishment of this work, and to Dr. Gunter Strube from MunEDA® Company to provide essential references for the accomplishment of this work.

References

1. Schmid, A., Leblebici, Y.: Robust circuit and system design methodologies for nanometer-scale devices and single-electron transistors. In: IEEE Transactions on Very Large Scale Integration (VLSI) Systems, v: 12, n: 11, pp. 1156–1166. IEEE Solid-State Circuits Society (Nov 2014). https://doi.org/10.1109/TVLSI.2004.836292
2. Redant, S., Marec, R., Baguena, L., Liegeon, E., Soucarre, J., Van Thielen, B., Beeckman, G., Ribeiro, P., Fernandez-Leon, A., Glass, B.: Radiation test results on first silicon in the design against radiation effects (DARE) library. IEEE Trans. Nucl. Sci. **52**(5), pp. 1550–1554 (2005). https://doi.org/10.1109/TNS.2005.855818
3. Kinget, P.: Integrated GHz voltage controlled oscillators. In: Analog Circuit Design, pp. 353–381 (1999)
4. Jovanovic, G., Stojcev, M., Stamenkovic, Z.: A CMOS voltage controlled ring oscillator with improved frequency stability. Sci. Publ. State Univ. Novi Pazar Ser. A: App. Math. Inf. Mech. **2**(1), 1–9 (2010)
5. Alioto, M., Palumbo, G.: Oscillation frequency in CML and ESCL ring oscillators. IEEE Trans. Circ. Syst. I: Fundam. Theory Appl. **48**(2), pp. 210–214 (2001). https://doi.org/10.1109/81.904885
6. Razavi, B.: A 2-GHz 1.6-mW phase-locked loop. IEEE J. Solid-State Circ. **32**(5), pp. 730–735 (1997). https://doi.org/10.1109/4.568843

7. Park, C.-H., Kim, O., Kim, B.: A 1.8-GHz self-calibrated phase-locked loop with precise I/Q matching. IEEE J. Solid-State Circ. **36**(5), pp. 777–783 (2001). https://doi.org/10.1109/4.918915
8. Sun, L., Kwasniewski, T.A.: A 1.25-GHz 0.35μm monolithic CMOS PLL based on a multiphase ring oscillator. IEEE J. Solid-State Circ. **36**(6), pp. 910–916 (2001). https://doi.org/10.1109/4.924853
9. Savoj, J., Razavi, B.: A 10-Gb/s CMOS clock and data recovery circuit with a half-rate linear phase detector. IEEE J. Solid-State Circ. **36**(5), pp. 761–768: 5 (2001). https://doi.org/10.1109/4.918913
10. Ken Yang, C.-K., Farjad-Rad, R., Horowitz, M.A.: A 0.5-μmCMOS 4.0-Gbit/s serial link transceiver with data recovery using oversampling. IEEE J. Solid-State Circ. **33**(5), pp. 713–722 (1998). https://doi.org/10.1109/4.668986
11. Jovanovic, G., Stojcev, M.: A method for improvement stability of a CMOS voltage controlled ring oscillator. In: International Scientific Conference on Information, Communication and Energy Systems and Technologies (ICEST), vol. 2, pp. 715–718 (2007)
12. Jalil, J., Reaz, M.B.I., Arif Sobhan Bhuiyan, M., Rahman, L.F., Chang, T.G.: Designing a Ring VCO for RFID Transponders in 0.18μm CMOS process. Sci. World J. (2014)
13. Tong, T., Wenhua, Z., Hvolgaard, J.M., Larsen, T.: A 0.18μm CMOS Low Power Ring VCO with 1GHz tuning range for 3-5 GHz FM-UWB applications. In: 10th IEEE Singapore International Conference Communications System, pp. 1–5 (Nov 2006). https://doi.org/10.1109/ICCS.2006.301371
14. Zhang, C., Li, Z., Fang, J., Zhao, J., Guo, Y., Chen, J.: A novel high-speed CMOS fully-differentical ring VCO. In: 12th IEEE International Conference on Solid-State and Integrated Circuit Technology (ICSICT), pp. 1–3. (2014). https://doi.org/10.1109/ICSICT.2014.7021580
15. Hajimiri, A., Limotyrakis, S., Lee, T.H.: Jitter and Pahse noise in ring oscillators. IEEE J. Solid-State Circ. **34**(6), 790–804 (1999)
16. Toh, Y., McNeill, J.A.: Single-ended to differential converter for multiple-stage single-ended ring oscillators. IEEE J. Solid-State Circ. **38**(1), pp. 141–145 (2005). https://doi.org/10.1109/JSSC.2002.806262
17. Chen, Z.-Z., Lee, T.-C.: The design and analysis of dual-delay-path ring oscillators. IEEE Trans. Circ. Syst. I: Regul. Pap. **58**(3), pp. 470–478 (Nov 2010). https://doi.org/10.1109/TCSI.2010.2072390
18. Pinto, J.P.C.: Quadrature generators based on ring oscillators and shift registers. Dissertation, New university of Lisboa (2015)
19. Dornelas, H. Schmidt, A., Strube, G., Fabris, E.: New technology migration methodology for analog IC design. In: 28th Symposium on Integrated Circuits and Systems Design (2015)
20. *WiCkeD Manual*
21. Pinto Jr., A.M., Iano, Y., Manera, L.T., Souza, R.R.N.: An optimization tool-based design strategy applied to divide-by-2 circuits with unbalanced loads. Int. J. Electr. Comput. Energ. Electron. Commun. Eng. **9**(6), 408–412 (2015)
22. Pinto, Jr., A.M., Souza, R.R.N., Iano, Y., Manera, L.T.: WiCkeD tool-based design method for divide-by-2 circuits with multiple loads. In: 2015 European Microelectronics Packaging Conference (EMPC), pp. 1–5 (2015)
23. Pelgrom, M.J.M., Duinmaijer, A.C.J.: Matching properties of MOS transistors. IEEE J. Solid-State Circ. **24**(5), 1433–1439 (1989)
24. Lakshmikumar, K.R., Hadaway, R.A., Copeland, M.A.: Characterisation and modeling of mismatch in MOS transistors for precision analog design. IEEE J. Solid-State Circ. **21**(6), pp. 1057–1066. (1986). https://doi.org/10.1109/JSSC.1986.1052648
25. Conti, M., Crippa, P., Orcioni, S., Turchetti, C.: Parametric yield formulation of MOS IC's affected by mismatch effect. IEEE Trans. Comput.-Aided Des. Integr. Circ. Syst. **18**(5), pp. 582–596 (1999) (IEEE—Council on Electronic Design Automation). https://doi.org/10.1109/43.759074
26. Michael, C., Ismail, M.: Statistical modeling of device mismatch for analog MOS integrated circuits. IEEE J. Solid-State Circ. **27**(2), pp. 154–166 (1992). https://doi.org/10.1109/4.127338

27. Pelgrom, M.J.M., Tuinhout, H.P., Vertregt, M.: Transistor matching in analog CMOS applications. In: Electron Devices Meeting, 1998. IEDM '98. Technical Digest., International. pp. 915–918 (1998) https://doi.org/10.1109/IEDM.1998.746503
28. Rout, P.K., Acharya, D.P.: Design of CMOS ring oscillator using CMODE. In: 2011 International Conference on Energy, Automation and Signal, pp. 1–6 (2011). https://doi.org/10.1109/ICEAS.2011.6147142
29. Pokharel, R.K., Nizhnik, O., Tomar, A., Lingala, S. Kanaya, H., Yoshida, K.: Wide tuning range CMOS quadrature ring oscillator of best figure of merit. In: 2009 European Microwave Integrated Circuits Conference (EuMIC), pp. 172–175 (2009)
30. Pinto, Jr, A.M.: Oscilador controlado Por tensão com estrutura em anel, com critérios de confiabilidade aos efeitos da radiação. Dissertation, Campinas State University (2017)

Computational Performance of an Model for Wireless Telecommunication Systems with Discrete Events and Multipath Rayleigh

Reinaldo Padilha, Yuzo Iano, Edson Moschim, Ana Carolina Borges Monteiro and Hermes José Loschi

Abstract With the objective of improving the transmission of content in wireless telecommunication systems, in simulation environment is proposed a precoding process of bits based in the application of discrete events in the signal before of the modulation process. The proposal brings a different approach of usual technical, in which the signal transmission on the channel is realized in the discrete domain with the implementation of discrete entities in the process of bit generation. The present work implements a model based on discrete events applied at a low level of abstraction in a wireless telecommunication system named hybrid method, being used the Simulink simulation environment of the MATLAB® software. In the simulation are considered advanced modulation format for signal transmission in an AWGN channel, and the studies are performed in different physical machines. The results show improvement of 9 to 30% in memory utilization, as also best computational performance as to simulation time. The extension of the results of this study has a strong impact on the methods performed in higher layers, being able to improve them even more, since this proposal acts on the bits.

Keywords Methodologies · Discrete events · Simulation · Precoding

---

R. Padilha (✉) · Y. Iano · E. Moschim · A. C. B. Monteiro · H. J. Loschi
School of Electrical and Computer Engineering (FEEC), University of Campinas—UNICAMP, Av. Albert Einstein – 400, Barão Geraldo, Campinas, SP, Brazil
e-mail: padilha@decom.fee.unicamp.com

Y. Iano
e-mail: yuzo@decom.fee.unicamp.com

E. Moschim
e-mail: moschim@dsif.fee.unicamp.br

A. C. B. Monteiro
e-mail: monteiro@decom.fee.unicamp.com

H. J. Loschi
e-mail: eng.hermes.loschi@ieee.org

Y. Iano et al. (eds.), *Proceedings of the 3rd Brazilian Technology Symposium*,
https://doi.org/10.1007/978-3-319-93112-8_20

1 Introduction

Computer simulations are strong tools that support the best knowledge of how a wireless telecommunication system is operating. In the simulation environment, the designer has the flexibility to implement different types of system architecture to analyze different layers, such as physical, transport, transmission and higher layer, improving and validating the system for different applications [1–6].

Through simulation, real systems can be studied in a dynamic and flexible manner, allowing changes of various aspects can be easily evaluated without a physical experimental setup. This flexibility allows adjust project parameters, aiming improve system performance as a whole and minimizing costs [2].

Some of the advantages of simulations are related to the freedom of the designer to model a system, could model coherent concepts compared to reality as well as propose new methodologies.

The power of the received signal is a factor that is directly linked to the possible data rates stable in a mobile communications system across the transmission. The major challenge related to transmission in these systems falls on the multipath effect in these channels.

Multipath fading is the result of propagating multiple versions of signals transmitted across different paths before they reach the receiver.

The attenuations suffered by the signal due to the effects of the multipath fading can be represented with the statistical distribution of Rayleigh.

The term Discrete Event is mainly used in the sense to denote the modeling that suggests representing the system as a sequence of operations performed on entities (transactions) of certain types such as data packets, bits, among others. These entities are discrete items of interest in a discrete event simulation, your meaning depends on what is being modeled and the type of system [1, 2].

This technique is usually used to model concepts with a high level of abstraction, such as patients in a hospital, clients in a queue, emails on a server, flow of vehicles, manufacturing enterprise, transmission of data packets in telecommunications systems, among others [2–11].

In this paper, a hybrid model for wireless telecommunication systems was made and implemented using an AWGN channel and advanced modulation format DBPSK in simulation environment, with the objective of to increase the transmission capacity of information content through of the channel and to compensate for the additional complexity of modulation across the multipath for better performance related to memory consumption.

Where a bit treatment with discrete events methodology was modeled in the step of bit generation, being the differential of this work the use of discrete events applied in a low level of abstraction. The results show better computational performance related to memory utilization of the simulation model.

The present work is organized as follows: Sect. 2 discusses traditional simulation models, showing the modeling of transmission channel AWGN. Section 3 presents and describes the proposed framework of this paper, based on the hybrid model with discrete event methodology. Section 4 presents the results and, finally, in Sect. 5, the conclusions are presented as also the potential of the search.

2 Traditional Simulation Model

A model widely used due to its simplicity and mathematical treatment, and what applies to a large set of physical channels, is the Additive White Gaussian Noise channel model, AWGN, which introduces in the transmitted signals a noise modeled. Statistically as a white gaussian additive process [12–14].

However, the mobile wireless channel is susceptible to a number of impediments including multipath, fading, shadowing, noise among other interferences. In such a way that such deficiencies can cause an enormous degradation in the performance of the system [15, 16, 17, 12, 18, 14].

Rayleigh fading is considered as a reasonable and ideal model for heavily built urban environments for radio signals with the propagation of signals in a means where there is no dominant propagation along a line of sight between the transmitter and the receiver [12, 14, 18].

Generally, wireless networks differ mainly in their physical layer, wherein the transmission of data electromagnetic waves are used that propagate through space, where are reflected, dispersed, attenuated, among other interferences. Thus, for this type of transmission an efficient carrier frequency data modulation is required [14–18].

The DBPSK modulation (Differential Binary Phase Shift Keying) eliminates phase ambiguity and the need for phase acquisition and tracking, resulting in advantages in the reduced cost of energy. In modulation is employed in a non-coherent way to solve the need for a coherent reference signal at the receiver. Thus, the input binary sequence is first differentially encoded and then modulated using a BPSK modulator. In demodulation, there is no need to know about the initial state of the bit, simplifying synchronization [12–14, 17].

In BPSK (Binary Phase- Shift Keying), one phase represents the binary 1 and the other phase represents the binary 0, and as the digital input signal changes state, the phase of the output signal changes between two angles separated by 180° [12–17].

The model presented in this section aim to display an AWGN transmission channel with DBPSK modulation. For this, was used the Simulink simulation environment of the MATLAB® software in its version 8.3 of 64 bits (2014a).

In the model, in Fig. 1, the signals corresponding to bits 0 and 1 are generated and then modulated in DBPSK, passing through a multipath Rayleigh fading channel with Jakes model with Doppler shift defined at 0.01 Hz, as also inserted a block incorporated which has a math function 1/u.

Such a function is required to track the time-variability channel where the receiver implementation ordinarily incorporates an automatic gain control (AGC), after following for an AWGN channel according to the parameters specified as sample time of 1 s, power input signal of 1 W, initial seed in the generator of 37 and in the channel

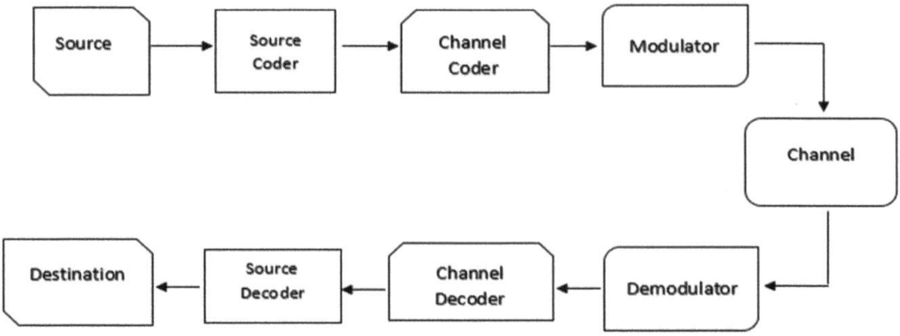

Fig. 1 Traditional model

of 67, Eb/No of 0 to 14 dB. Then the signal is demodulated in order to perform the bit error rate (BER) of the channel. The values obtained referring the BER are sent to the MATLAB® workspace, for further processing and generating of the signal BER graph.

3 Proposed Framework

The modeling according to discrete events is identical with that shown in previous section. Differentiating that in this model, was added the discrete events process of precoding, consisting of the treatment performed on the signal corresponding to bit 0, being converted into discrete entities, and forwarded for a FIFO queue with infinite capacity. Moreover, there is no limit of capacity and retention in the block, storing entities in the First-In-First-Out sequence, ordering the bits following really your order of arrival, and thus driving to a server, which have configuration of service time equal to the simulation time.

Where the differential of this work is in the use of discrete events applied in such low level of abstraction, being the bit generation. After the signal passes through the server, is converted back to its original format respecting the original format and data type specified and maintaining the sampling period respectively. Thus, the signal is modulated in DBPSK and inserted into the AWGN channel, and then demodulated for the purposes of calculating the BER of the signal. These relative values to this BER are also sent to the MATLAB® workspace, for further processing and generating of the signal BER graph.

The model presented in Fig. 2, incorporates the traditional modeling with a proposal presented, as also highlights the part modeled according to the technique of discrete events, in blue, as previously described.

And in Fig. 3, using 10,000 s of simulation time, was placed the flows of transmission of the DBPSK signal in relation to hybrid model (below) and traditional

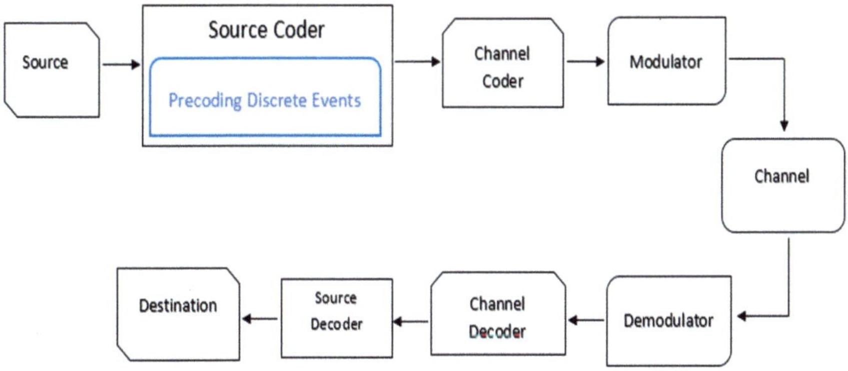

Fig. 2 Hybrid model

model (top) for better viewing and comparison, noting that both methodologies also generated the same result.

Also used was the scatterplot command, which is a scatter diagram or constellation diagram, to view the constellation of a modulated digital signal, being useful for comparing the performance of one system with another.

In Fig. 4 is shown the results for visualization of the constellations in 5, 10 e 15 dB, according to the hybrid model (below) and traditional model (top).

4 Results

In this section, the results will be showed where were performed 5 sequential simulations with the models presented previously, on physical machines with different hardware configuration, being an Intel Core i5 processor and 8 GB RAM, and another with an Intel Core i3 processor and 4 GB RAM.

Was used the `sldiagnostics` function, that displays diagnostic information about the modeling system in Simulink, calculating the sum all of the memory consumption processes used in the model in simulation, by the `ProcessMemUsage` parameter, which counts the amount of memory utilized in each phase of the model, during the entire simulation, displaying the total amount in MB, according presented in the Figs. 5 and 6.

Also was analyzed the first simulation of both models, because it is in the first that the variables are allocated and the memory reserved for the execution of the model, having a better performance as shown in Table 1 and related with Figs. 7 and 8.

To analyze the relationship between the simulation methodology and the impact on the physical layer of the channel, scripts were made in the MATLAB® workspace for processing of the graph relative to BER, that allows analyzing the performance of bit error rate (BER) in communications systems.

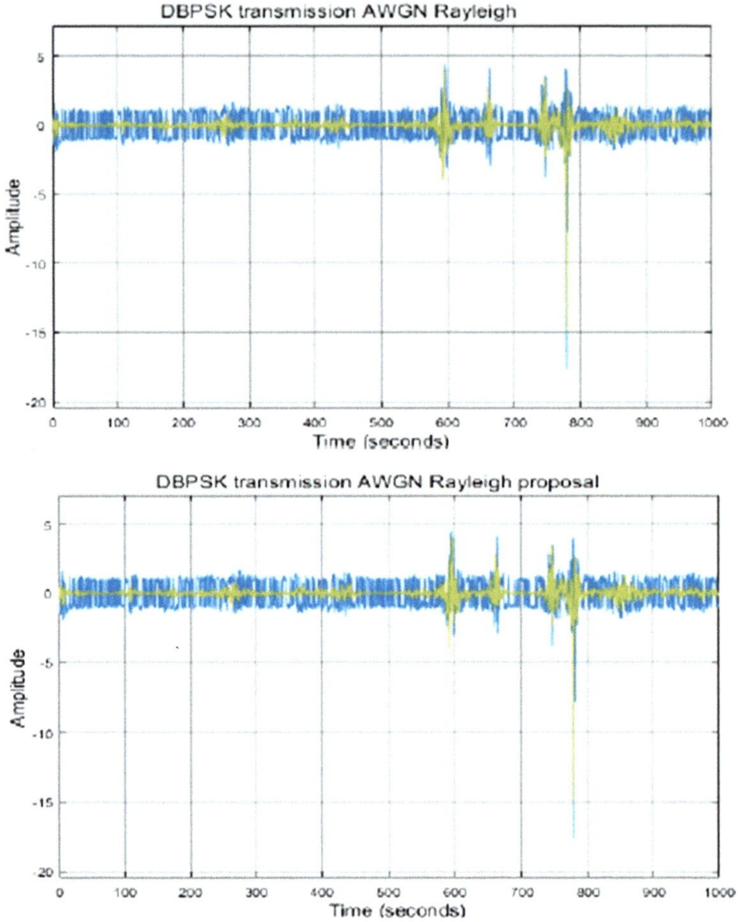

Fig. 3 Transmission flow DQPSK

In Fig. 9, is displayed the performance of the models according to simulation methodologies under study, along with a transmission with noise ranging from 0 to 60 dB.

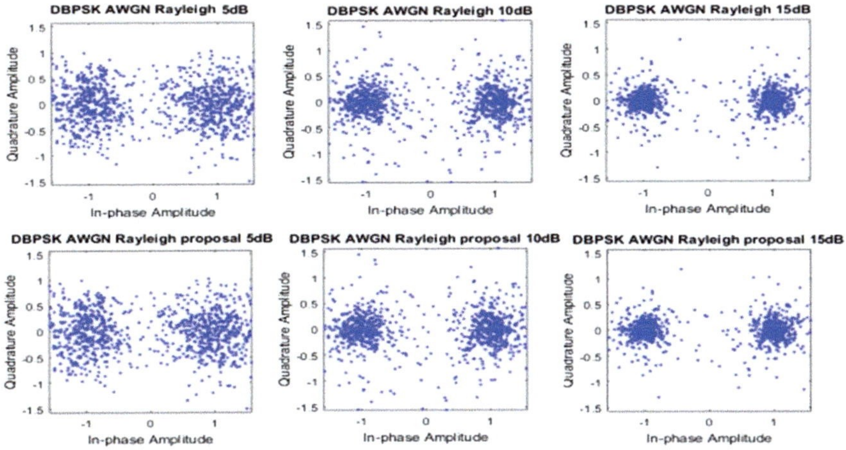

Fig. 4 Simulated DQPSK constellation

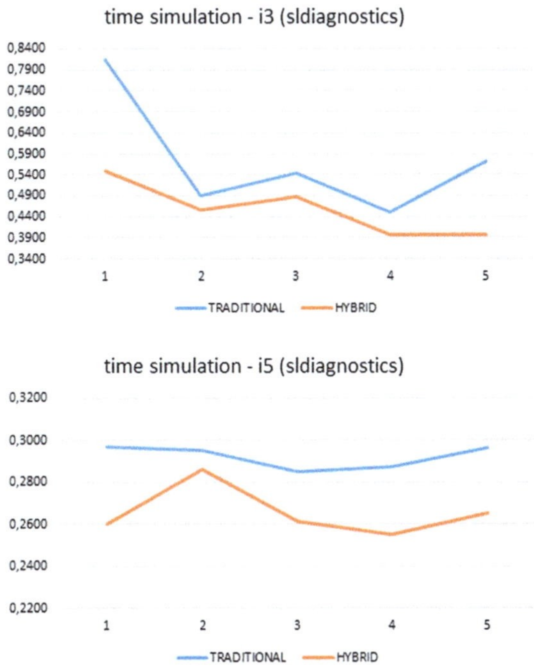

Fig. 5 Time simulation with `sldiagnostics`

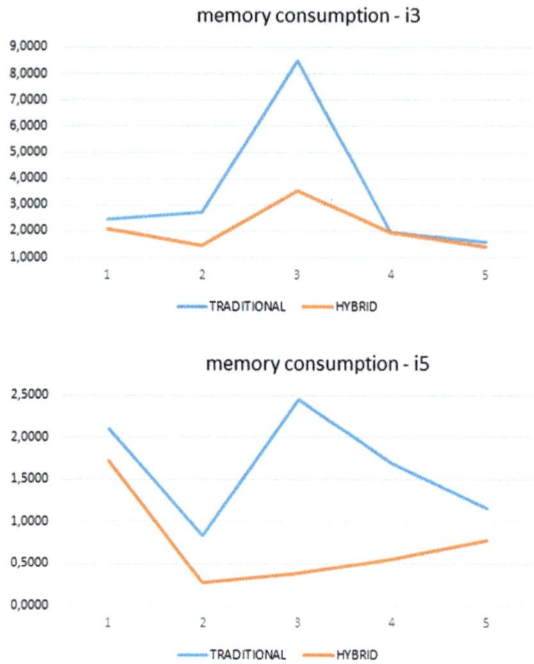

Fig. 6 Memory consumption simulation

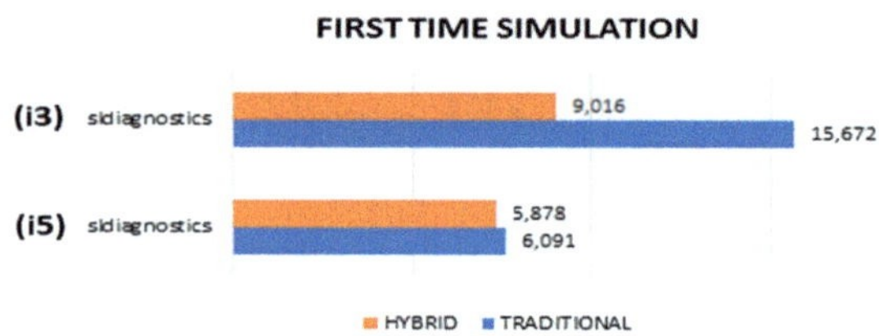

Fig. 7 First time simulation

Table 1 Computational improvement

Memory consumption		
Machines	**i3**	**i5**
Modelo DBPSK Rayleigh	29,72%	9,36%

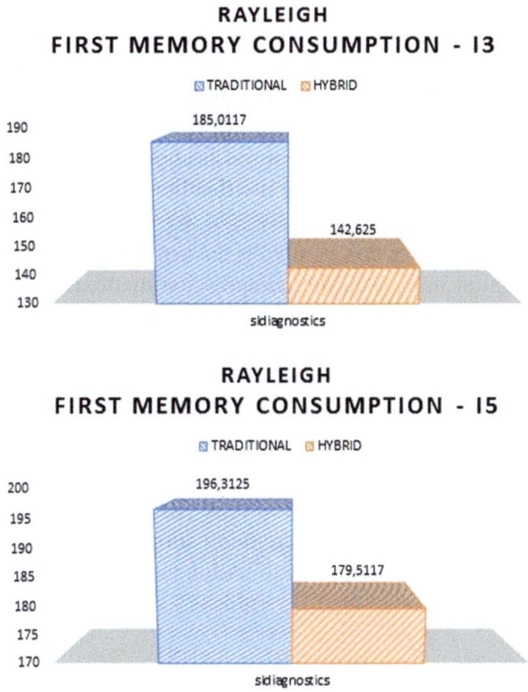

Fig. 8 First memory consumption simulation

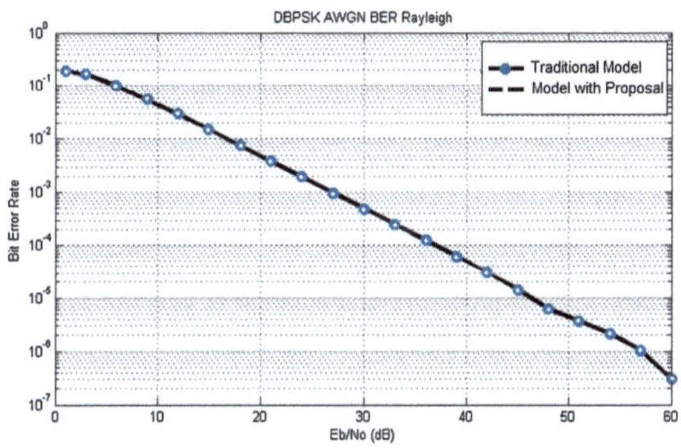

Fig. 9 BER comparative between the models

5 Conclusions

In all scenarios analyzed, the simulation model of the system with discrete event methodology on both different hardware configurations, evaluated on memory consumption, obtained better results, when compared with the model with the traditional methodology, either in its first simulation or along the sequence of 5 simulations.

Thus, the use of discrete events applied in a low level of abstraction such as bit generation in wireless telecommunication system model performed a treatment of the bits before of the modulation process, functioning as a pre- coding process differentiated.

Another important factor that the proposed framework achieves is the compensation obtained concerning to the complexity that the multipath fading, Rayleigh in this work, generates the modulation.

Better information compression is a byproduct of this work since it acts on the bits. It has a substantial impact on the compression methods performed in higher layers (e.g., Digital Dolby Plus Audio, MPEG-4 AVC/H.264, etc.) of a broadcasting system.

The purpose of this research, was the development of simulation models of wireless telecommunication systems, taking a different approach from what is normally done and applying a concept of a methodology at a lower abstraction level than it is normally used.

And looking at the Non-Orthogonal Multiple Access (NOMA) context, used in 5G, consisting of the near future in telecommunications, to compensate for the additional complexity brought by the multipath fading techniques in the applied modulation formats.

References

1. Digital Modulation in Communications Systems. An Introduction, Agilent Technologies (2001)
2. Padilha, R., Martins, B.I., Moschim, E.: Discrete event simulation and dynamical systems: a study of art. BTSym'16, Campinas, SP—Brasil, December (2016)
3. Pereira, F.T., Takano, A.M., Leal, F., Pinho, F.A.: Aplicação Da Simulação A Eventos Discretos Em Um Ambiente Hospitalar Visando A Melhoria No Processo De Atendimento. XLVSBPO, Natal, RN –Brasil (2013)
4. Sharda, B., Bury, J.S.: A Discrete event simulation model for reliability modeling of a chemical plant. Winter Simulation Conference (2008)
5. Hu, W., Sarjoughian, H.S.: Discrete-event simulation of network systems using distributed object computing. SPECTS'05 (2005)
6. Sasaki, N.K., Moschim, E.: Simulação de Sistemas de Comunicação Óptica Baseada em Simulação a Eventos Discretos. Universidade Estadual de Campinas. Campinas, SP—Brasil. July (2007)
7. Pissinelli, J.G., Risso, L.A., Picanco, S.R.A., Ignacio, A.S.P., Silva, L.A.: Modelo De Simulação De Eventos Discretos Para Análise De Fluxo De Veículos. ENEGEP, Fortaleza, CE—Brasil (2015)
8. Rangel, J.J.A., Costa, J.V.S., Laurindo, Q.M.G., Peixoto, T.A., Matias, I.O.: Análise do fluxo de operações em um servidor de e-mail através de simulação a eventos discretos com o software livre Ururau. Produto and Produção **17**(1), 1–12 (2016)
9. Gomes, E.N., Fernandes, M.S.R., Campos, C.A.V., Viana, A.C.: Um Mecanismo de Remoção de Mensagens Obsoletas para as Redes Tolerantes a Atrasos e Interrupções. CSBC (2012)
10. Godoy, E.P., Lopes W.C., Sousa, R.V.: Porto, A.J.V.: Modelagem E Simulação De Redes De Comunicação Baseadas No Protocolo Can—Controller Area Network. Revista SBA: Controle & Automação, **21**(4), (2010)
11. Forrester, J.W.: Industrial dynamics—after the first decade. Manage. Sci. **14**(7):398–415 (1968)

12. Freeman, R.L.: Fundamentals of Telecommunications. Wiley (1999)
13. Freeman, R.L.: Telecommunication System Engineering, 4th edn. Wiley (2004)
14. Proakis, J.G.: Digital Communications, 5rd edn, McGraw-Hill (2008)
15. Tozer, E.P.: Broadcast Engineer's Reference Book. Focal Press (2012)
16. Whitaker, C.J.: Standard Handbook of Broadcast Engineering. McGraw-Hill (2005)
17. Couch II, L.W.: Digital and Analog Communication Systems. 8th Prentice Hall (2013)
18. Freeman, R.L.: Telecommunication System Engineering. 4th Wiley (2004)

Electrical Power Monitoring of Low-Power Devices Using a Smart Meter

Rachel Batalha de Lima and Edgard Luciano Oliveira da Silva

Abstract Integration of sensors and actuators via the Internet can optimize energy consumption. Hence, IoT devices will integrate several types of energy consuming things and connect them to the energy supply company to balance power generation and usage. Users will interact with their appliances, or handle them using a cloud-based interface. This article describes an intelligent electric energy meter relying on the Esp8266 microcontroller with the NodeMCU firmware as an IoT gateway to forward in conjunction with current and voltage sensors connected to the ThingSpeak analysis platform. The resulting values were used as a way of timely controlling the consumption of low-power appliances.

Keywords Esp8266 · Electric power control · Internet of things · Thinkspeak Sensors

1 Introduction

The electrical energy brings many benefits to society's development, boosting the economy and improving population's quality of life. However, it is necessary to be aware of the energy consumption, production, and distribution in Brazil because of some factors as the depletion of natural resources, environmental impacts, and investments on new power plants' constructions. A solution to this is the implementation of efficient electrical energy usage through public politics to aware people and some research development.

The number of consumers in the residential electrical sector has been growing every year [1], as shown in Table 1, which consequently impacts on the increase of electrical consumption in this sector. Consumers are separated according to the

R. B. de Lima (✉) · E. L. O. da Silva
University of the State of Amazonas, Manaus, AM 69050-010, Brazil
e-mail: rbdl.eng@uea.edu.br

E. L. O. da Silva
e-mail: elsilva@uea.edu.br

Y. Iano et al. (eds.), *Proceedings of the 3rd Brazilian Technology Symposium*,
https://doi.org/10.1007/978-3-319-93112-8_21

Table 1 Number of consumers per year

	2012	2013	2014	2015	2016
Residential	61.697	63.862	66.007	67.746	69.277

voltage level they use. The rule says that the higher the consumption, the higher the voltage level: Group A—Class A1 (230 kV or more), Class A2 (88–138 kV), Class A3 (69 kV) and Class A4 (2.3–25 kV); Group B—Low Voltage (Less than 2, 3 kV) and Residential (110–220 V) [2].

Consumers, even when having an impact on the consumption of electricity, are not aware of the real-time quantity of kilowatt-hours spent by small attitudes like unnecessary lighting, the usage of electric shower, standby appliances mode, among others. The price paid by the final consumer, about the delivered service, is given monthly by the concessionaire based on its consumption and the price of a kilowatt-hour. A meter quantifies all this energy, in its different routes to the final consumer.

Taking in consideration the power measurement, the concessionaires cannot obtain the real-time levels of current and voltage, nor do they have the control of the technical losses during the transmission, nor the theft of energy, since their monitoring of the meters is done only once a month. It affects the improvement of consumption's analysis and the knowledge of the real quality of the service being delivered.

In this context, this paper contributes to the development of an effective and low-cost Smart Meter, using IoT concepts. It does the comparison and analysis between the developed project and preexisting solutions in the market, presenting advantages and disadvantages of the proposed methodology. IoT support comes from the ThingSpeak platform which is an open source Internet of Things (IoT) API that stores and saves data from things via the HTTP protocol over the Internet or using a Local Area Network.

The paper is organized as follows: Sect. 2 presents the necessary theoretical basis for the comprehension of this paper. Section 3 points out some of the main related projects about the subject. Section 4 describes the approach used in the proposed meter development. Section 5 exposes the experiments and results obtained, and finally, Sect. 6 presents the conclusions and suggestions for further projects.

2 Theoretical Basis

2.1 Electrical Quantities

Energy is the capacity of something to do work, what means, to generate force in a particular body, substance or physical system [3]. To obtain a better understanding on the subject, it is necessary to know some technical concepts that involve this theme. These are: Current (A), Voltage (V), and Power (W).

The electrical voltage is the difference in electric potential between two points. Its unit of measurement is the volt (V) and is usually represented in the equations and circuits by the letters U or V. This voltage is supplied to the circuit by a generator. In electronic circuits, generators are usually batteries, which transform chemical energy into electric. There are also other types of generator as mechanical, solar, thermal, and magnetic generators.

The electric current can be defined as the flux of electrons in a conductor when exposed to a difference in electric potential controlled by a generator, the unit of measurement is Ampère (A), is usually used the letter I to represent it in electrical circuits and equations.

Power is defined as the energy rate (U) that is transformed or transferred by time, is equal to the product of the current in the element by the voltage at its terminals. It is measured in units of joules/second, also known as watts [4].

2.2 *Power Meters*

The operation principle of the electromechanical energy meter is the same as the induction motor. Its rotation, along with the disk that has an axis connected to an auger, causes registers' rotation which will provide the energy consumption reading. Consumption is a measurement performed monthly. The readings must be subtracted from the previous month to have the actual consumption [5].

Electronic meters read the variables through short intervals of time samples using Analog-to-Digital (A/D) converters. The accuracy class of the equipment is determined by the number of bits in the A/D converters and processing precision, allowing them to be more accurate than the electromechanical meters. To perform equipment readings, a communication protocol is used between the meter and the equipment used by a reader to store the read registers, or through network communication protocols, sending the data directly to concessionaire's database [6].

The implementation of electronic meters has been progressively deployed due to the high cost of new equipment. In 2012, the Brazilian Electricity Regulatory Agency (in Portuguese, Agência Nacional de Energia Elétrica, ANEEL) regulated a resolution for electronic energy metering systems for consumer units of Group B (residential, rural and other classes, except low income and public lighting) [7].

The IoT can help energy management because it offers systems to automatically collect and act on energy and power-related data with the goal to improve the efficiency, economy, reliability, and sustainability of both the generation and the distribution of electricity. Using advanced meter devices coupled to the Internet backbone can allow collecting data from end-user, and manage energy distribution.

2.3 *Esp8266*

The ESP8266 is a 32-bit microcontroller that includes a Tensilica L106 microprocessor core with standard frequency ranging from 80 to 160 MHz. As communication interfaces, there are serial interfaces synchronous, SPI, I2C, and I2S; asynchronous USART and a Wi-Fi interface, which can operate in the AP mode, as a client, in an *ad hoc* or Wi-Fi Direct [8] mode.

The NodeMCU module used in this project is a development board that adds an ESP8266 chip, has development options in Lua language or with the same Arduino IDE. It has two buttons: flash, used in recording firmware, and RST (Reset). On the same side, we have the micro-USB connector for power and connection to the computer. On the opposite side, the ESP-12E and its built-in antenna already welded on the board. On the sides are the GPIO pins, external power, and communication.

3 Related Projects

A residential electricity demand meter with remote access was presented in [9] using an ACS712 current sensor, a voltage sensor module and a W5100 Ethernet shield for communication with the web application in conjunction with the Arduino UNO for measuring voltage and current, obtaining a high cost of R$ 320.50.

As exemplified in [10], an ASC12 30 A current sensor was used, a voltage sensor 127/220 V with Arduino UNO getting the voltage and current values, and the PLX-DAQ software to collect the data in a spreadsheet, presenting a total cost payment of R$ 200.00.

To develop a prototype for an electric energy measurement [11] presents a prototype at the cost of $108.00. Consisting of a non-invasive current sensor, a voltage filter, and a Bluetooth sensor to send data to the computer.

This work differs from previous ones due to the replacement of the Arduino Uno by a NodeMCU, which has higher processing power. It also brings a prototype with the most affordable price and makes use of IoT concepts, so the user obtains a better experience with the software usage.

4 Proposed Approach

This paper proposes the development of Smart Meter for low-power devices consisting of two phases, as shown in Fig. 4

1. Hardware development (current gauge circuit); and
2. Communication between the ThingSpeak platform and the hardware for analysis of IoT data.

In the hardware development phase, the prototype of the circuit for the current acquisition was made using the non-invasive sensor SCT 013-30A.

The ESP8266 NodeMCU module communicates with the ThingSpeak platform over the HTTP network protocol, sending data every 15 s (Fig. 1).

4.1 *Current Gauge Circuit*

The current gauge circuit is shown in Fig. 2. It presents the NodeMCU in conjunction with a 4-channel ADS1115 Digital-to-Analog converter for reading the analog ports, a 330-ohm resistor, and a P2 connector for acquiring the current sensor signal.

5 Experiments and Results

The experiments carried out in this paper had the objective of monitoring the current using the SCT 013-30A sensor in conjunction with the ThingSpeak platform.

The first experiment was performed on a lamp with the following characteristics 280 mA and 110–127 V as seen in Fig. 3.

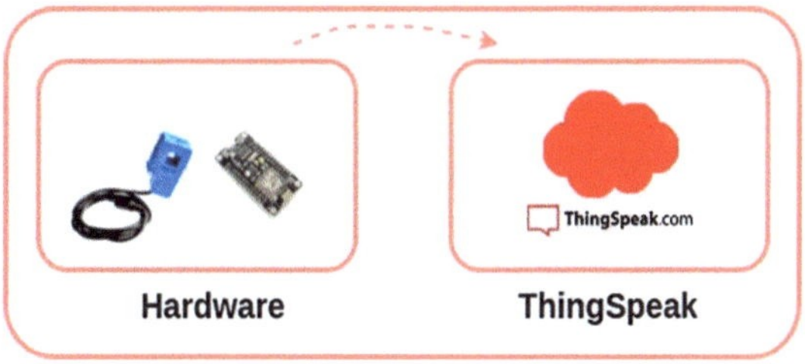

Fig. 1 The proposed framework for the Smart Meter

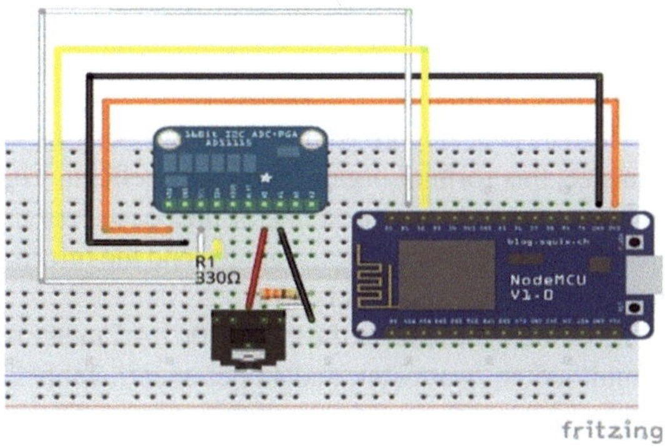

Fig. 2 Current gauge circuit

Fig. 3 Measurements on a lamp with the Smart Meter

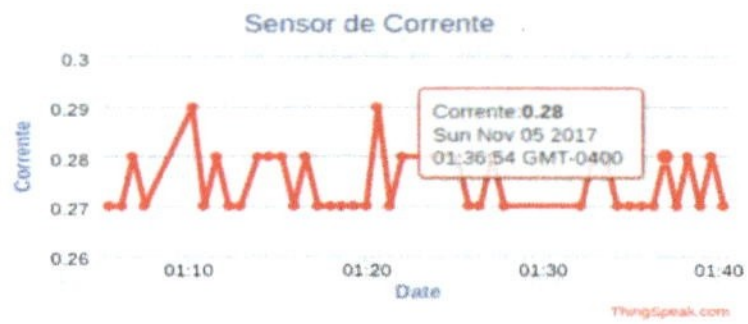

Fig. 4 Chart with current sensor data generated by the ThingSpeak platform

Table 2 Values of current and power

	Current (A)	Power (Watts)
1	0.27	33.46
2	0.27	33.46
3	0.28	34.72
4	0.27	33.46
5	0.29	35.96
6	*0.27*	33.46
7	0.28	34.72
8	0.27	33.46
9	*0.27*	33.46
10	*0.28*	34.72

The data go to the ThingSpeak platform which serves as data storage on the web, displaying and storing it as charts as shown in Fig. 4. The browser displays the data, and it is possible to view the information from any device that has Internet access and browser available.

The data are categorized into ThingSpeak in the form of channels. Due to the large data size, this form of organization is beneficial. The HTTP protocol is used to send the data, and in this request one of the passed parameters is a unique key (linked to the ThingSpeak channel you want to send the information to). The minimum interval between two submissions is 15 s.

To validate the results, the PZEM-061 AC meter was used, which has the functions of multimeter, wattmeter, and amperemeter. Table 2 shows the resulting current and power values.

Regarding cost, compared to other microcontrollers, the approach using Arduino UNO takes advantages because it can achieve same power processing with a lower market cost.

6 Conclusion and Further Projects

In this paper was presented the monitoring of electric power in low-power devices through a Smart Meter intended to be used according to the IoT paradigm. The results are satisfactory for a first version. There was a variation in the power values due to the use of an approximated voltage value which was 127 V.

As further projects will be implemented the usage of a voltage sensor module to have more accurate measurements, the two-wattmeter method has also been studied to achieve greater accuracy. Thus, the present article aims to evolve the Smart Meter for three-phase installations by coupling it to the residential electric grid, so the user will be able to obtain real-time control of their electric consumption.

References

1. Energy Research Company—EPE, Statistical Yearbook of Electrical Energy 2017, http://www.epe.gov.br/AnuarioEstatisticodeEnergiaEletrica/Forms/Anurio.aspx. Last Accessed 23 Sept 2017
2. Brazilian Association of Large Industrial Energy Consumers and Free Consumers—ABRACE, An Introduction to the Electrical Sector, http://www.abrace.org.br/wpcontent/uploads/2015/12/manualenergiaeletrica.pdf. Last Accessed 24 Sept 2017
3. Alexander, C.K., Sadiku, M.N.O.: Fundamentals of Electric Circuits. McGraw-Hill, Boston (2009)
4. Nilsson, J.W. R., Nilsson, S. A. J.W., Riedel, T.: Electric Circuits, 10th. Prentice Hall (2014)
5. Creder, H., T.: Electrical Installations, LTC Editors (2007)
6. da Silva, M.R.: Modeling and analysis of the useful Life (meters) of Induction type Meters of active Electric Energy. M.S Thesis, Unesp (2010)
7. National electric energy agency—Aneel. public consultations 2017, http://www2.aneel.gov.br/aplicacoes/consultapublica/. Last Accessed 23 Sept 2017
8. Oliveira, Sergio: Things internet with ESP8266, Arduino and Raspberry PI. Novatec, São Paulo (2017)
9. de Paula, G.J.: Residential electric power demand meter with remote access, 2013
10. Marcelino, J.E.C., Costa, T.C.C., Morais, P.R.R.: Prototype of a smart electric meter: applied to residential consumer units. V SINGEP (2016)
11. Silva, I.A.D.: Intelligent voltage-current sensor with smartphone monitoring and control online (2016)

Applied Instrumentation: Strain Measurements Using Arduino and Strain Gauge

Elton Fernandes dos Santos, Vlademir de Jesus Silva Oliveira, Wagner de Almeida Ferreira and Julio César Beltrame Benatti

Abstract Strain gauges are devices used to measure strain on objects. The measured value is proportional to the variation of its electrical resistance, and this method is widely used in circuits for data acquisition. Many methods of analog signal conditioning are found in the literature, but the conversion and storage are dependent on the technology employed. In this paper, a study of a data acquisition module to perform mechanical tests on metal bars is presented. The system consists of a signal conditioning circuit compatible with the Arduino platform, where the collected data can be processed and stored. This work covers the steps by step of signal conditioning, such as amplification, and analog-to-digital conversion. The circuits employed are explained, and the detail about software and hardware technology are commented. In the end, experimental tests using the module are presented where a discussion about the test results are given. The method affords changes in the setup, which can be useful in similar tests.

Keywords Difference amplifier · Arduino · Strain gauge · Data acquisition

1 Introduction

Proper planning of a structural project requires accurate structure information. Extensometry, which is the technique used for the experimental analysis of tensions and deformations in mechanical structures, is the main technique used for this purpose

E. F. dos Santos · V. J. S. Oliveira (✉) · W. A. Ferreira · J. C. B. Benatti
Universidade do Estado de Mato Grosso, Jd. Imperial Sinop Ingás 3001, Brazil
e-mail: vlademir.oliveira@unemat.br

E. F. dos Santos
e-mail: eltonfernando90@gmail.com

W. A. Ferreira
e-mail: wagnervt.almeida@gmail.com

J. C. B. Benatti
e-mail: benatti@unemat.br

Y. Iano et al. (eds.), *Proceedings of the 3rd Brazilian Technology Symposium*,
https://doi.org/10.1007/978-3-319-93112-8_22

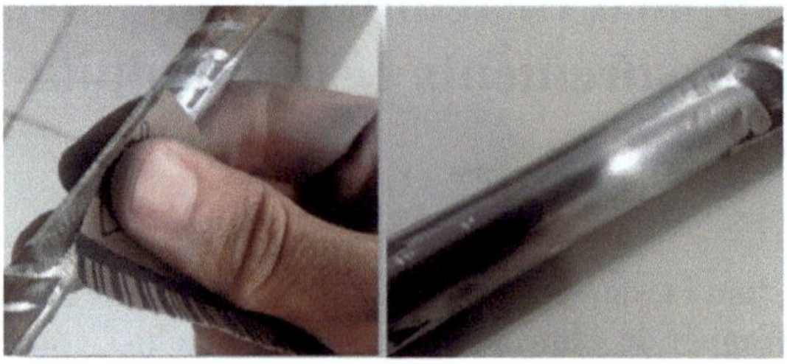

Fig. 1 Surface preparation

[1–4]. Although there are other methods, the use of strain gauge is one of the most versatile [5]. This work presents a model for signal processing using extensometry, in which it has been implemented in a module compatible with the Arduino platform and a set up of a test carried out with standard steel bar (500 MPa) used in reinforced concrete structures.

2 Background

2.1 Strain Gauge Features

The preparation of the surface and fixation of the strain gauge is a very important step, so all manufacturer's recommendations should be carefully checked. If this step is conducted improperly, the results will be compromised [6]. The objective is that the strain gauge accurately reproduces the small deformations, so the surface must be uniform and properly prepared, as in the Fig. 1. The fixation can be done with cyanoacrylate at the base of the sensor. The surface must be completely clean, acetone works well for this purpose.

The strain gauge is designed to measure only longitudinal deformations. For this reason, it is important to identify in advance the direction of the deformations that you want to measure. However, for specific measurements, there are combinations of extensometers that can be angularly arranged to allow measurement in various directions. The size is determined by the characteristic of the material used in the test, in heterogeneous materials a larger area is required for the resistance variation to be significant. The choice of the appropriate sensor also depends on the conditions that affect it during the process. The most relevant information is expressed in code, which combined appear in the sensor name [7]. The model used was the BF-350-3-AA from AGS.

- BF: characteristics related to the sensor manufacturing material.
- 350: Ohmic resistance.
- 3-AA: geometric feature.

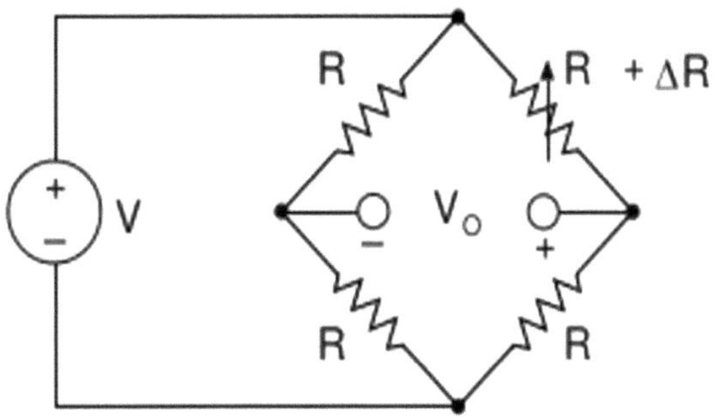

Fig. 2 A quarter-bridge configuration

The strain gauges are temperature sensitive, so depending on the application, compensation methods must be used.

2.2 *Signal Conditioning*

The Wheatstone bridge is employed in many applications, especially in small signal measurements. A bridge can be composed of four components, resistors or sensors. It can be called a quarter, half, or full bridge when the resistors are replaced by one, two, or four sensors respectively. In Fig. 2 there is a quarter-bridge example. There are several types of orientation settings depending on the type of measurement. The system sensitivity is proportional to the number of sensors in the bridge [8].

Assuming that the resistors are equal and ideal and that the sensor has a variation ΔR, the potential difference V_o can be deducted by calculating the voltages adopting the source negative terminal as ground, resulting in the Eqs. (1) and (2)

$$V_{o-} = \left(\frac{VR}{2R}\right) = \frac{V}{2}, \tag{1}$$

and

$$V_{o+} = \left(\frac{VR}{2R + \Delta R}\right) = \left(\frac{V}{2 + \Delta R/R}\right). \tag{2}$$

The output V_o is given by the difference between the Eqs. (2) and (1)

$$V_o = -\frac{V}{2}\left(\frac{\Delta R/R}{2 + \Delta R/R}\right). \tag{3}$$

Operational amplifiers are integrated circuits that can be configured according to the needs of the designer, and have interesting features such as high input impedance and very high gain, enabling them for a wide range of applications [9]. One of the

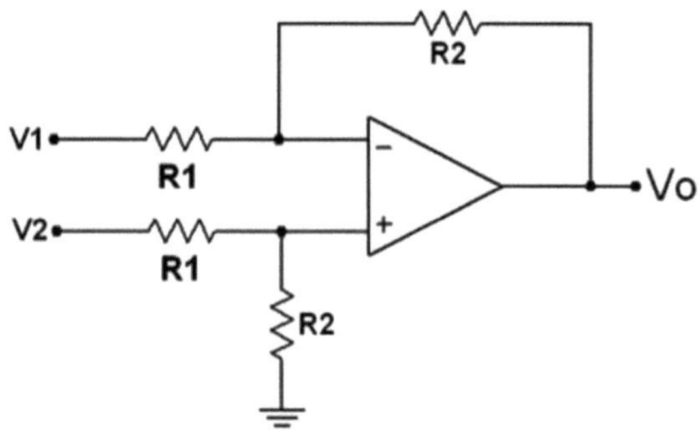

Fig. 3 Difference amplifier configuration

most used configurations in instrumentation is the difference amplifier, which can be seen in Fig. 3.

The circuit output is a result of the input voltage difference multiplied by a gain, as shown in Eq. (8). But first, let us define

$$V_+ = \frac{V_2 R_2}{R_1 + R_2}, \tag{4}$$

$$\frac{V_1 - V_+}{R_1} = \frac{V_+ - V_o}{R_2}, \tag{5}$$

and

$$V_o = V_+\left(\frac{R_2}{R_1} + 1\right) - \frac{V_1 R_2}{R_1}. \tag{6}$$

Substituting (4) into (6) one can get

$$V_o = \left(\frac{V_2 R_2}{R_2 + R_1}\right)\left(\frac{R_2 + R_1}{R_1}\right) - \frac{V_1 R_2}{R_1}, \tag{7}$$

and

$$V_o = \frac{R_2}{R_1}(V_2 - V_1). \tag{8}$$

Assuming that inputs V_1 and V_2 are both subject to the noise signal, then this unwanted signal would be amplified at the output, but this circuit has a characteristic called the Common-Mode Rejection Ratio (CMRR), which is the ability to attenuate identical signals in the entries [9]. Thus, the difference amplifier, using precision resistors, only amplifies differential signals coming from the input.

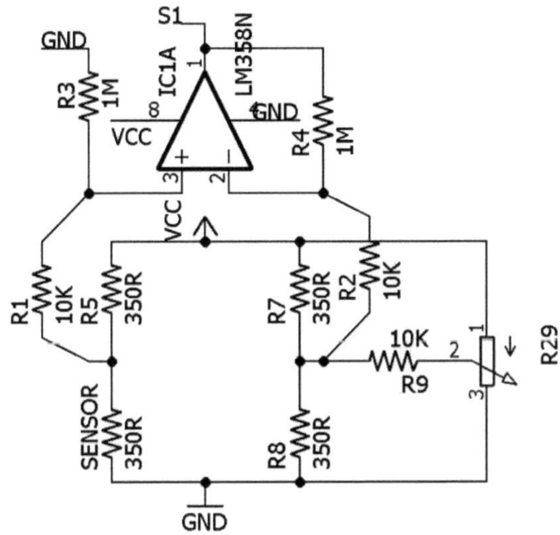

Fig. 4 Sensor signal conditioning circuit

2.3 *Circuit Design*

The circuit has the function of amplifying the differential signal from the sensor. It is the connection of a difference amplifier to the bridge poles. Potentiometers are coupled to avoid offset between the bridge poles. In the case of Fig. 4, one of the bridge resistors must be replaced by the sensor.

The amplifier gain must be adjusted according to the application. Although theoretical methods can determine the gain, it may not be satisfactory to measure, and since the tests are usually destructive, a calibration procedure should be performed. In this work, a load press has been used to adjust the strain. Figure 5 shows the experimental setup.

The sensors are attached to the steel bar to measure strain (compression or tension), which is subjected to a specific load. The Printed Circuit Board (PCB) has potentiometers to nulling the gain at each sensor change. In order to avoid slipping, a few sheets of paper may be placed between the bar and the press table, the calibration procedure must be performed until the maximum load is reached that will be used in the load test.

2.4 *Signal Processing*

The analog-to-digital conversion of a physical phenomenon can be made by acquiring a portion of the analog signal using a sensor and representing it in binary form by the converter. There are ADC (Analog-to-Digital Converter) with 10–16 bits and sample rate of Mega-Samples per Second (MSPS) or Giga-Samples per Second (GSPS),

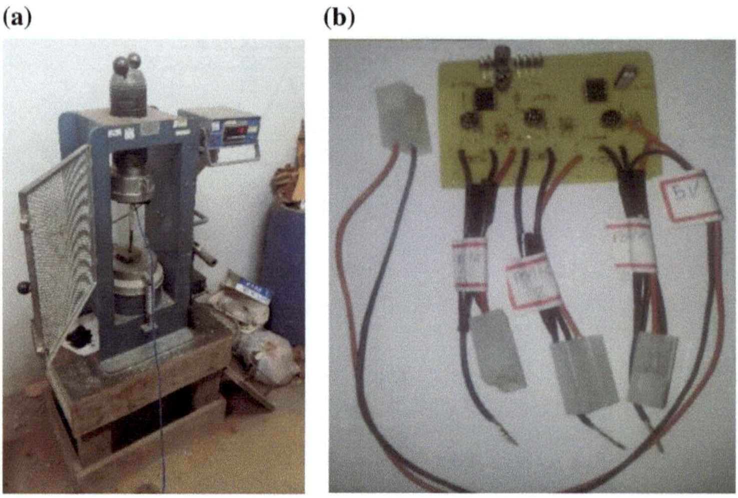

Fig. 5 Experimental set up used in tests. **a** Hydraulic press. **b** PCB and cables

which can be a single integrated circuit or a microcontroller module. The design specification depends on the type of application. In this design has been employed Arduino board. Arduino has become popular because is easy to program and has a lot of information available on the Internet.

Although almost all information can be converted to digital, there are some restrictions in the process, which must be remembered. The Arduino has a 10-bit integrated converter which on a scale from 0 to 5 V will represent 1024 positions. Therefore the lowest value that can be read is 5/1024, or 4.9 mV. The reference voltage can be changed to a value less than 5 V to improve resolution, but the range decreases [10]. Another factor is the quantization process, according to [10] there is an absolute error of 1/2 LSB, in this case, 4.9 mV/2 = 2.45 mV due to the finite resolution of the converter.

The sampling frequency is subject to the response time of the analogRead() function, which is about 100 μs. However, this is a particularity of the Arduino platform where the Prescaler division factor is fixed and set to 128. The pre-scaler of the AVR microcontroller can be reconfigured by changing the first 3 bits of the ADCSRA register, using its assembly language [11].

3 Data Acquisition System

The Arduino SD.h library has been used for storage on SD card, then a text file is created, where the data is organized into two columns format and later imported into Excel®. The Arduino receives the analog values from the sensors and converts them to

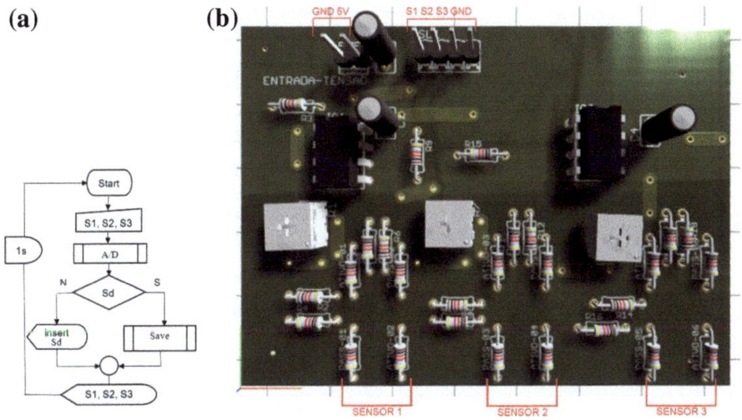

Fig. 6 Characteristics of the data acquisition system. **a** Arduino algorithm flowchart. **b** PCB generated in the 3D BRD Viewer online software

digital. After conversion, the Arduino SD recorder saves the data in a format that can be processed later. The algorithm developed for data acquisition is shown in Fig. 6a as a flowchart. Figure 6b shows the PCB that perform the analog processing. The board has support for up to three simultaneous sensors, i.e., three signal conditioning circuits are required. In this design, two LM358 has been used with decoupling capacitors connected to the sources, and 1% tolerance resistors. The output pins S1, S2, and S3 are connected to the analog pins of the Arduino. The sensors replace the resistors connected to the Wheatstone bridge. The circuit can be used for any combination of the bridge, but mainly one-fourth and half. The experiments were conducted using an Ethernet cable and three strain gauges. The Ethernet cable has four twisted wires. In this way, they allow the connection to three sensors sharing the same GND.

4 Results

In Fig. 7, the graph of voltage versus strain for the experimental test with a maximum load of 56 kN is presented. The saturation voltage of the amplifier depends on the output load, in this particular case the saturation occurred at 0 and 3.5 V. Thus, the voltage range shown in the graph of Fig. 7 (0–0.8 V) can be increased if a new gain be determined using the Eq. (9).

$$G = \frac{3.5 \times 100}{0.8} = 437.5, \tag{9}$$

where 3.5 is the saturation voltage, 100 the amplifier gain and 0.8 the range obtained in the test. It is advisable to round the gain values down to ensure that the amplifier

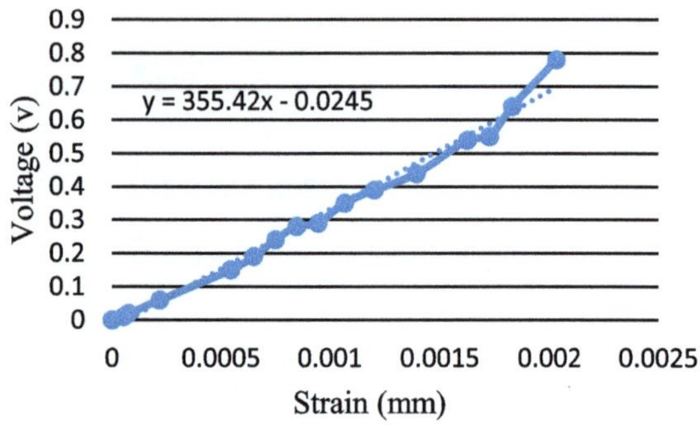

Fig. 7 Voltage versus strain for the experimental test with maximum load of 56 kN

does not operate in saturation. Therefore, to improve the range of the test using the standard steel bar (500 MPa), assuming that the load does not exceed 56 kN, the ratio of R4/R2 and R3/R1 should be changed to approximately 437.5.

The module developed in this work has been used to perform lattice structures tests using extensometry methods. In this test, three sensors were placed in some points of the structure in order to compare the measurements with other methods. The result has been compared with the calculation using the ideal model and simulation in Ftool 3.0 software. The measured results showed similarity with the simulation, demonstrating the efficacy of the method.

5 Conclusions

The proposed system presents two interesting points, which are a low cost and a clear method of application. The potentiometers coupled to the Wheatstone bridge allow the initial measurement value to be adjusted, so it is possible to carry out measurements in tension (increasing voltage scale) or compression (decreasing voltage scale) tests, only making changes in the initial value of the measurements.

The difference amplifier allows adjustment of input signal range of the A/D converter, by adjusting the gain. An undesirable effect has been observed in the resolution, as residual forces cause oscillations in the output. The half bridge configuration using a 90° extensometer helps to inhibit these effects.

References

1. Kettenbeil, C., Ravichandran, G.: Experimental investigation of the dynamic behavior of meta-concrete. Int. J. Impact Eng **111**, 199–207 (2018)
2. Lee, B., Mulheron, M.: Measurement of bar strain during pull-out tests: use of electrical resistance gauge methods under large displacement. Mag. Concr. Res. **67**(10), 523–531 (2015)

3. Lu, X., Li, G., Pei, H., Su, G., Wu, J., Amenuvor, A.C.: Experimental investigations on load transfer of PHC piles in highway foundation using FBG sensing technology. Int. J. Geomech. **17**(6), 1–12 (2017)
4. Nie, R., Leng, W., Yang, Q., Chen, Y.: An improved instrumentation method for PHC piles. Proc. Inst. Civil Eng.: Geotech. Eng. **169**(6), 494–508 (2016)
5. Pallas-Areny, R., Webster, J.G.: Sensors and Signal Conditioning. Wiley, New York (2001)
6. Excel Sensores.: Preparação da superfície e colagem do extensômetro na peça. 1.ed. São Paulo (2004)
7. AGS-TECH: Coding System of Strain Gauges, Albuquerque novo México. In: https://www.elecrow.com/download/Coding%20System%20of%20Strain%20Gauges-AGS-TECH%20Version.pdf, last accessed 15 Oct 2017
8. National Instruments - White-Paper.: Measuring Strain with Strain Gages. In: http://www.ni.com/white-paper/3642/en/, last accessed 21 Feb 2018
9. Franco, S.: Design with Operational Amplifiers and Analog Integrated Circuits. McGraw-Hill, New York (2014)
10. ATMEL.: AVR120: Characterization and Calibration of the ADC on an AVR, application note. In: http://www.atmel.com/Images/Atmel-2559-Characterization-and-Calibration-of-the-ADC-on-an-AVR_ApplicationNote_AVR120.pdf, last accessed 15 Nov 2017
11. ATMEL.: Atmel 8-bit microcontroller with 4/8/16/32KBYTES in-system programmable flash, datasheet. In: http://www.atmel.com/images/Atmel-8271-8-bit-AVR-Microcontroller-ATmega48A-48PA-88A-88PA-168A-168PA-328-328P_datasheet_Complete.pdf, last accessed 15 Nov 2017

Overview About Radiation–Matter Interaction Mechanisms and Mitigation Techniques

R. N. S. Raphael, L. E. Seixas Jr., Agord M. Pinto Jr., S. A. Bascopé, L. T. Manera, S. Finco and S. P. Gimenez

Abstract This work presents a general overview about the origin and manifestation of ionizing radiation-induced effects over CMOS technology-based semiconductor structures. From the characterization of radiation–matter interaction mechanisms and effects, this work summarizes the set of design strategies described in the literature for radiation hardened electronic implementation, considering system level, block level, and device level approaches. Additionally, a final case study is presented characterizing the Total Ionizing Dose TID tolerant operation of the Diamond MOSFET transistor as an innovative alternative of non-standard MOSFET's layout for space applications.

Keywords Cosmic radiation effects · Displacement damage · Total ionization dose · Single event effects · Radiation hardening techniques · Diamond MOSFET

R. N. S. Raphael (✉) · L. E. Seixas Jr. · S. A. Bascopé · S. Finco
Center of Information Technology Renato Archer (CTI), Campinas, Brazil
e-mail: raphael.souza@cti.gov.br

L. E. Seixas Jr.
e-mail: luis.seixas@cti.gov.br

S. A. Bascopé
e-mail: saul.bascope@cti.gov.br

S. Finco
e-mail: saulo.finco@cti.gov.br

A. M. Pinto Jr. · L. T. Manera
University of Campinas (Unicamp), Campinas, Brazil
e-mail: agord@engineer.com

L. T. Manera
e-mail: manera@fee.unicamp.br

S. P. Gimenez
FEI University Center, São Bernardo do Campo, Brazil
e-mail: spgimenez@fei.edu.br

Y. Iano et al. (eds.), *Proceedings of the 3rd Brazilian Technology Symposium*,
https://doi.org/10.1007/978-3-319-93112-8_23

1 Introduction

The development of electronic components and circuits associated with advanced mechanic systems allowed to explore the space above people's heads. Nevertheless, the launch of satellites over the atmosphere introduced to the science knowledge cosmic effects not observed at sea level. Much of the verified cosmic radiation is filtered by the atmosphere, but above it, satellite circuits are prone to suffer radiation interferences from a variety of sources (solar particles, galactic cosmic rays, and magnetosphere) [1]. As a result, the set of radiation-induced effects from the radiation–matter interaction implies on the performance degradation of electronic devices through transient and fixed states.

Additionally, the CMOS technology downscaling contributes with the rising sensitivity of electronic devices (even for terrestrial applications) to radiation-induced performance degradation from transient effects, especially considering structures for low-power- and high-frequency applications [2–4]. This scenery establishes a research and development demand toward the proposition of effective solutions for mitigating radiation effects, according to required levels [5–7] for each application.

From this application context, this work presents a general overview about radiation effects on CMOS-based electronic devices. Thus, Sect. 2 presents a brief description about ionizing radiation, Sect. 3 presents a theoretical description about the physical effects from the interaction of radiation with semiconductors, and Sect. 4 summarizes the application of mitigation techniques for radiation hardened electronic design at system, topology, and device level, in conformity with the nature of radiation-induced degrading effect. At the end of this section, a case study is presented considering the device level TID robustness performance of the Diamond MOSFET.

2 Radiation: Origin and Structure

Considering the origin criterion, cosmic radiation can be categorized as extragalactic, galactic, solar, interplanetary, or magnetosphere [1]. Produced by charged particles, electromagnetic radiation is physically described under a wave approach as a self-propagating synchronized oscillation of transverse electric and magnetic fields that propagate at speed of light in the vacuum. Under a corpuscular approach, it is composed by mass less particles or photons whose energy level is associated with the frequency. Considering the interaction with semiconductor structures, significant effects are verified from high energy (high frequency) radiation (X-rays and gamma-rays) [1]. Particulate radiation is composed by matter units and elementary particles whose energy level is associated with mass and electric charge (protons, neutrons, electrons, positrons, heavy ions, and alpha particles) [1]. The main physical features of the radiation structure are summarized through the Table 1, considering

Table 1 Radiation structure: mass and electric charge

Electric charge	Electromagnetic radiation	Particulate radiation
Neutral	X-rays	Neutrons
	gamma-rays	
Charged	–	Protons
		Electrons/positrons
		Heavy ions
		Alpha particles

two categorization criteria: **mass** (electromagnetic or particulate nature) and **electric charge**.

3 IC Radiation Effects

Through direct or indirect physical mechanisms, the interaction radiation–matter involves different processes of energy transfer: atom displacement, nuclear reactions, decomposition of molecules, and excitation/ionization of atoms [1]. The resulting effects are associated with the radiation features (mass, electric charge, energy level and incidence angle), and with the nature of the reference material (atom number and density). Thus, the verified set of interaction mechanisms is usually described considering three physical effects [6–8]: (1) **Single Event Effects** (**SEE**), (2) **Total Ionizing Dose** (**TID**), and (3) **Displacement Damage** (**DD**).

The general features associated with each class of effect are summarized according to the Table 2, considering the time features of the interaction (categorized as **single** or **cumulative**), and the nature of the event (involving processes of **ionization** or **displacement**).

Table 2 Radiation–matter interaction physical mechanisms

Features	Interaction		
	Ionization		Displacement
	Temporary	Fixed	Fixed
Cumulative effects	–	TID	DD
Single event effects	Soft errors	Hard errors	–

3.1 Single Event Effects (SEE)

Single Event Effects (**SEE**) represent a general class of random physical effects derived from the single incidence of high energy ionizing particles (protons, alpha particles, heavy ions) over semiconductor structures. According to the literature [9–11], a general classification can be defined considering the nature of the effects.

- **Hard Errors**: permanent damages from destructive effects.
- **Soft Errors**: temporary damages from non-destructive effects.

As an example, from the incidence of a high energy ionizing particle over the NMOS structure through the drain, according to Fig. 1, a general sequence of physical events can be described for the generation of a parasitic current pulse, as illustrated considering the double exponential-based model in Fig. 1a [12]. Thus, considering the mass, electric charge, energy level and incidence θ angle of the particle, different penetration paths can be established inside the semiconductor structure after collision, for characterizing the **charge deposition** step [10]. In this case, a set of electron–hole pairs is composed around the resulting penetration path, generating a distortion in the depletion layer of the drain-substrate junction around a funneling region, as illustrated in Fig. 1b. As a result, a sequence of 2 physical effects establishes in the drain a two components-based current transient for characterizing the **charge transport** step [10]: (1) **drift current** generated by the electrons flow from the strong electric field in the depletion region of the reverse-biased PN junction (drain-substrate junction), according to Fig. 1c, and (2) **diffusion current** generated from the scattering of the remaining electrons in the substrate, according to Fig. 1d.

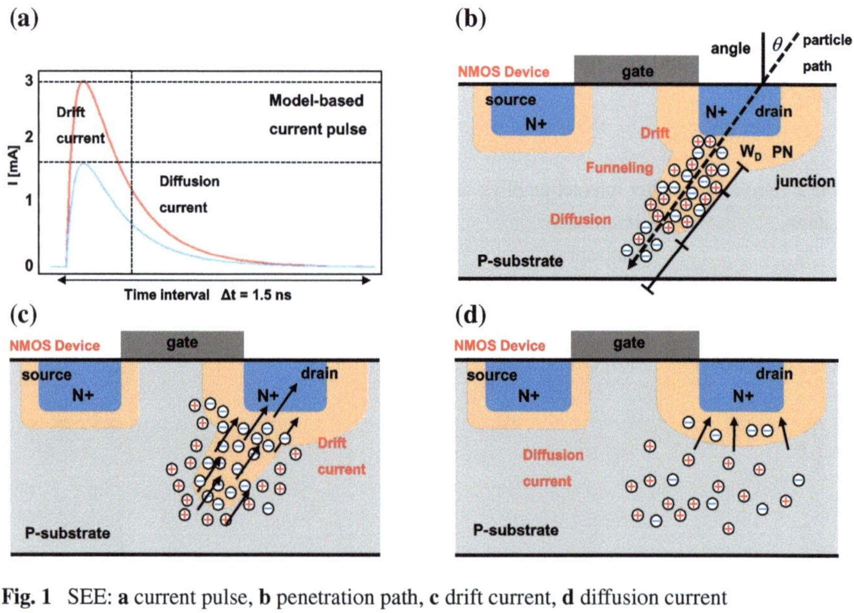

Fig. 1 SEE: **a** current pulse, **b** penetration path, **c** drift current, **d** diffusion current

Considering the digital domain, the resulting current transient is a spurious current pulse characterizing a **Single Event Transient** (**SET**) effect (transient disturbance on combinational structures) or a **Single Event Upset** (**SEU**) effect (static disturbance on memory units). Additionally, high intensity current pulses may generate permanent damages in the semiconductor structure, like for example **Single Event Gate Rupture** (**SEGR**). Thus, the set of SEE-based transient or permanent effects characterizes the **charge collection** step [10].

3.2 Total Ionizing Dose (TID)

Total Ionizing Dose (**TID**) represents a general class of ionizing-based cumulative physical effects derived from the continuous exposure of dielectric structures to ionizing radiation incidence [6–8].

Thus, from the distributed incidence of radiation over a NMOS transistor through the gate, according to Fig. 2, a sequence of resulting events can be described as follows: (a) incidence of ionizing radiation over the gate, (b) composition of a resulting set of electron–hole pairs in the gate oxide region, (c) migration of electrons (higher mobility) through the channel electric field-induced drift current and (d) migration of holes (lower mobility) toward the traps in the Si/SiO_2 interface region.

As a cumulative effect, the trapping of holes impacts on the electrical properties of the device, generating variations in many performance parameters: threshold voltage V_{TH}, current leakage, carrier's mobility, transconductance parameter, subthreshold slope, noise level, and input capacitance.

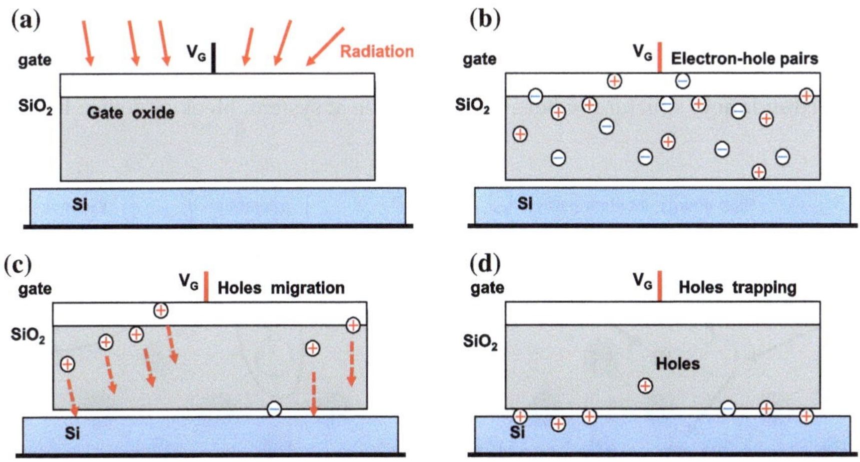

Fig. 2 TID: **a** radiation, **b** electron–hole pair, **c** holes migration, **d** holes trapping

3.3 Displacement Damage (DD)

Displacement Damage (**DD**) is a cumulative effect-based event generated from mechanical processes through the collision of particulate radiation (protons, neutrons, heavy ions, alpha particles or high energy electrons) with the silicon lattice, with the resulting displacement of atoms from the original position [13], as illustrated through Fig. 3.

As a permanent effect resulting from high energy incident particles, according to Fig. 3a, the resulting distortion on the silicon lattice can be characterized through the presence of **vacancies** (indicating previous positions of displaced atoms) and **interstitial spaces** (indicating new positions after displacement), according to Fig. 3b.

4 Radiation Hardening-Based Design Techniques

A general criterion is usually defined for categorization of radiation hardening-based techniques [6, 14]

- **Radiation Hardening by Process** (**RHBP**): implementation based on manufacturing process.
- **Radiation Hardening by Design** (**RHBD**): implementation based on design techniques.

In this context, a significant number of works has been proposed for describing RHBD-based design techniques for electronic structures implementation [14–18]. A general reference for robustness and reliability can be established from the following premises [19]:

- **Simplicity**: small number of components.
- **Robustness**: components insensitivity to variations in the operation conditions.
- **Redundancy**: structure or behavior replication at system, block or device level.

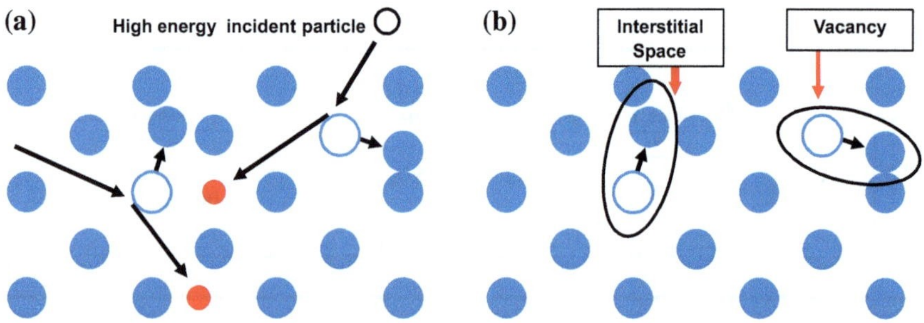

Fig. 3 DD: **a** incident particle in the lattice, **b** distortion (vacancies and interstitial spaces)

Thus, from the referred design premises, the set of proposed mitigation techniques for radiation effects can be summarized according to the general block diagram [14] in Fig. 4. In this case, considering the nature of the physical effect and the particularities of the circuit or system for implementation, the design strategy to be applied involves the definition and sizing of structures in different design stages (schematic and layout), in different redundancy [17] domains (hardware, software, time, and information) [14], according to Fig. 4a, and in different hierarchy levels (system, architecture, topology and device), according to Fig. 4b. On the other hand, the application of different radiation hardening design strategies usually establishes a trade-off between the obtained radiation tolerance and the system level reference performance parameters as area, power consumption and noise level.

Considering the referred block diagram, the next subsections present a general description about radiation effects mitigation techniques in different hierarchy levels for hardware redundancy-based applications.

4.1 System Level Techniques

System level methods for hardware redundancy-based design include different strategies for replication, [14, 17] considering fault detection-based structures as **Dual Modular Redundancy** (**DMR**), and fault masking-based structures as **Triple Modular Redundancy** (**TMR**) [19–23].

As a primary concept for fault masking-based reliability, Triple Modular Redundancy TMR represents a particular structure of modular redundancy derived from the triplication of a given module. In this context, considered as black box, this module

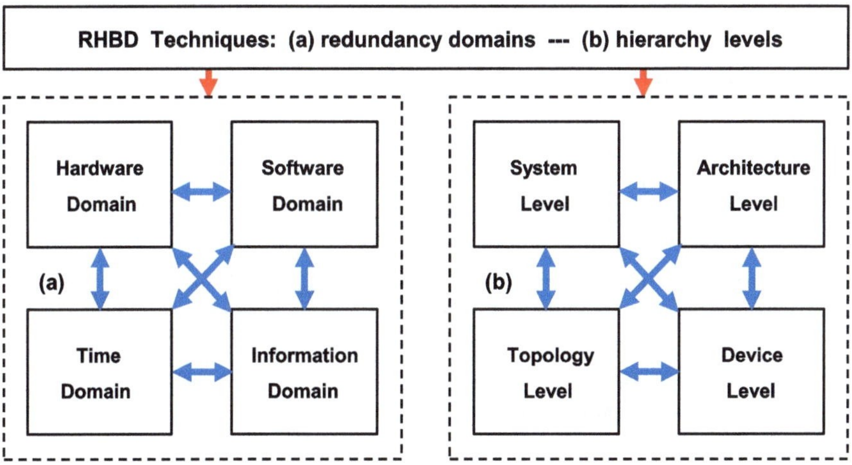

Fig. 4 RHBD design techniques: **a** redundancy domains, **b** hierarchy levels

may represent different hierarchy levels in digital domain (logic gates, combinational structures, or a complex digital system) or analog domain (delay cells in a ring oscillator). Thus, the collection of output data is provided from a voter circuit through the reproduction of the output generated from most of the modules, according to Fig. 5.

In this case, the resulting TMR network generates a correct output (without failures) even if one of the modules at most fails, assuming an uncorrelated operation between the modules [19].

From the analysis of the fault masking-based performance of a TMR network [24], the reliability model of the resulting redundant system R_S, disposed through TMR architecture, can be expressed as a function of the module reliability R_M, considering the voter reliability $R_V = 1$ (perfect voter) [19, 23, 24], according to the equation in (1).

$$R_S(R_M) = 3\,R_M^2 - 2\,R_M^3 \tag{1}$$

Thus, considering Fig. 6a, the resulting R_S variation profile is characterized with respect to R_M (in red), comparatively to an average straight line (in blue). In this case, the intersection point at P_E indicates an equivalent reliability level, and the subsequent data points indicate the R_M range with an effective reliability gain. Thus, an additional parameter can be defined for describing the effective average reliability gain G_R, according to the model in (2).

$$G_R(R_M) = \frac{R_S}{R_M} = 3\,R_M - 2\,R_M^2 \tag{2}$$

As result from this parameter, an additional variation profile can be characterized according to Fig. 6b, indicating a maximum point in the curve at P_M (R_M = 0.75, G_R = 1.125) for a specific reliability R_M = 0.75.

Alternative approaches for TMR-based systems have been proposed considering the impact over the area, power consumption and design complexity and, in this context, architectural variants from the reference TMR structure include proposals

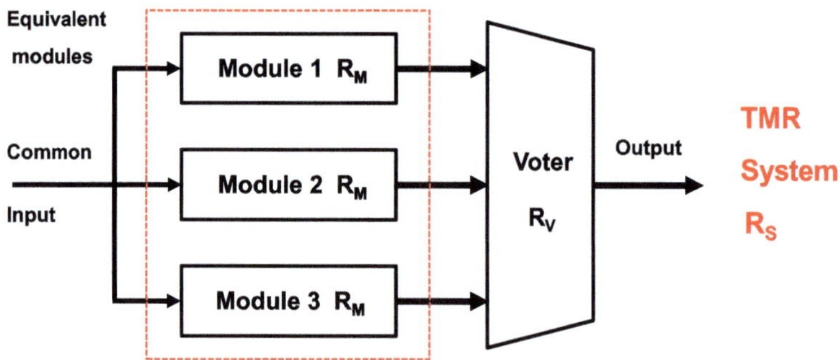

Fig. 5 Triple Modular Redundancy (TMR): basic conceptual representation

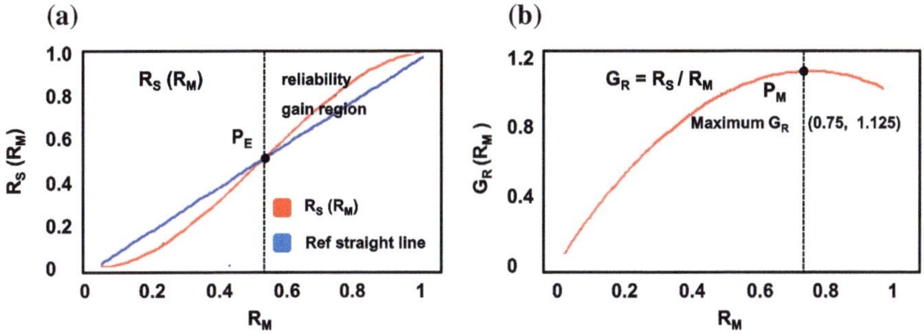

Fig. 6 TMR-based curves: **a** system reliability R_S, **b** corresponding reliability gain G_R

as **Cascaded TMR** [25, 26], **Hybrid Modular Redundancy** (**HMR**) [27], and **Hierarchical TMR** [28].

4.2 *Block Level Techniques*

Considering hardware redundancy-based techniques [17], block level strategies for SEE tolerant structures involves the replication of components according to different patterns. Toward this end, different approaches have been reported through the literature involving **Quadded Logic** [29, 30], **Interwoven Logic** [31], and **Dotted Logic** [32].

Additionally, design techniques for TID tolerant structures include topology strategies for reduction of threshold voltage variation ΔV_{TH} effects. As an example, Fig. 7a illustrates a standard CMOS inverter and Fig. 7b characterizes the corresponding transfer curve considering the normal inverter operation (green line), an intermediary distortion state (blue line) and a higher distortion level after irradiation (red line). Resulting from V_{TH} reduction for NMOS devices, V_{TH} increase in module for PMOS devices, and increase in the current leakage I_{LEAK}, TID-induced distortion effects in the transfer curve of logic gates imply in a reduction of the output rail voltage level and variation in the switching point (Fig. 7b) [18].

A components replication-based strategy is illustrated from Fig. 8a, as a topology proposal for an inverter logic gate implementation [18]. This building feature provides additional components for compensating the threshold voltage variation in the both NMOS and PMOS transistors [18], for allowing a transfer curve with lower distortion levels, as indicated in Fig. 8b. On the other hand, this topology pattern implies on a comparatively higher input capacitance and silicon area and reduced speed performance, considering devices with the same sizing.

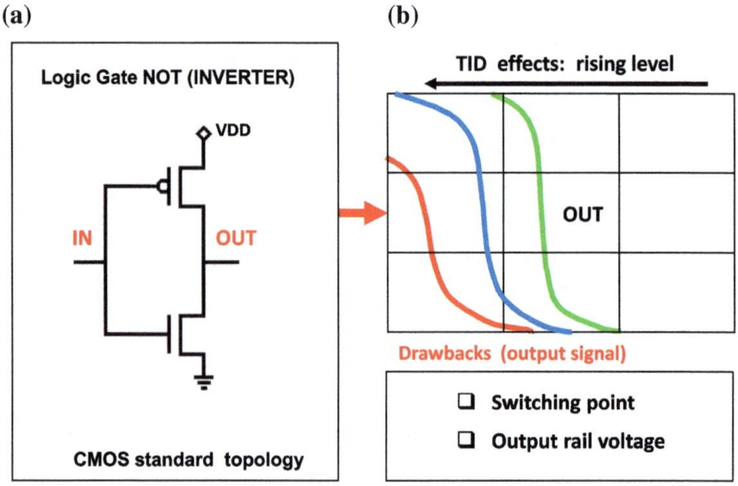

Fig. 7 Standard CMOS inverter: **a** conventional topology, **b** distorted transfer curves

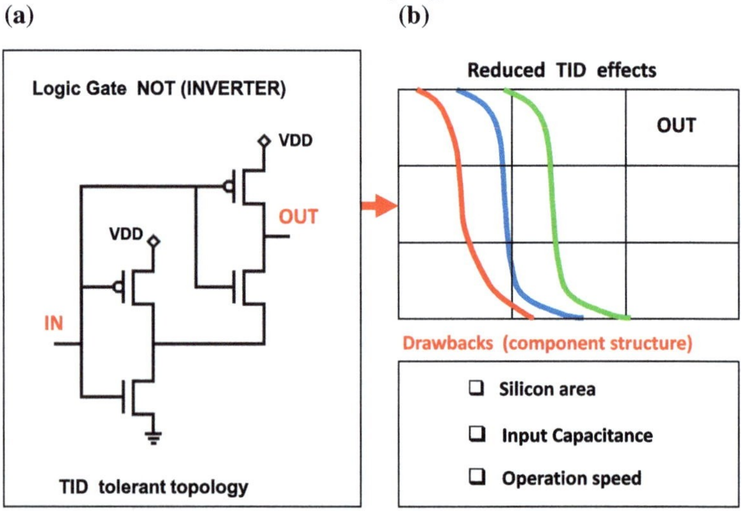

Fig. 8 Radiation hardened CMOS inverter: **a** topology with redundancy, **b** resulting curves

4.3 Device Level Techniques

From device-based hierarchy level, design strategies for radiation hardening usually involves a proper selection of (1) device geometry or **layout style** and (2) **sizing techniques** [33].

In this context, considering the first method, TID-based mitigation effects can be obtained through the current leakage reduction by applying enclosed layout device geometries.

Resulting from the continuous incidence of ionizing radiation over the MOS gate structure, the presence of positive charges or holes accumulated in oxide traps (thin gate oxide and thick field oxide) and Si/SiO_2 interface traps generates current leakage I_{LEAK} by attracting negative carriers and creating a source-drain parasitic path in parallel with the device channel.

Considering the conventional rectangular layout transistor or **Conventional MOSFET** (**CM**) represented in top view in Fig. 9a, the thick field oxide in touch with both source and drain creates an inversion layer in the nearby region, generating two main leakage paths: intra-transistor and inter-transistor. Thus, from the representation of the CM side view, Fig. 9b illustrates the trapping of positive charges in the **Bird's Beak Region** (**BBR**), at the transition between the thin gate oxide and the thick field oxide, and the resulting migration of negative carriers (for composing the parasitic path).

In this context, considering the building features of **Enclosed Layout Transistors** (**ELT**) as devices with edgeless structure or annular gate shape [34–36], as represented in Fig. 10, the BBR involves a common voltage node, allowing a reduction in the edge current leakage. On the other hand, the drawbacks of this edgeless structure include the drain-source asymmetry (considering the area surrounded through the gate), increased capacitance and area, difficult to implement smaller levels of W/L ratio without enlarging L, and non-availability for application in sub-100 nm technologies.

Additionally, a more advanced approach for current leakage reduction through device level TID tolerant structures considers non-standard gate geometries without requiring additional costs in the manufacturing process, by applying octagonal [37] and hexagonal (Diamond) patterns [38–42]. The application of the referred gate geometries implies on additional effects in the device operation as the **DEactivation** of **PArasitic MOSFETs** in the **Bird's Beak Regions** (**BBR**) **Effect** (**DEPAMBBRE**) [41, 42], allowing an improved and stable electrical performance for a wider temperature range [43–45].

Thus, for a given channel width W, a comparative illustration is indicated through Fig. 11, considering the conventional layout transistor or **Conventional** rectangular **MOSFET** (**CM**) with a channel length L, according to Fig. 11a, and the **Diamond**

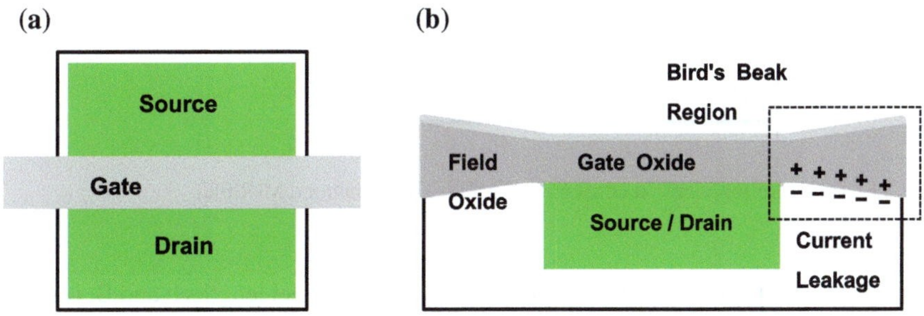

Fig. 9 Conventional transistor layout: **a** top view, **b** side view

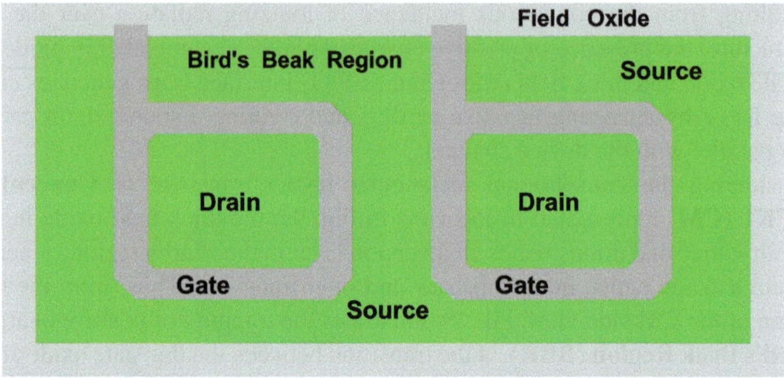

Fig. 10 Enclosed Layout Transistor (ELT): top view representation

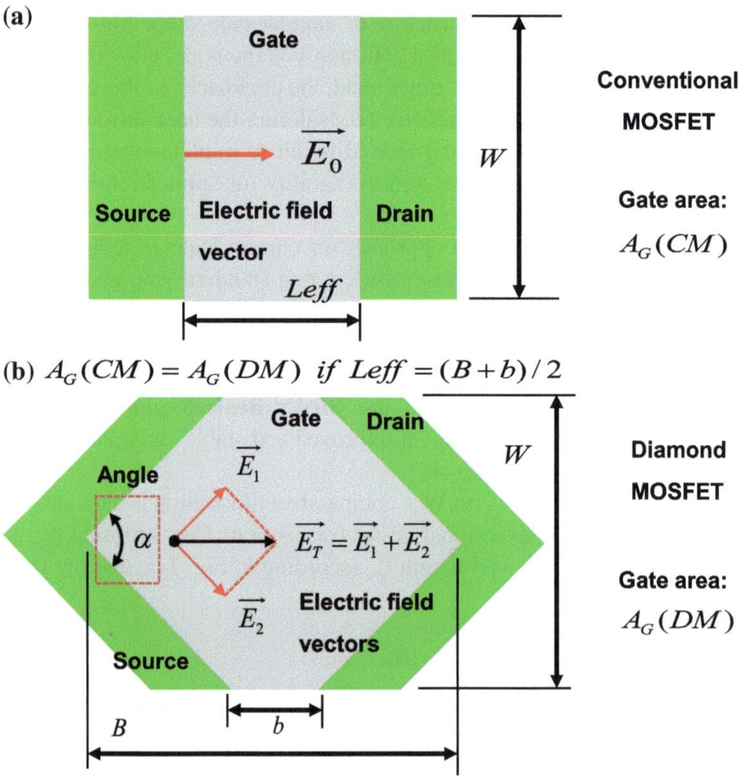

Fig. 11 Transistor top views: **a** conventional MOSFET, **b** diamond MOSFET

MOSFET (**DM**) [46], with the shortest and longest channel lengths b and B, respectively, according to Fig. 11b.

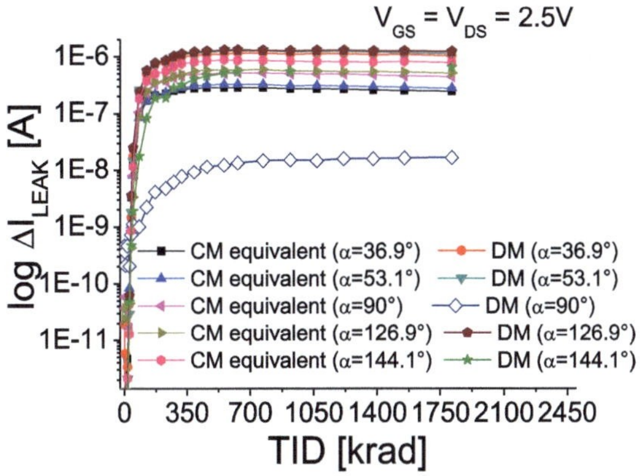

Fig. 12 Current leakage ΔI_{LEAK}: variation profile as a function of the applied radiation doses

In this context, considering the devices in biased condition, E_0 indicates the vector representation for the **Longitudinal Electric Field** (**LEF**) in the CM transistor (Fig. 11a), and E_T indicates the resulting LEF in the DM transistor, considering the vector components E_1 and E_2 for a given α angle of the hexagonal gate geometry (Fig. 11b). In this case, for a common state of biasing and sizing condition (channel width W and aspect ratio W/L) for the both devices, the resulting DM LEF tends to be comparatively higher than the CM counterpart, considering the **Longitudinal Corner Effect** (**LCE**) [41, 42].

For performance verification effects from a set of DM devices with different angles (α = 36.9°, 53.1°, 90°, 126.9° and 144.1°) and a given bias condition ($V_{DS} = V_{GS} = 2.5$ V), Fig. 12 characterizes the current leakage variation ΔI_{LEAK}, considering a ^{60}Co (standard) radiation source for a reference dose variation range. Thus, the corresponding curves indicate a current leakage reduction in DM devices with the increase in the α angle and, in this case, the lowest ΔI_{LEAK} level was obtained for α = 90° due to the DEPAMBBRE Effect [41, 42].

5 Conclusions

Driven through the semiconductors market expansion, CMOS technology downscaling has represented one of the main bottlenecks for the electronic development, considering the new paradigms of operation for each technology node: reduced area and cost, decreased power, and higher performance. On the other hand, this process implies in a set of degrading factors as short channel effects, increased noise levels, process variation and radiation effects sensitivity.

Thus, from a radiation effects sensitivity-based approach, this paper presented a general description about the concepts involving the radiation–matter interaction, the set of interaction patterns (SEE, TID and DD) and resulting effects, and a general overview about mitigation techniques in different hierarchy levels (system, topology and device).

Through the application of RHBD-based techniques, several mitigation strategies can be applied in different domains: hardware, software, time, and information. In this context, each method for radiation hardened design can be applied for comprising a given radiation-induced effect (SEE or TID), in a given hierarchy level and according to the requirements of each application.

Different approaches for hardware redundancy-based techniques has been proposed and validated comparatively to TMR-based faulting masking techniques for mitigation of SET-induced transient effects. Additionally, device level techniques have been described considering alternative non-standard gate geometries for MOS structures implementation, for allowing the improvement of TID robustness performance.

Finally, considering the analyzed case study, the application of Diamond MOSFET DM ($\alpha = 90°$) proved to be an innovative and effective solution for improving the TID tolerance performance for operation of electronic systems in space applications.

References

1. Stassinopoulos, E.G., Raymond, J.P.: The space radiation environment for electronics. Proc. IEEE, pp. 1423–1442 (1988)
2. Johnston, A.H., Radiation effects in advanced microelectronics technologies. IEEE Trans. Nucl. Sci. (1998)
3. Johnston, A.H.: Scaling and technology issues for soft error rates. In: 4th Annual Research Conference on Reliability (2000)
4. Koons, H.C., et al.: The impact of the space environment on space systems. In: 6th Spacecrat Charging Technology Conference (2000)
5. Hughes, H.L., Benedetto, J.M.: Radiation effects and hardening of MOS technology: devices and circuits. IEEE Trans. Nucl. Sci. (2003)
6. Camplani, A., Shojaii, S., Shrimali, H., Stabile, A., Liberali, V.: CMOS IC radiation hardening by design. Facta Univ.—Ser.: Electron. Energ., pp. 251–258 (2014)
7. Bezhenova, V., Michalowska-Forsyth, A.M.: Effects of ionizing radiation on integrated circuits. Elektrotech. Informationstechnik, pp. 39–42 (2016)
8. Maurer, R.H., Fraeman, M.E., Martin, M.N., Roth, D.R.: Harsh environments: space radiation environment, effects, and mitigation. In: Johns Hopkins APL Technical Digest (2008)
9. Karnick, T., Hazucha, P., Patel, J.: Characterization of soft errors caused by single event upsets in CMOS processes. IEEE Trans. Dependable Secure Comput. (2004)
10. Munteanu, D., Autran, J.-L.: Modeling and simulation of single-event effects in digital devices and ICs. IEEE Trans. Nucl. Sci. (2008)
11. Adell, P., et al.: Analysis of single-event transients in analog circuits. IEEE Trans. Nucl. Sci. (2000)
12. Messenger, G.: Collection of charges on junction nodes from ion tracks. IEEE Trans. Nucl. Sci. (1982)
13. Srour, J.S., Marshall, C., Marshall, P.W.: Review of displacement damage effects on silicon devices. IEEE Trans. Nucl. Sci. (2003)
14. Dubrova, E.: Fault Tolerant Design: An Introduction. Sweden Kluwer Academic Publishers (2008)

15. Nicolaidis, M.: Design of soft error mitigation. IEEE Trans. Device Mater. Reliab. (2005)
16. Todd, B., Uznanski, S.: Radiation risks and mitigation in electronic systems. In: CAS-CERN Accelerator School: Power Converters (2015)
17. Gurudatt Kulkarni, M.G., Wani, M.L.: Review on redundancy in electronics. Int. J. Eng. Comput. Sci. (2013)
18. Pan, D., Li, H.W., Wilamowski, B.M.: A radiation-hard phase-locked loop. In: International Symposium on Industrial Electronics ISIE (2003)
19. Lyons, R.E., Vanderkulk, W.: The use of triple modular redundancy to improve computer reliability. IBM J. (1962)
20. Abraham, J.A., Siewiorek, D.P.: An algorithm for the accurate reliability evaluation of triple modular redundant networks. IEEE Trans. Comput. (1974)
21. Teifel, J.: Self-voting dual modular redundancy circuits for single-event-transient mitigation. IEEE Trans. Nucl. Sci. (2008)
22. Engelman, C., Ong, H., Scott, S.L.: The case of modular redundancy in large-scale high performance computing systems. In: 27th IASTED International Conference on Parallel and Distributed Computing and Networks (2009)
23. Wakerly, J.F.: Microcomputer reliability improvement using triple-modular redundancy. Proc. IEEE (1976)
24. Mine, H., Hitayama, K.: Reliability analysis and optimal redundancy for majority-voted logic circuits. IEEE Trans. Reliab. (1981)
25. Hamamatsu, M., Tsuchyia, T., Kikuno, T.: On the reliability of cascaded TMR systems. In: Pacific Rim International Symposium on Dependable Computing (2010)
26. Rahman, M.H., Rafique, S., Alam, M.S.: A fault tolerant voter circuit for triple modular redundant system. J. Electr. Electron. Eng. (2017)
27. Adeosun, O.O., Ismaila, W.O., Omotoso, I.O., Adeosun, T.H.: Hybrid modular redundancy network for critical systems. Int. J. Emerg. Technol. Adv. Eng. IJETAE (2012
28. Alagoz, B.B.: Hierarchical triple-modular redundancy (H-TMR) network for digital systems. In: OncuBilim Algorithm And Systems Labs (2008)
29. Han, J.: A Fault-Tolerant Technique using quadded logic and quadded transistors. IEEE Trans. VLSI Syst. (2014)
30. Grace Vimala, E., Nagavalli, V.: Design of quadded logic and quadded transistor using low power consumption. Int. J. Adv. Res. Electr. Electron. Instrum. Eng. (2016)
31. Han, J., Jonker, P.: A study on fault-tolerant circuits using redundancy. In: Multiconference (2003)
32. Freeman, H.A., Metze, G.: Fault-tolerant computers using "Dotted Logic—Redundancy Techniques". IEEE Trans. Comput. (1972)
33. Beiu, V., Ibrahim, W., Beg, A., Tache, M.: On sizing transistors for threshold voltage variations (2012)
34. Stabile, A., Liberali, V., Calligaro, C.: Design of a rad-hard library of digital cells for space applications. In: 15th International Conference on Electronics, Circuits and Systems ICECS (2008)
35. Snoeys, W.J., Gutierrez, T.A.P., Anelli, G.: A new NMOS layout structure for radiation tolerance. IEEE Trans. Nucl. Sci. (2002)
36. Nowlin, R.N., McEndree, S.R., Wilson, A.L., Alexander, D.R.: A new total-dose-induced parasitic effect in enclosed-geometry transistors. IEEE Trans. Nucl. Sci. (2005)
37. Gimenez, S.P., Galembeck, E.H.S., Renaux, C., Flandre, D.: Impact of using the octogonal layout for SOI MOSFETs in a high temperature environment. IEEE Trans. Dev. Mater. Reliab. (2015)
38. Seixas, L.E., Gonzalez, O.L., Souza, R.R.N., Finco, S., Vaz, R.G., da Silva, G.A., Gimenez, S.P.: Improving MOSFETs tolerance through diamond layout style. IEEE Trans. Dev. Mater. Reliab. (2017)
39. Seixas, L.E., Finco, S., Silveira, M.A.G, Medina, N.H., Gimenez, S.P.: Study of proton radiation effects among diamond and rectangular gate MOSFETs layouts. Mater. Res. Exp. (2017)

40. Gimenez, S.P., Correia, M.M., Neto, E.D., Silva, C.R.: An innovative ellipsoidal layout style to further boost the electrical performance to MOSFETs. IEEE Electr. Dev. Lett. (2015)
41. Gimenez, S.P.: Layout Techniques for MOSFETs (Synthesis Lectures on Emerging Engineering Technologies). Morgan & Claypoole Books (2016)
42. Gimenez, S.P.: Diamond MOSFET: an innovative layout to improve performance of ICs. Solid-State Electr. (2010)
43. Gimenez, S.P., Alati, D.M.: Electrical behavior of the diamond layout style for MOSFETs in X-rays ionizing radiation environments. Microelectron. Eng. (2015)
44. Gimenez, S.P., et al.: Diamond layout style impact on SOI MOSFET in high temperature environment. Microelectron. Reliab. (2015)
45. Gimenez, S.P., Flandre, D. et al.: Using diamond layout style to boost MOSFET frequency response of analogue IC. Electron. Lett. (2014)
46. Gimenez, S.P. et al.: Compact diamond MOSFET model accounting for PAMDLE applicable down 150 nm node. Electron. Lett. (2014)

Biopotential Amplification System Developed for Surface Electromyography Using Dry Electrodes

Alex Toshio Kakizaki, Marcos Henrique Mamoru Otsuka Hamanaka, Vinícius do Lago Pimentel, Carlos Alexandre Ferri and Antônio Augusto Fasolo Quevedo

Abstract Herein we show the results of dry electrodes based on two types of electrodes chromium-gold and gold and the integration to a real-time sEMG collection system. The system uses a PXI chassis with a National Instruments FPGA module that performs a manual selection of amplification gains. A real-time PXI module was utilized with LabVIEW to allow the use of EMG signal decomposition techniques in the study of motion intention identification, prosthesis movements, and neuromuscular integrity analysis.

Keywords Surface electromyography · Dry electrodes · LabVIEW

1 Introduction

Electromyography (EMG) is a biopotential acquisition technique from muscle contraction [1]. The analyses of this EMG signal are used in the diagnosis of neuromuscular diseases, gait analysis, and activation of prostheses. The EMG signal can be used in problems called direct and inverse [2]. The inverse problem is used to recover the signals from the source. The decomposition of the surface EMG signal is a theme of the signal processing area. The unsupervised processing using the blind sources separation technique aims to recover the signals that constitute the MUAPs (Motor Unit Action Potential) from the mixture of the signals of interest without knowledge of the mixing system and the electrical activity of the motor unit (MUs) [1].

A. T. Kakizaki (✉) · M. H. M. O. Hamanaka · V. do Lago Pimentel
Information Technology Center Renato Archer (CTI Renato Archer), Rod. Dom Pedro I, Km 143.6, Campinas, Brazil
e-mail: alex.kakizaki@cti.gov.br

M. H. M. O. Hamanaka
e-mail: marcos.hamanaka@cti.gov.br

A. T. Kakizaki · C. A. Ferri · A. A. F. Quevedo
University of Campinas (UNICAMP), Cidade Universitária "Zeferino Vaz, Campinas, SP, Brazil

Y. Iano et al. (eds.), *Proceedings of the 3rd Brazilian Technology Symposium*,
https://doi.org/10.1007/978-3-319-93112-8_24

Muscle contraction is the result of recruiting three types of muscle fibers. Each fiber has a characteristic related to the speed of contraction and resistance to fatigue [3]. Table 1 has qualitative information summarized for each type of fiber.

The method of acquisition of biopotentials resulting from muscle contraction can be performed by invasive or non-invasive electrodes. In the invasive electrodes, the examination is performed through contraction using a thread or needle inserted into the muscle group. The method is painful and the risk of infection is higher if compared to the non-invasive method. The non-invasive uses surface electrodes and it is more comfortable eliminating the pain of the invasive method. The surface electromyography is the most indicated for activities that require strength, rapid muscle contraction and monitoring of muscular electrical activity simultaneously. Table 2 shows a comparison between invasive and non-invasive methods.

Therefore, surface electromyography depends on the use of electrodes that can detect the action potentials on the surface of the skin. There are two types of electrodes applied to the surface electromyography: wet and dry. The most common is the wet that you normally use Ag/AgCl electrodes with gel to obtain the stability in skin-electrode contact. The dry electrode has the advantage of not needing the gel [2, 4, 5]. The disadvantage is the high contact impedance which can be minimized by increasing the contact area [2].

In the present work, we developed dry electrodes to be used in the collection of surface EMG signals. The electrodes were fabricated by depositing metal on a flexible substrate (Kapton®). Two types of electrodes have been tested: Cr/Au and Au films. The electrodes were integrated to an EMG system operating in real-time to test and validate the electrodes. The system has embedded algorithms that will be used for further signal processing, such as surface EMG decomposition.

Table 1 Qualitative information on the velocity of contraction and resistance to fatigue

Type of fiber	Information related to contraction velocity and fatigability
I	Slow contraction and fatigue resistant
IIa	Rapid contraction and have a certain resistance to fatigue
IIb	Fast contraction and fatigue more quickly than others

Table 2 Comparison of invasive and non-invasive methods to signal acquisition

Acquisition type	Cons	Pros
Invasive	Painful, risk of infection and time-consuming	A greater amplitude and spatial selectivity
Non-invasive	A spatiotemporal mixture, high contact impedance and lower amplitude of the signal	More comfortable, quick and simple preparation of the skin

2 Methods

Our system is based on a signal conditioning circuit to be used with the dry electrodes. Using the RC model [6, 7], the system was designed and developed based on a resistor of 1 MΩ in parallel with a capacitor of 10 nF as circuit input requirements. The circuit has two stages of amplification: the first stage has a fixed gain and the second stage has a programmable gain selected via LabVIEW. The circuitry has seven differential channels and its total gains vary from 10 to 10,000 V/V.

2.1 *Dry Electrodes*

The electrode was fabricated by thermal deposition process at the Information Technology Center Renato Archer. Kapton® is a flexible substrate used in the manufacture of dry electrodes. The metals Cr and Au were thermal evaporated on the substrate.

The substrate was cut in cleanroom and washed with Isoprapanol. After that, the substrate was dried at the furnace at 50 °C for, approximately, two hours.

The next step includes to fix the substrate at sample holder, fix sample holder at the planetary system, insertion of metal (Cr or Au) on tungsten boat and a close vacuum chamber. The vacuum was made until pressure achieves 10^{-7} Torr to start deposition process. During the process, the temperature inside the chamber was around 194 °C. The thicknesses of the Cr and Au films were respectively 5 and 200 nm.

2.2 *Measurement System*

The measuring system was designed and developed to operate with dry electrodes (Fig. 1). The circuit has a first amplification stage with fixed gain, the second stage of amplification with programmable gain and a band-pass filter [1]. Gain control can be adjusted through a dedicated program in LabVIEW Real-Time that commands the FPGA module model PXI-7851R (National Instruments). The LabVIEW RT was used due to low latency data transmission through Gigabit Ethernet protocol between chassis PXI and computer.

The first stage of amplification uses the INA129 instrumentation amplifier [8]. Gain is set by using an external resistor that provides a gain of 20 dB. The INA129 has an input buffer for the impedance to be high [9]. In common mode, the typical impedance for the INA129 is 10^{11} Ω in parallel with 9 pF. Another feature of the amplifier is the common-mode rejection rate (CMRR). For the gain used in our system, the rejection rate is 106 dB which assists in the quality of the signal to be acquired.

The second stage of amplification is based on LM1973 [10] in the feedback of operational amplifier. In this configuration, the circuitry assembled will operate as a

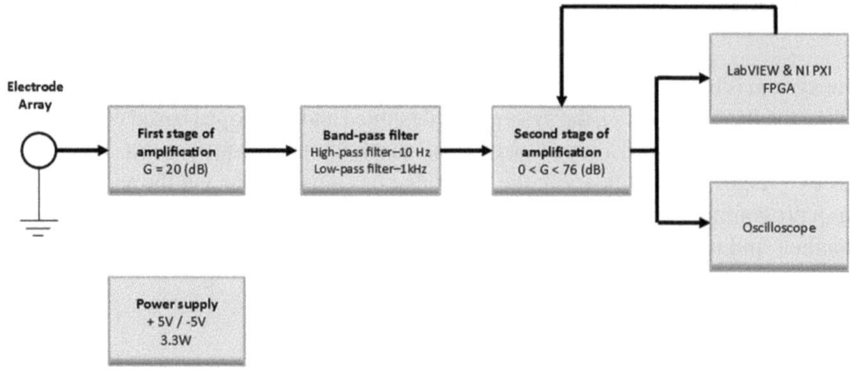

Fig. 1 Block diagram of the measurement system (modified of [1])

programmable gain amplifier. In addition to the amplification, it is possible to control the gain according to the needs [10]. The digital control was implemented using NI LabVIEW and the NI PXI-7851R FPGA module. The gain control is in open loop.

2.3 *Test Protocol*

The test protocol includes three steps

i. The frequency response of circuit through the relation between output and input signal as shown in Fig. 3. A sinusoidal waveform was applied to front-end biopotential amplifier and the output signal was collected with an oscilloscope. The procedure was repeated for each gain;
ii. Commercial wet electrode test using Noraxon system and developed system. The signals were decomposed into frequency through FFT MATLAB® script for comparison;
iii. Fabricated electrode validation using the developed system and commercial electrode like standard. The signals were decomposed into the frequency for response comparison.

The test with wet electrode using Noraxon system and the developed system were performed following the sequence: Tests only on the right forearm, cleaning the skin for electrodes contact, positioning electrodes with a distance of 40 mm from each other and connection of electrodes with the circuit.

3 Results

The acquisition system can be seen in Fig. 2. Observe the main information of the setup and the image of the equipment.

The gains are programmable via LabVIEW. The frequency response of the acquisition system was measured for five gains: 5, 10, 20, 100 and 501 V/V as shown in Fig. 3.

The biopotential amplification system developed was validated. Summarizing the tests a Noraxon system with Ag/AgCl electrodes was used as a standard to validate

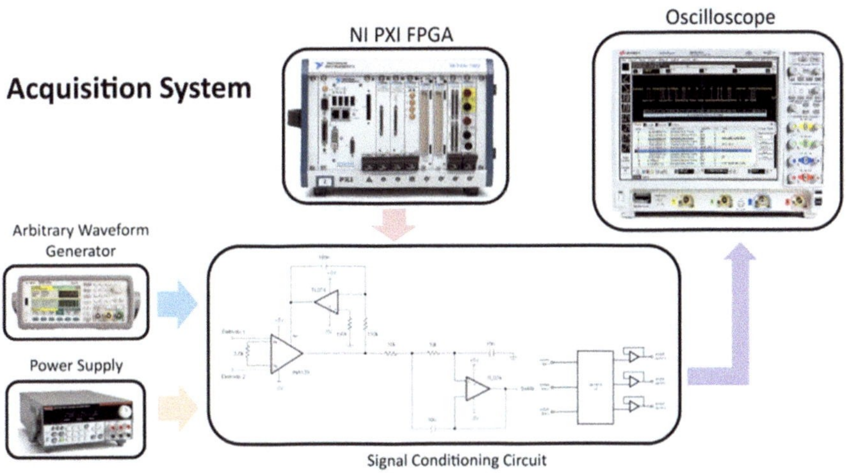

Fig. 2 Block diagram of acquisition system [2]

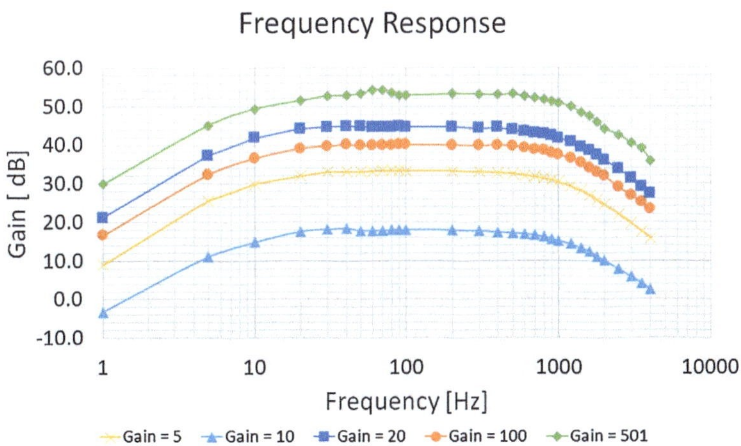

Fig. 3 The frequency response of the two-stage amplifier for five gains: 5, 10, 20, 100 and 501 V/V

the developed system. The first test was performed with the Ag/AgCl electrodes and the gel using Noraxon system and the developed system. The second test was performed without the gel using the developed system. The preliminary results of these tests are published in Brazilian Technology Symposium 2017 (BTSym'17).

Several tests were performed with the dry electrodes: The tape test to verify the adhesion of the metallic film on the substrate. The other tests were performed comparing the commercial electrode with those fabricated using the developed system (Fig. 4). Some test results that validated the two types of fabricated electrodes are plotted below.

The raw data collected with the oscilloscope for four contractions with, approximately, the one-second duration can be seen in Fig. 5.

A comparison between amplitude spectrum from wet Ag/AgCl electrode and fabricated dry Cr/Au electrode for: one contraction during one second (Fig. 6); one contraction during four seconds (Fig. 7); and, four contractions during one second

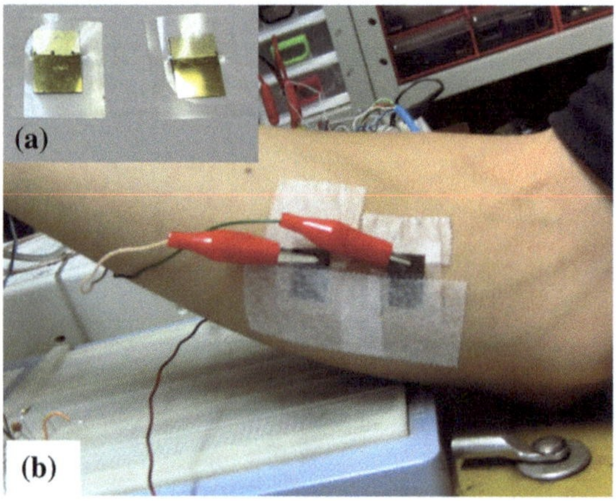

Fig. 4 Electrodes used in the tests (**a**) Detail of Kapton® with Cr/Au film - dry electrode with a sheet of polyester to limit skin-metal contact area, (**b**) The interelectrode distance was 40 mm and the contact area with skin was 78.5 mm^2

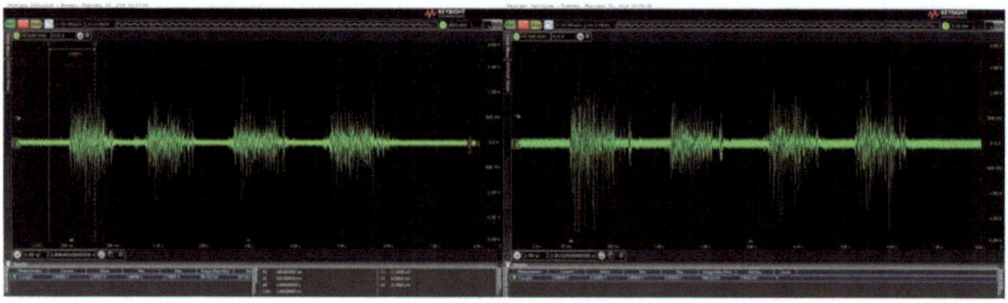

Fig. 5 RAW surface electromyography signal collected with (left) wet commercial Ag/AgCl electrode and (right) dry fabricated Cr/Au electrode

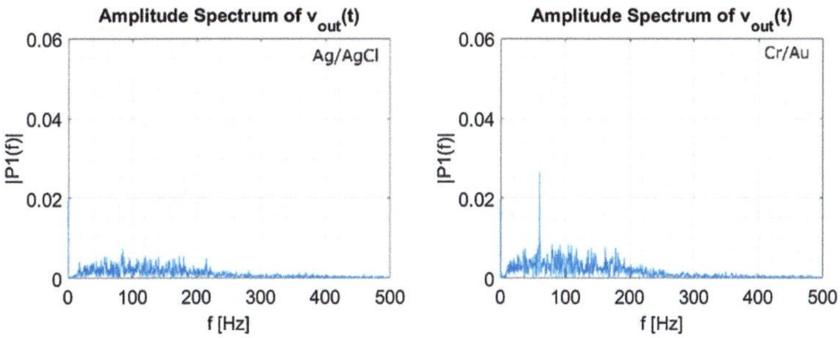

Fig. 6 Comparison between amplitude spectrums from wet Ag/AgCl electrode and fabricated dry Cr/Au electrode utilizing the developed system. The results correspond to one contraction during one second

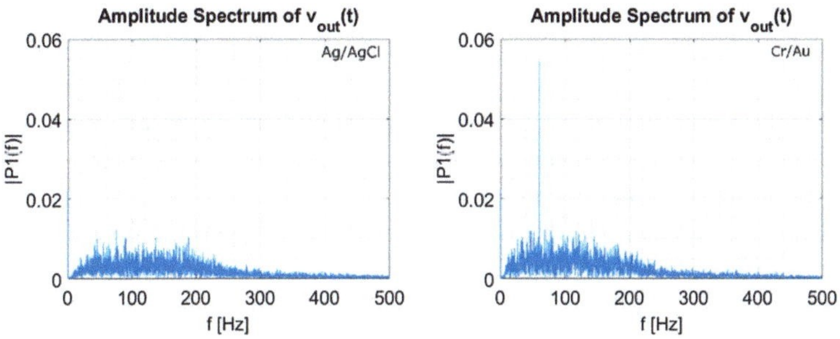

Fig. 7 Comparison between amplitude spectrums from wet Ag/AgCl electrode and fabricated dry Cr/Au electrode utilizing the developed system. The results correspond to one contraction during four seconds

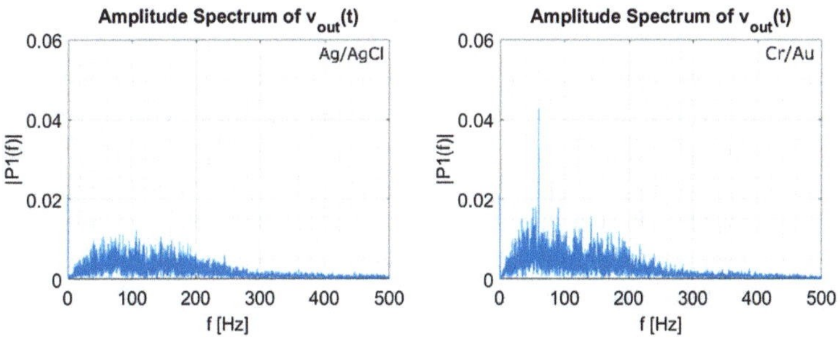

Fig. 8 Comparison between amplitude spectrums from wet Ag/AgCl electrode and fabricated dry Cr/Au electrode utilizing the developed system. The results correspond to four contractions during one second and interval equal to one second

(Fig. 8) are illustrated. Figure 9 shows an amplitude spectrum for wet Ag/AgCl electrode, dry Cr/Au, and dry Au electrode.

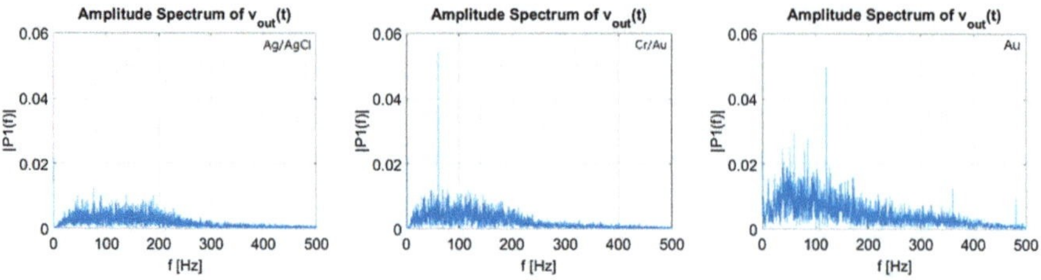

Fig. 9 Comparison between amplitude spectrums from commercial wet Ag/AgCl electrode, fabricated dry Cr/Au electrode, and fabricated dry Au electrode utilizing the developed system. The results correspond to one contraction during four seconds

4 Discussion

The results validate the two types of electrodes fabricated Cr/Au and Au. The composition of the electrode influences the results even if the electrical connections are on the surface of the electrode and does not have contact with the metal film below the surface of the covering metal.

The frequency analysis of dry electrodes shows noises in the 60 Hz (including harmonics) that need to be investigated.

The results show that the system can be used to design an array of dry electrodes for signal processing application.

5 Conclusion

The developed measurement system was validated by comparison with standard commercial electrodes. The dry electrodes of Cr/Au and Au presented similar signals to the commercial electrode and therefore, can be used in the biopotential amplification system developed. In addition, the deposited metal films presented good adhesion to the flexible substrate. These preliminary results demonstrate the possibility to create conditions to improve the actual standard exams such as increase the points of data collection in the muscle contractions.

Acknowledgements We are thankful to Dr. Tiago P. Almeida for making the system for EMG recording used in this work available. We thanks the Coordination of the Open Laboratory (Colab) of the CTI Renato Archer, a branch of the Ministry of Science, Technology, Innovations and Communications of Brazil where the experiments were executed in a partial or integral.

References

1. Almeida, T.P.: Decomposição de Sinais Eletromiográficos de Superfície Misturados Linearmente Utilizando Análise de Componentes Independentes. Master's thesis, University of Campinas, Campinas, SP (2012)
2. Merletti, R., Parker, P.: Electromyography. Hoboken. Wiley, NJ (2004). https://doi.org/10.1002/0471678384
3. Silverthorn, D.U.: Fisiologia Humana: uma abordagem integrada. Editora Manole Ltda, Barueri, SP (2003)
4. Baek, J.Y., Na, J.H., Choi J.M., Park, K.S., Lee, S.H.: Flexible polymeric dry electrodes for the long-term monitoring of ECG. Sens. Actuators A, 423–429 (2007). https://doi.org/10.1016/j.sna.2007.11.019
5. Myers, A.C., Huang, H., Zhu, Y.: Wearable silver nanowire dry electrodes for electrophysiological sensing. R. Soc. Chem. **5**, 11627–11632 (2015). https://doi.org/10.1039/c4ra15101a
6. Chi, Y.M., Jung, T.-P., Cauwenberghs, G.: Dry-contact and noncontact biopotential electrodes: methodological review. IEEE Rev. Biomed. Eng. **3**, 106–119 (2010). https://doi.org/10.1109/RBME.2010.2084078
7. Meziane, N., Webster, J.G., Attari, M., Nimunkar, A.J.: Dry electrodes for electrocardiography. Physiol. Meas. **34**, R47–R69 (2013). https://doi.org/10.1088/0967-3334/34/9/r47
8. Datasheet INA129, http://www.ti.com/lit/ds/symlink/ina129.pdf
9. Webster, J.G.: Medical Instrumentation: Application and Design 4th edn. Hoboken, NJ, Wiley (2010)
10. Datasheet LM1973, http://www.ti.com/lit/ds/symlink/lm1973.pdf

E-Street for Prevention of Falls of the Elderly an Urban Virtual Environment for Human–Computer Interaction from Lower Limb Movements

Alexandre Brandão, Diego Dias, Iago Alvarenga, Glesio Paiva, Luis Trevelin, Karina Gramany-Say and Gabriela Castellano

Abstract Virtual reality (VR) interfaces have been widely applied in different areas of knowledge, due to the popularization of smartphones and the low cost of immersive devices (VR glasses). In this context, we present a VR solution that reproduces an urban virtual environment (called e-Street), aimed at preventing falls in the elderly, improving spatial orientation/navigation and stimulating the movement of the lower limbs (stationary march) for patients in the physical and neurofunctional recovery process. The user controls the virtual environment with body movements of the lower limbs from ultrasound sensors (positioned in the ankle region) connected to an Arduino board, which communicates via Bluetooth with e-Street (running on a smartphone, inserted in VR glasses). e-Street can be used as (physically active) digital entertainment and provides accessibility to adapted physical activity, considering that the input movement is personalized for each user.

Keywords Virtual reality · Urban environment · Physical therapy · Gesture interaction · Assistive technology · Neurofunctional recovery

---

A. Brandão (✉) · G. Castellano
Institute of Physics Gleb Wataghin, University of Campinas\UNICAMP, Campinas, SP, Brazil
e-mail: abrandao@ifi.unicamp.br

A. Brandão · G. Castellano
Brazilian Institute of Neuroscience and Neurotechnology\BRAINN, Campinas, SP, Brazil

D. Dias · I. Alvarenga
Computer Science Department, Federal University of Sao Joao Del Rei, Sao Joao Del Rei, MG, Brazil

G. Paiva · L. Trevelin
Computer Science Dept, Federal University of Sao Carlos\UFSCar, Sao Carlos, SP, Brazil

K. Gramany-Say
Gerontology Department, UFSCar, Sao Carlos, SP, Brazil

Y. Iano et al. (eds.), *Proceedings of the 3rd Brazilian Technology Symposium*,
https://doi.org/10.1007/978-3-319-93112-8_25

1 Introduction

Human–Computer Interaction (HCI) allows communication between people and computer systems in a variety of ways, including body movements (or gesture interaction). HCI based on body movements can occur in a physically active manner and therefore achieve benefits related to physical activity and health promotion. This approach can contribute to physical and neurofunctional recovery processes in patients undergoing physical therapy and can simulate routine situations aka Activities of Daily Living (ADL) in order to prevent mechanical injuries, such as frequent falls in the elderly.

Falls are a major cause of mortality and morbidity in people over 65; approximately 30–40% of people from this age group will suffer at least one fall. These falls can cause minor and even serious injuries in many cases, with a real risk of impairment to the independence of these individuals [1]. The elderly can maintain their body balance when they are focused on performing a specific activity. However, they may not present the same balance when performing multiple tasks simultaneously. Thus, when there is a division of attention in the execution of two or more activities, motor performance and balance are impaired in these older people.

The estimated incidence of falls in Brazil is 30% of the elderly who fall at least once a year, and constitute 7% of accidental deaths in the elderly with more than 75 years, which is the sixth largest cause of death among the elderly. Older people who suffer falls have a functional decline, and about 5% of falls result in fractures and require complementary medical care [2]. The use of Virtual Reality (VR) interfaces enables the simulation of a patient's routine and can compose trainings for motor and neurofunctional therapies in controlled environments. Currently, the lower cost of the equipment needed to experience VR allows more professionals (kinesiology experts) to present this type of technology for the benefit of a larger portion of the population.

In this work, the e-Street software is presented, with its development tool (Unity3D) as well as the gestural interaction device (Arduino and ultrasound sensors) also developed in this multidisciplinary study. This article is organized as follows. First, an overview of the computational tools is given in Sect. 2. Then, Sect. 3 presents the functional and usability requirements. Section 4 presents a discussion considering other VR systems applied to the health areas, and the main conclusions are drawn.

2 Materials and Methods

In this work, the Unity3D [3] game engine (development platform) and the Natural User Interface (NUI) [4, 5] concept was used, which allow natural means of interacting with digital systems, including interaction through physical activity, inherent to human behavior. The study was approved by the Ethics Committee on Human Research of the institution (UNICAMP), with application approval number CAAE

35771314.4.0000.5404, and software registration at the Instituto Nacional de Propriedade Industrial (INPI) of Brazil, with approval number BR 51 2017 000419-6.

2.1 *Unity-3D: E-Street Software Interface*

The software e-Street was modeled on the Unity-3D game engine, widely used for the development of video games (consoles) and applications for mobile devices (smartphones and tablets). The Unity Technologies had their initial launch in 2005 and their stable release October 2017.

Concepts of Computer Graphics, Virtual Reality, Human–Computer Interaction, and Programming were used during the development and animation of the urban virtual environment, providing the user with an "urban walk" experience in a highly immersive environment.

2.2 *Natural User Interface (NUI): Interaction Device*

NUI is used by designers and developers when they refer to interaction with the computer in an effectively invisible way, or that becomes invisible during learning, after successive interactions. Usually, computer interfaces use artificial control devices, in which the learning of commands is accomplished by buttons, such as in the alphanumeric keyboard or video game control, in which case the learning curve tends to be high.

A NUI only requires the user to be able to interact with the environment through previously known interactions, such as gesture interaction and voice commands. This type of natural interface also requires learning, but the user is not presented to a new device or way of using it, that is, the communication through gestures is inherent to the human being, and the learning curve tends to be easier.

In this context, an interaction device (based on an Arduino board and ultrasound sensors) was built, allowing the user to control the virtual environment from body movements, specifically, isolated plantar flexion (ankle joint) or plantar flexion associated with flexion of the hip and the knee joints.

3 Results

A software called e-Street (Fig. 1) was modeled (Unity-3D), to simulate an urban environment oriented for navigation from movements that reproduce the stationary march (simulated walk without spatial displacement), which require spatial orientation and balance from the user. The e-Street software runs on an Android smartphone within Google Cardboard [6].

Fig. 1 Top view of the e-Street software (prepared by the authors)

3.1 Requirements

To control the virtual environment, a device was built from sonar sensors physically connected (wired) to a controller board (Arduino). The input signals to the Arduino consist of the height difference (ΔS, considering the initial position S0) of the sonar. Via Bluetooth, the Arduino sends the signal to the software (e-Street) that is running on a smartphone (inserted in VR glasses), which allows control of the navigation (by gesture interaction) in that environment by the user. Thus, system control (Fig. 2) is performed from ultrasound sensors attached to the distal position of the lower limb (near the ankle joint).

The sensor holders were printed on a 3D printer, with Polylactic acid (PLA) plastic and optimized to occupy the smallest space in the user's body (portable), considering the body anatomy.

Each time the user performs the movement of raising the feet from the floor alternately, simulating a walk (without spatial displacement), a forward shift is assigned within the e-Street environment. Thus, it is possible to explore the VR environment from real body movements (Fig. 3).

In Fig. 3, it is possible to visualize one of the several cars that are permanently transiting in e-Street and the crosswalk (highlighted in green), that is responsive to the user, that is, the vehicles stop when the user accesses the crosswalk.

4 Discussion and Conclusion

VR can stimulate several sensory systems of the human body, including visual perception, which facilitates the input and output of information to the brain. VR can be used in conjunction with other therapeutic interventions to increase the complexity

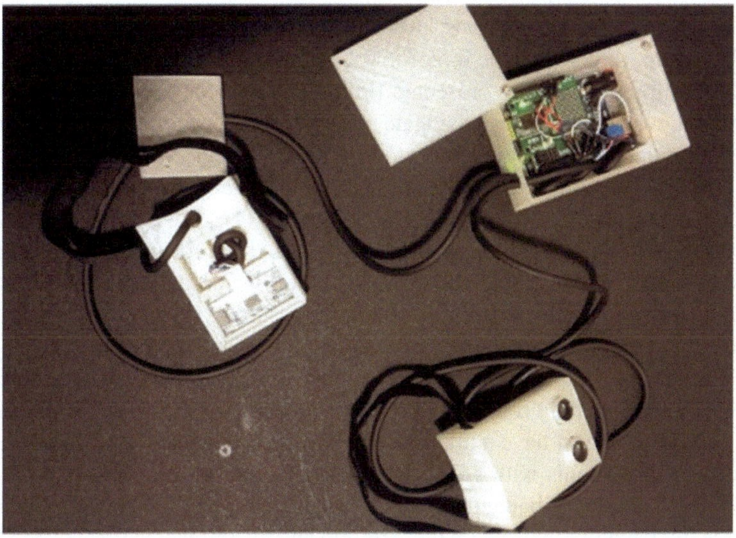

Fig. 2 Arduino board (above and right in the image) and connected ultrasound sensors (prepared by the authors)

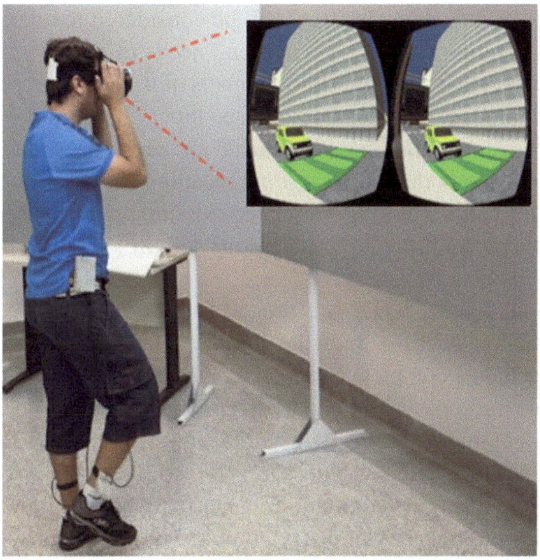

Fig. 3 The user controlling e-Street (prepared by the authors)

of the requested tasks during the rehabilitation process [7]. Moreover, with VR, the rehabilitation process can be done in a controlled environment and bring the complexity of the physical world [8], ensuring the safety of the patients and even reducing fall risk [9].

Given the above considerations, this work aimed to develop the first prototype of a VR system, directed to the motor and cognitive rehabilitation of patients with lower limb impairments. The presented prototype, e-Street, includes a rich virtual environment, where the user can navigate in a controlled fashion, without leaving the rehabilitation unit. Also, the idea was to make this system as low cost as possible, so that it could be acquired and used by an extensive range of people, and even in public health care units.

However, being a prototype, the system still has room for improvements. In the following, we discuss the limitations of the presented device, and the possible changes that can be made to it in order to ensure that the resulting system can be safely and accurately applied to the rehabilitation process of impaired patients.

The limitation of the ultrasound sensor is the signal noise. In order not to lose its reference, that is, the ground, lower limb movement is restricted to the sagittal plane (y-axis), and must have a minimum amplitude to be accurately detected. This is a problem for patients with greater restriction of movement of the lower limbs. Therefore, future work is being directed to the development of an updated version of the device, which will replace both the Arduino and the ultrasonic sensors by inertial units capable of capturing movement in any plane/axis. This will allow the development of a wearable device that adapts to user functionality with greater precision and is able to reliably reproduce the desired movements.

For this, the following characteristics will be considered in relation to NUI [10]: 1. User-centered design: provides the changing needs of the interface for different users; 2. Multichannel: captures multiple channels of user input to perform the interaction with the computer. The considered user input channels are the body parts, such as the ankle. 3. Interaction based on behavior: recognizes the user's body language, that is, movements that express some meaning (gestural interaction). These NUI characteristics can be used to simulate real situations and benefit motor and neurofunctional therapies through multisensory stimuli [11], which is proposed by e-Street.

Other limitations of the device are related to the created virtual environment, where user interaction is still somewhat limited. The next research goals in this direction are to include a complete residential environment in the simulation, called e-House; to enable multiuser functionality (social interaction); and to allow manual interaction (from computer vision) with virtual objects during e-Street navigation, and with virtual furniture within e-House.

Despite the above-mentioned limitations, when compared to the VR apps available on Google Play [12], e-Street has unique features and is directly associated with motor stimuli focused on physical therapy. Table 1 presents the capabilities (motor skills) for interaction with e-Street.

Considered a non-invasive technology, e-Street can be used to reproduce ADLs from immersive virtual experiences with the user, assisted by trained professionals (kinesiology experts) and in a safe and controlled environment. It can, therefore, be used for motor rehabilitation of gait and fall prevention in older adults.

More studies are necessary to indicate the use of e-Street in the clinical environment and demonstrate that HCI, in a physically active way, can be used safely in

Table 1 Caption missing

Capabilities/Motor skills	e-Street + Arduino board + Ultrasound sensor	Android apps on google play
Lower limbs	Yes	No
Upper limbs	No	No
Trunk rotation	Yes	Yes
Orthostatic position	Yes	Yes
Sitting position	Yes	Yes
Range of motion	Medium	No
Motor coordination	High	Low
Balance	High	Medium

motor and neurofunctional recovery processes. In addition, the investigation must provide evidence that this type of interaction could be mitigating the damages related to sedentarism and associated diseases. Our future research goals include these issues.

Acknowledgements The authors would like to acknowledge the contribution of Marcelo Fernandes de Oliveira and Jorge Vicente Lopes da Silva from the Information Technology Center (CTI – Renato Archer), who helped to develop the support (3D printing) for the ultrasonic sensors and the Arduino board. This study was supported by grants from Sao Paulo Research Foundation (FAPESP number 2015/23830-4, linked to BRAINN RIDC number 2013/07559-3).

References

1. Ambrose, A.F., Paul, G., Hausdorff, J.M.: Risk factors for falls among older adults: a review of the literature. Maturitas **75**, 51–61 (2013)
2. Antes, D.L., D'orsi, E., Benedetti, T.R.B.: Circumstances and consequences of falls among the older adults in Florianopolis. Epi Floripa Aging 2009*. Rev. Bras. Epidemiol. **16**(2), 469–481 (2013). São Paulo
3. Unity–Game Engine, https://unity3d.com/pt/. Accessed Nov 2017
4. Wigdor, D., Wixon, D.: Brave NUI World. MKP Inc., San Francisco (2011)
5. Glonek, G., Pietruszka, M.: Natural user interface: review. J. Appl. Comput. Sci. **20**(2), 27–45 (2012)
6. Google Cardboard, https://vr.google.com/cardboard. Accessed Nov 2017
7. Da Silva Cameirão, M., Bermúdez i Badia, S., Duarte, E., Verschure, P.: Virtual reality based rehabilitation speeds up functional recovery of the upper extremities after stroke: a randomized controlled pilot study in the acute phase of stroke using the rehabilitation gaming system. Restorative Neurol. Neurosci. **29**, 287–298 (2011)
8. Keshner, E.A.: Virtual reality and physical rehabilitation: a new toy or a new research and rehabilitation tool? J. NeuroEng. Rehabil. **1**(8), 1–2 (2004)
9. Mirelman, A., Rochester, L., Reelick, M., Nieuwhof, F., Pelosin, E., Abbruzzese, G., Dockx, K., Nieuwboer, A., Hausdorff, J.M.: V-TIME: a treadmill training program augmented by virtual reality to decrease fall risk in older adults: study design of a randomized controlled trial. BioMed. Cent. Neurol. **13**(15), 1–12 (2013)

10. Liu, W.: Natural user interface-next mainstream product user interface. In: IEEE 11th International Conference on Comp-Aided Industrial Design & Conceptual Design (CAIDCD), vol. 2, pp. 203–205 (2010)
11. Rose, F.D., Brooks, B.M., Rizzo, A.A.: Virtual reality in brain damage rehabilitation: review. CyberPsychology Behav. **8**(3), 241–262 (2005)
12. Google Play–VR, https://play.google.com/store/search?q=VR&c=apps. Accessed Nov 2017

A Time Series Analysis Applied to the Generation of Energy at the Santo Antonio Hydroelectric Plant Located in the State of Rondonia in Brazil in the Year 2016

Nickolas Bastiani Cassiano, Joao Gabriel Ribeiro, Giovane Maia do Vale and Vlademir de Jesus Silva Oliveira

Abstract The purpose of this work is to use a time series methodology, or more specifically, the Multivariate Vector Error Correction Model (VECM), to analyze the impacts of the variables average hourly flow rate of water and hourly reservoir level upstream, about the generation of hourly energy in the Brazilian hydroelectric plant of Santo Antonio located in the state of Rondonia in the year 2016. The model estimated the presence of positive impacts of the explanatory variables on the power generation of the plant in the short and long term. Analogous results were diagnosed in the prediction and impulse response functions of the series used.

Keywords Environmental impacts · Power generation · Time series
Multivariate vector error-correction model

1 Introduction

The Santo Antonio hydroelectric plant, located in the city of Porto Velho capital of Rondonia, at the bed the Madeira River, has been active since 2008 in its partial

N. B. Cassiano · J. G. Ribeiro (✉) · G. M. do Vale · V. de Jesus Silva Oliveira
Electrical Engineering of the State University of Mato Grosso (UNEMAT), Sinop-MT, Brazil
e-mail: jgr06041980@gmail.com

N. B. Cassiano
e-mail: nickolasbastiani.18@gmail.com

G. M. do Vale
e-mail: giovanemvale@gmail.com

V. de Jesus Silva Oliveira
e-mail: vlademir.oliveira@unemat.br

Y. Iano et al. (eds.), *Proceedings of the 3rd Brazilian Technology Symposium*,
https://doi.org/10.1007/978-3-319-93112-8_26

capacity, and its construction work was completed in 2016. The plant is a water thread, that is, it stops accumulating large water reservoirs, using types of turbines to take advantage of the flow of the river. So it has been operating with 50 turbines of the bulb type, having today 3568 MW of installed power, and capacity to generate estimated consumption to 45 million Brazilians.

Due to the importance of this plant for Brazil, an analysis of this hydroelectric was initiated, through the use of time series methodologies. This study evaluated the series of hourly electric power generation of the plant in the year 2016. Then, it investigated the explanatory behavior of this series, starting from the series average hourly flow rate water, which is the amount of flow arriving, and hourly level of the reservoir upstream, being the amount in the direction of the source of the river this year. According to author [34], they have the role of input variables in an energy generation model.

Section 2 describes the materials and methods used. Section 3 shows results and discusses important issues. Finally, conclusions are presented in Sect. 4.

2 Materials and Methods

Data collection took place together with the Santo Antonio hydroelectric plant. In a second step occurs the composition of a database is organized, with the input variables hourly energy generation (ge_t) in MWh, the average hourly flow rate of water (vam_t) in m^3/s, and hourly level of the reservoir (nrm_t) in m, throughout the year 2016.

In this way, one can conjecture that the generation of energy will be a function of these two variables, plus the random stochastic component (e_t) as follows:

$$ge_t = f(vam_t, nrm_t) + e_t, \tag{1}$$

where, $t\epsilon[01/01/2016; 30/12/2016] \equiv [1; 8753]$ are the observations used of all the series involved in the study.

Then, appropriate time series analysis procedures were then applied as follows the sequence below:

(1) Exploratory analysis of the series under study;
(2) Dickey–Fuller stationary series tests;
(3) Selection criteria for the best lag;
(4) Johansen Cointegration Tests;
(5) Vector Model Error Correction (VECM);
(6) Series forecasts; and
(7) Impulse Functions Answers.

The implementation of the techniques occurred in the in the R-Studio software. This manuscript concludes with the analyses of the hypotheses investigated.

For more details on the methodologies of these classic time series techniques, it is advisable to consult the following texts: [14, 23, 24, 26, 32, 33, 35].

3 Results and Discussion

From the data collected and referring to the variables under study was evaluated the descriptive behavior of the series of daily vectors ge_t, vam_t and nrm_t, containing each of the 8753 hourly observations, measured from the 1st to the 24th hour, during the year of 2016 on the dates from 01/01/2016 to 12/30/2016, at the Santo Antonio plant in Rondonia. Table 1 shows the behavior of the series.

In the year 2016, the average hourly power generation of the Santo Antônio hydroelectric power plant was 1351.94 MWh, an amount from an average hourly flow rate of water of 14,105 m^3/s and a level of the average reservoir time of 70.48 m with a low Coefficient of Variation (CV) of 0.08% as expected. It is also observed that the variable that most varies with the CV is the average hourly flow rate of water with 72.44%, it is also verified that in this year the hydroelectric plant guaranteed homogeneity of 69.44% in its generation of average energy, with a total generated in this year of 11,833,560 MWh.

The behaviors can be seen in Figs. 1, 2, and 3.

It can be seen that the series of hourly energy generation remains stable until July, and from then until September there is a fall, and later it is growing until December. As for the series of hourly flow, it shows increasing behavior until March, and from there it decreases until September and then begins to grow slowly until December. On the other hand, the series level of the reservoir is stable between the lower and upper limits of 70.4 and 70.6 m approximately during the hours of the year 2016. It is also observed some similarity of the series of generation of energy and tributary flow hours, from July to December.

Then, the Dickey–Fuller stationarity tests were carried out to evaluate if the series had similar stabilization behaviors when differentiated.

It is verified by the Dickey–Fuller test of Table 2, that the series under study have similar stabilization behavior in their respective lower and upper limits, as well

Table 1 Detail of the main relative descriptive measures of the time series under study in 2016

Statistics	ge_t (MWh)	vam_t (m^3/s)	nrm_t (m)
Minimum	7.00	1806	69.94
Maximum	2215	42,138	70.62
Average	1351.94	14,105	70.48
Standard deviation	413.20	10,218	0.06
CV (%)	30.56	72.44	0.08
Total sum	11,833,560	12,3461,800	616,917
Observations	8753	8753	8753

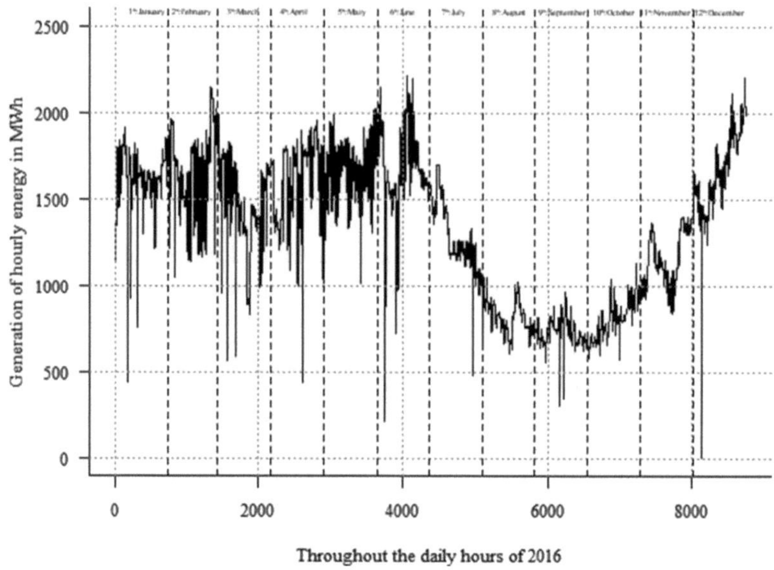

Fig. 1 The behavior of the Series of hourly energy generation in MWh (ge_t) in the year 2016

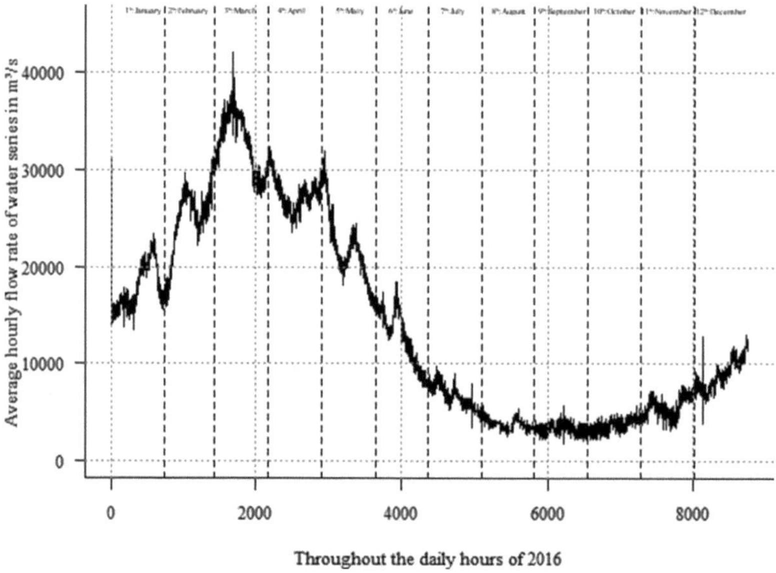

Fig. 2 The behavior of the average hourly flow rate of water series in m^3/s (vam_t) in 2016

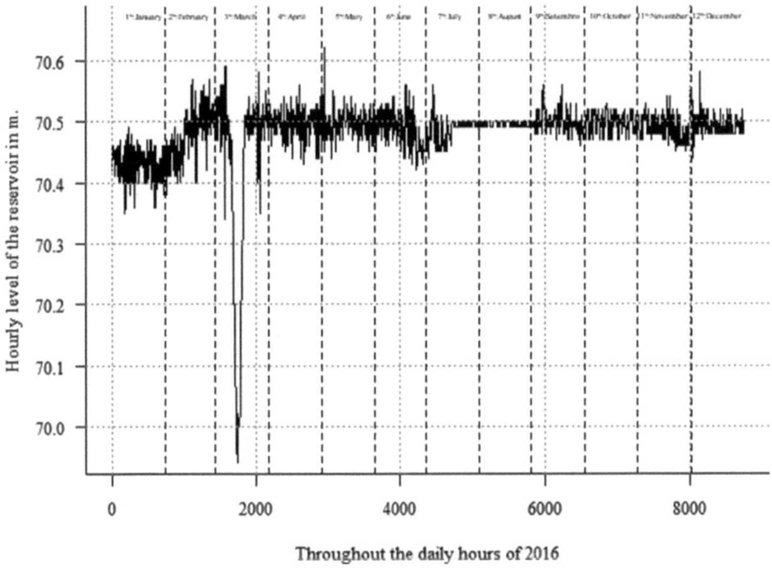

Fig. 3 The behavior of the hourly series of the reservoir in m (nrm_t) in the year 2016

Table 2 Dickey–Fuller (DF) stationarity tests for the series under study

	Statistics	DF—α (trust level)			
Series	τ_τ	1%	5%	10%	Integration I(n)
ge_t	−2.40	−2.58	−1.95	−1.62	–
Δge_t	−66.21	−2.58	−1.95	−1.62	I(1)
vam_t	−1.26	−2.58	−1.95	−1.62	–
Δvam_t	−104.93	−2.58	−1.95	−1.62	I(1)
nrm_t	0.053	−2.58	−1.95	−1.62	–
Δnrm_t	−46.80	−2.58	−1.95	−1.62	I(1)

H_0 Absence of stationarity

as the same order of integration I(1). Since the absolute values of the statistic are greater than the values set at 1, 5, and 10%, the series is in their first differences, and therefore, in this format of the differentiated series, we reject the null hypothesis of absence of stationarity.

Before estimating the number of cointegrating relationships between the series, we analyzed the best lag in the input matrix of the data composed of the vectors of the variables, and to start the methodology. The criteria and information used in this procedure are AIC(n), HQ(n), SC(n) and FPE(n), according to Table 3.

Table 3 shows that the lag indicated by most criteria was a value of 10, to be used in the Johansen cointegration test, and if there is any cointegration relation, the VECM is estimated, with the same value of lags.

Subsequently, we tried to analyze if the series have relations of cointegrating, which means, in other words, to say that they have similar trajectories throughout the hours of the year 2016. Table 4 presents the results obtained by the Johansen test.

In the first and second lines of Table 4, it can be observed that the statistic $J_{traço}$ presents absolute values higher than the indexes of comparisons of scientific errors of 1, 5, and 10%. This behavior indicates that the null hypothesis of at least 0 and 1 cointegration relationship will be rejected, so these results indicate that there is at least 1 cointegrated relationship in the series under study. However, nothing can be said about the presence of two cointegrating relations in the series in use, by the analysis of the third line from Table 4.

Hence, due to the Dickey–Fuller test, the series is stationary in the first difference (all are integrated of order 1, I(1)). Furthermore, there is at least one cointegration relation between the series under study by the Test of Johansen, we can estimate the VECM, with the lag 10, indicated by the information criteria in Table 3.

From the proposed VECM model, the short-term impacts on hourly energy generation at the Santo Antonio plant in Rondonia were estimated by the Ordinary Least Squares (OLS) method, as indicated in Table 5.

Table 3 Selection of the optimal lag according to the information criteria of AIC(n), HQ(n), SC(n) and FPE(n)

Information criteria	Statistics	Lag
AIC(n)	8.81	10
HQ(n)	8.84	10
SC(n)	8.89	9
FPE(n)	6761.35	10

Table 4 Cointegration tests for the series, ge_t, vam_t and nrm_t (Johansen test)

Hypotheses Tested	Statistics $J_{traço}$	1%	5%	10%
H_0: 0 cointegration relationships	94.42	28.71	31.52	37.22
H_1: 1 cointegration relationships	41.50	15.66	17.95	25.52
H_2: 2 cointegration relationships	0.45	6.50	8.18	11.65

H0: n presence of cointegration relationships compared to
H1: r > n presence of cointegration relationships

Table 5 Short-term analysis of the VECM model with the response variable (ge_t), explained by its own lags and also by the other variables (vam_t, nrm_t)

Variable response ge_t			
Explanatory variables	Coefficients	Test-t	p-value
Long term	−6.31E−03	−4.80	1.62E−06
Constant	2.32E+03	4.80	1.62E−06
ge_{t-1}	3.25E−01	30.37	2.00E−16
vam_{t-1}	−5.28E−03	−2.38	0.02
nrm_{t-1}	3.19E+02	1.74	0.08
ge_{t-2}	−1.85E−01	−16.39	2.00E−16
vam_{t-2}	−2.87E−03	−1.22	0.22
nrm_{t-2}	3.52E+02	1.65	0.10
ge_{t-3}	−4.53E−03	−0.39	0.69
vam_{t-3}	−2.63E−03	−1.11	0.27
nrm_{t-3}	−8.16E+01	−0.38	0.71
ge_{t-4}	−2.24E−02	−1.95	0.05
vam_{t-4}	−3.58E−03	−1.50	0.13
nrm_{t-4}	3.53E+02	1.63	0.11
ge_{t-5}	−5.53E−02	−4.82	1.45E−06
vam_{t-5}	3.47E−03	1.45	0.15
nrm_{t-5}	−5.48E+02	−2.53	0.01
ge_{t-6}	−5.63E−02	−4.91	9.52E−07
vam_{t-6}	8.14E−03	3.40	6.88E−04
nrm_{t-6}	−5.57E+01	−0.26	0.80
ge_{t-7}	−3.20E−02	−2.77	5.55E−03
vam_{t-7}	8.80E−03	3.68	2.37E−04
nrm_{t-7}	8.54E+01	0.40	0.69
ge_{t-8}	−3.58E−02	−3.15	1.63E−03
vam_{t-8}	1.68E−03	0.73	0.47
nrm_{t-8}	5.50E+02	2.58	9.87E−03
ge_{t-9}	−6.33E−02	−5.84	5.43E−09
vam_{t-9}	5.20E−04	0.48	0.63
nrm_{t-9}	−1.87E+01	−0.11	0.92

$F_{calculated} = 46.67$ and p-value = 2.20E−16
$R^2 = 13.15\%$
Coefficients Significant: ge_t:08, vam_t:03, nrm_t:04

In the short term, it is found that the variable ge_t in its hourly lags, in most of the coefficients negatively impacts on the hourly energy generation response variable of 2016, and of these effects, only eight are statistically significant. For the variable vam_t in its time lags, in most of the coefficients, it produces positive effects in the response variable, but only 3 coefficients have statistical significance. In the case of the lagged variable nrm_t, it is observed that four estimated coefficients indicate a positive effect on the response variable, and of the total of 4 coefficients present statistical significance.

It is observed that by the test t of the table, the long-term component and the constant component have statistical significance. From the long-term component we obtain the estimated cointegration vector $\hat{\beta} = (1.00; -0.02; 5200.56)^t$. And from this vector, we can remove the long-term effects (ge_{lp}) on energy generation, according to

$$\begin{aligned} ge_{lp} = & -(6.31E-03)ge_{t-10} + (1.26E-04)vam_{t-10} \\ & + 32.82nrm_{t-10}. \end{aligned} \tag{2}$$

It is observed that the variables vam_{t-10} and nrm_{t-10}, indicate an increase of the generation of energy in the long term (ge_{lp}), a fact that is expected since they are variables of great importance for the success of the generation of energy.

In the model estimated, it was noticed the presence of problems and serial autocorrelation, and heteroscedasticity and the abnormality of the residuals, but still we chose to estimate the predictions and functions impulse responses, to evaluate the behavior of the series in the future and the impacts of the explanatory variables (vam_t and nrm_t) on the response variable (ge_t).

Figures 4, 5 and 6 indicate the forecasting behavior of each of the series, based on data for the year 2016.

It is detected that the series of hourly energy generation (ge_t) and hourly reservoir level (nrm_t) in the Santo Antonio plant, remains at the average levels in its forecast. However, the average hourly flow rate of water (vam_t) indicated a slight decay.

The estimates of the impulse response functions were performed to evaluate the effects of the variables and under the response variable, and these impacts can be observed in Fig. 7.

It is verified that the impacts of unitary shocks of the variable vam_t cause increases in the variable ge_t, which in a certain way was expected since the variable average hourly flow rate of water influences positively in the hourly generation of energy of the plant of Santo Antonio, corroborating with the long-term analyzes, in which the variable vam_t has a positive impact on ge_t. Analogous fact occurs with the unitary shocks of nrm_t, which also provoke positive results in ge_t, a correlated result can be observed in the long-term analyzes, for this variable.

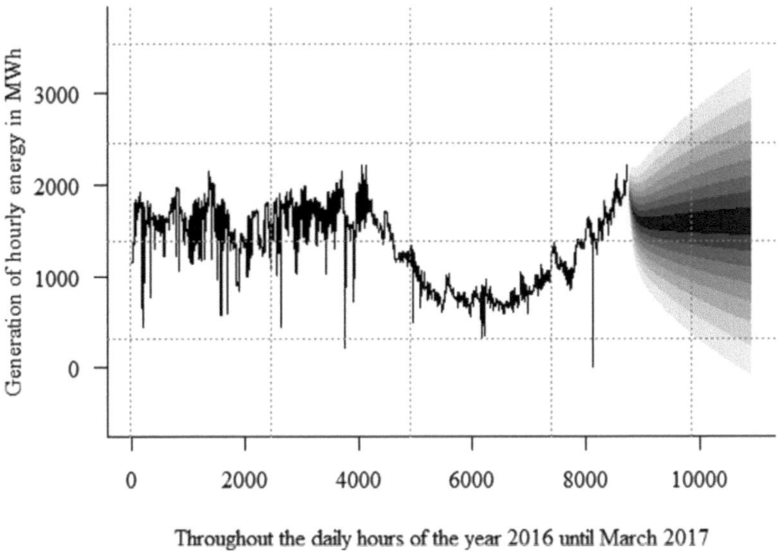

Fig. 4 Prediction of the series hourly generation of energy in MWh (ge_t) until March 2017, with confidence interval with 1% error

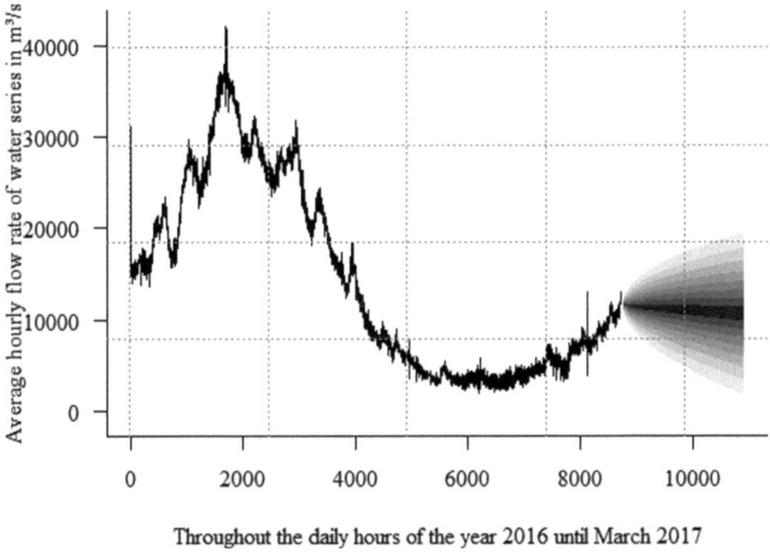

Fig. 5 Prediction of the average hourly flow rate of water series in m^3/s (vam_t) until March 2017, with confidence interval with 1% error

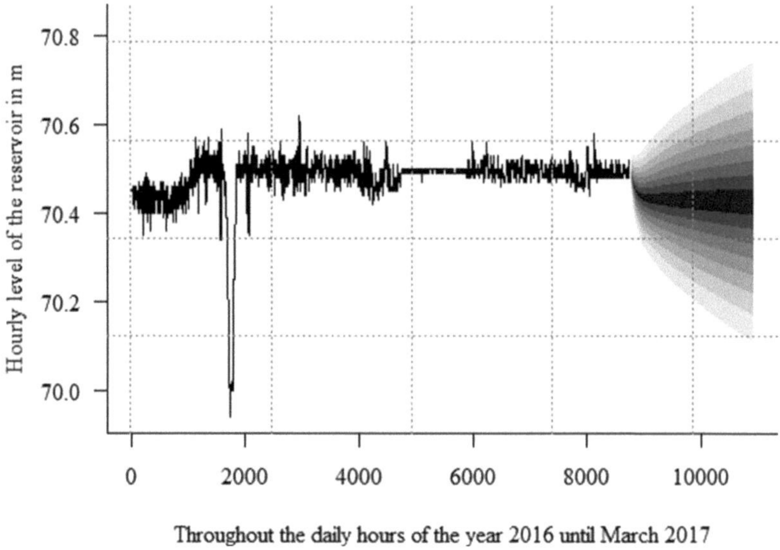

Fig. 6 Prediction of the hour series of the reservoir in m (nrm_t) until March 2017, with confidence interval with 1% error

4 Conclusions

The short-term analysis in 2016 indicated some positive lagged components of the variables ge_t, vam_t and nrm_t, and the total of significant components obtained in the estimates of these same variables was 08, 03 and 04 respectively.

In contrast, the long-term analysis revealed more expressive results, in which the components vam_t and nrm_t, positively impact on the hourly energy generation ge_t, these results were also achieved in the impulse functions built, and demonstrate an increase in the generation of hourly energy ge_t, each unitary shock of the variables vam_t and nrm_t.

Finally, the forecasting behavior of the series was evaluated, in which it is observed that the average hourly generation ge_t and nrm_t, remain constant in the average levels, already the variable vam_t, presents a slight decay.

It is suggested for future works, the use of more explanatory variables of the hourly generation, and also the updating of these studies year by year, evaluating and detailing the behavior of the generation of energy that is something of extreme importance in the life of any human being.

It is also worth considering that the study can also be replicated for several hydroelectric plants that capture these same variables.

Acknowledgement First of all, I thank God, and all those who in one way or another contributed to the making of this work.

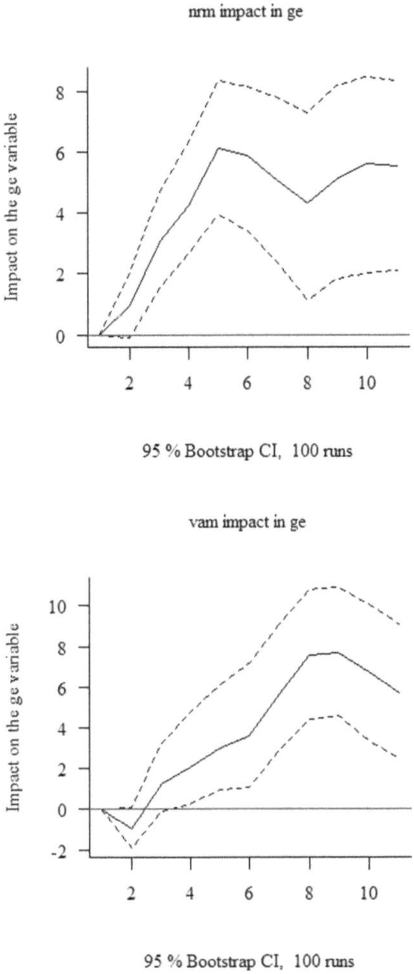

Fig. 7 The response to the unit shocks of the variables vam_t and nrm_t in ge_t

References

1. Akaike, H.: Likelihood of a model and information criteria. J. Econometrics. **16**, 3–14 (1980)
2. ANEEL.: Atlas de energia elétrica do Brasil/Agência Nacional de Energia Elétrica—ANEEL, pp. 1–153. Brasília (2002)
3. ANEEL.: Atlas de energia elétrica do Brasil/Agência Nacional de Energia Elétrica—ANEEL, 3ª edn., pp. 1–236. Brasília (2008)
4. Bermann, C.: Crise ambiental e as energias renováveis. Ciência e Cultura (online). v. 60, n.3. São Paulo (2008)
5. Bermann, C.: Energia no Brasil: Para que? Para quem? Crise e alternativas para um país sustentável. Editora Livraria da Física: FASE, São Paulo (2001)
6. Box, G.E.P., Pierce, D.A.: Distribution of residual autocorrelations in autoregressive integrated moving average time series models. J. Am. Stat. Assoc., pp. 1509–1526 (1970)

7. Box, G.E.P., Jenkins, G.M., Reinsel, G.: Time Series Analysis: Forecasting and Control, 3ª edn. Prentice Hall, Englewood Cliffs (1994)
8. Correia, E.F., et al.: Investimento em Subestações Aumenta Qualidade da Transmissão. Revista FURNAS Ano XXXI Nº 318 (2005)
9. Dickey, D.A., Fuller, W.A.: Distribution of the estimators for autoregressive time series with a unit root. J. Am. Stat. Assoc. (1979)
10. Enders, W.: Applied Econometric Times Series. Wiley Series in Probability and Statistics. Wiley, New York (2009)
11. Farias, H.P.: Função Resposta a Impulso e Decomposição da Variância do Erro de Previsão Aplicados às Principais Bolsas de Valores. Lavras, Minas Gerais (2008)
12. Filho, D.d.: Dimensionamento de Usinas Hidroelétricas Através de Técnicas de Otimização Evolutiva. São Carlos, SP (2003)
13. Goldemberg, J., Lucon, O.: Energia e meio ambiente no Brasil. Estudos Avançados (online) **21**(59), 7–20 (2007)
14. Gujarati, D.: Econometria Básica, 4edn. Campus, Rio de Janeiro (2006)
15. Hannan, E., Quinn, B.: The determination of the order of an autoregression. J. R. Stat. Soc. **B41**, 190–195 (1979)
16. Johansen, S.: Likelihood Based Inference in Cointegrated Vector Error Correction Models. Oxford University Press, Oxford (1994)
17. Johansen, S.: Statistical analysis of cointegration vectors. J. Econ. Dyn. Control **12**, 231–254 (1988)
18. Kennedy, P.: Manual de Econometria, 6.edn. Elsevier, Rio de Janeiro (2009)
19. Leão, Ruth P.S. GTD—Geração, Transmissão e Distribuição de Energia Elétrica. Capítulo 2: Tecnologias de Geração de Energia Elétrica: Geração Hidroelétrica. Universidade Federal do Ceará—UFC (2009)
20. Lutkepohl, H.: New Introduction to Multiple Time Series Analysis. Springer, Berlin (2006)
21. Mine, M., Tucci, C.E.M.: Gerenciamento da produção de energia e controle de inundação: Foz do Areia no rio Iguaçu. Rev. Bras. de Recursos Hídricos **7**(3), 85–107 (2002)
22. Modiano, E.M.: Elasticidade renda e preços da demanda de energia elétrica no Brasil. Departamento de Economia da PUC (Texto para Discussão, n. 68), Rio de Janeiro (1984)
23. Morettin, P.A.: Econometria financeira: um curso em séries temporais financeiras. E. Blucher, São Paulo (2008)
24. Morettin, P.A.: Toloi, C.M.C.: Análise de séries temporais, 2nd edn. E. Blucher, São Pauo (2006)
25. Passos, F.F.: Análise Temporal da Série de Consumo Residencial de Energia Elétrica no Brasil no Período de 1963 a 2012. Varginha (2015)
26. Pindyck, R.S., Rubinfeld, D.L.: Econometria: modelos e previsões, 4 edn. Campus, Rio de Janeiro (2004)
27. Quinn, B.: Order determination for multivariate autoregression. J. R. Stat. Soc. **B42**, 182–185 (1980)
28. Santos, A.J.: Modelos Vetoriais Auto-Regressivos com Transição Suave Estruturados por Árvores—STVAR-Tree. Rio de Janeiro (2009)
29. Schwarz, H.: Estimating the dimension of a model. Ann. Stat. **6**, 461–464 (1978)
30. Sims, C.A.: Macroeconomics and reality. Econometrica **48**(1), 1–48 (1980)
31. Siqueira, M.L., Cordeiro, Jr., H.H., Castelar, I.: A demanda por energia elétrica no Nordeste brasileiro após o racionamento de 2001–2002: previsões de longo prazo. Pesquisa e Planejamento Econômico, **36**(1), pp. 137–178 (2006)
32. Tsay, R.S.: Analysis of Financial Time Series, 2nd edn. Wiley, New York (2005)
33. Tsay, R.S.: Multivariate Analysis of Time Series with R and Financial Applications. Wiley, New York (2014)
34. Tucci, C.E.M.: Flood flow forecasting. WMO Bull. **52**(1), 46–52 (2003)
35. Zivot, E., Wang, J.: Modeling Financial Time Series with S-Plus, 2nd edn. Springer, New York (2005)

Optimization of Photocatalytic Degradation of Methyl Orange Using TiO_2

Gustavo Duran Iga, André Luis de Castro Peixoto and Ademir Geraldo Cavallari Costalonga

Abstract The environmental contamination is one of the most significant problems today, especially when it comes to natural waters. In Brazil, the textile industry consumes approximately 20 tons/year of dyes and about 20% are disposed as effluents. For the degradation of these organic and inorganic contaminants, one can use a process called photocatalytic oxidation, which uses semiconductors such as TiO_2. Many factors interfere with the photocatalytic activity. This study focuses on finding the best conditions for degradation of methyl orange, working on a variation of the TiO_2 quantity and the temperature of the solution. Following the experimental plan, was standardized constant at 2.0 the pH of the solution containing the dye with a buffer mixture of HCl/NaCl. This degradation was accompanied by the decrease of the peak absorption of the dye spectrum. The process was optimized, and in 60 min it was possible to achieve the degradation of up to 95% of the dye.

Keywords Photocatalysis · Titanium dioxide · Methyl orange Photodegradation · Environmental contamination · Dyes · Pollution by dyes

1 Introduction

Globalization and industrial expansion brought several non-biodegradable wastes, which can be harmful to the environment. Among these contaminants, the widely used dyes that pigment the industrial textile processes, had become a major risk for the environment and human health pose several risks. The variety of colors from

G. D. Iga (✉) · A. L. de Castro Peixoto · A. G. C. Costalonga
IFSP—Instituto Federal de Educação, Ciência e Tecnologia de São Paulo Câmpus Capivari, Avenida Doutor Ênio Pires de Camargo, 2971, Capivari SP 13360-000, Brazil
e-mail: gustavoiga@hotmail.com

A. L. de Castro Peixoto
e-mail: alcpeixoto@ifsp.edu.br

A. G. C. Costalonga
e-mail: ademirgcc@ifsp.edu.br

Y. Iano et al. (eds.), *Proceedings of the 3rd Brazilian Technology Symposium*,
https://doi.org/10.1007/978-3-319-93112-8_27

organic dyes occurs due to strong light absorption by organic molecules with the wavelength in the region of visible light. The dyes are substances known to provide vibrant colors to tissues and food. [1]

In Brazil, 20 ton/year of dyes are consumed by textile industry, of which 20% is disposed of as effluents. The main source of this loss happens due to the incomplete fixation of the dyes on fiber at the dyeing process [2, 3], turning into one of the worst wastes that can harm the environment.

This urgency poses an important technical challenge that involves solving environmental problems. The threat creates a demand, from environmental rules, to optimize the industrial processes themselves for proper disposal of waste generated. One of the methods for the degradation of contaminants is called an Advanced Oxidative Process (AOP).

1.1 Advanced Oxidative Process

AOP enables the transformation of organic contaminants into water, carbon dioxide and inorganic waste. These degradation mechanisms typically take effect by hydroxyl radicals [4]. Recent studies have shown the potential of the use of AOP in the removal of pharmaceuticals from wastewater [5].

There are two types of AOPs: heterogeneous, and homogeneous processes. They are defined by the presence or absence of solid catalysts, respectively [6].

Heterogeneous photocatalysis is a promising technique in the treatment of environmental pollution and, in the last decades, has been extensively studied due to its ability to completely oxidize organic molecules [7, 8], with low energy cost.

1.2 Photocatalytic Material

The first work that has the application of heterogeneous photocatalysis using TiO_2 was published in 1972 by Fujishima and Honda [9], experimenting a reaction to generate hydrogen and oxygen from the oxidation of water. They use as a catalyst the titanium dioxide in aqueous suspension irradiated by a source of light.

Heterogeneous photocatalysis based on titanium dioxide has been widely applied to a variety of harmful organic compounds. [10]. Due to its high reactivity, quickly oxidizes a wide range of pollutants and non-selectively [11]. TiO_2 is the most used semiconductor in photocatalysis [12] because it is an easily obtainable material, has low cost and high chemical stability [13].

1.3 Methyl Orange

The orange methyl dye was chosen as the pollutant model at this work because it is from the class of azo compounds. It is also known as an acid–base indicator, being yellow in the basic medium and orange in the acid medium [11].

Azo dyes are the most important chemical class of dyes in the textile industry, accounting for almost 70% of organic dyes on the market.

This class of compounds has an extended conjugation of the π orbitals, with two aromatic rings attached to the –N=N– group, originating the absorption of light in the visible region. Then, showing its typical color.

1.4 Photodegradation

Dai et al. [14] studied the photocatalytic activity of titanium dioxide nanoparticles, through the degradation of methyl orange. The degradation experiments were performed on photocatalytic reactors containing 20 ppm of dye concentration. The solutions were irradiated by a UV lamp with maximum emission at 365 nm. Next, with the catalyst concentration of 1.0 g/L, they varied pH between 1.5 and 5 and found higher efficiency at pH = 2. The degradation of the dye follows the kinetic law of pseudo-first order. After 45 min of illumination, 98% of the dye had been mineralized.

Guettai and Amar [11] studied the degradation of the methyl orange by heterogeneous photocatalysis, using the powder form of TiO_2 in as a catalyst. Two UV lamps of 15 W each were used, with an emission at the wavelength of 365 nm. The pH range studied ranged from 2 to 10, and the catalyst concentration was set at 0.8 g/L. They observed that dye concentrations of up to 50 mg/L degrade after 3 h of exposure, while higher concentrations of dye degrade at a slower rate. The best results were obtained in acidic pH (2, 3, and 5) after 5 h of irradiation, reaching 98.58% of the color loss.

Qaradawi and Salman [15] studied the photodegradation of the orange methyl dye using TiO_2 too. Ultraviolet radiation, to activate the catalyst, was obtained from sunlight. The values of pH 3, 5, 7, and 9 were studied. The irradiation time was 5 h. It was observed that the greatest degradation happened at pH = 3. No decay was observed in the concentration of the dye in the absence of light, neither in the absence of a catalyst.

Section 2 explains the proposed method. Results and a discussion are presented in Sect. 3. Conclusions appear in Sect. 4.

2 Method

This project was developed to reach the best conditions and optimize the heterogeneous photocatalysis with titanium dioxide. Methyl Orange, an acid–base indicator, was chosen to be the organic compound to be degraded. An 8 W germicide mercury lamp was installed at the center of a photoreactor to provide the UV radiation for the system. There are a wide variety of types and configurations of photochemical reactors that are studied and developed for specifics degradations, in this project used as a polyvinyl chloride one.

The system consists of a 500 mL vessel connected to a photoreactor and a pump that is used for effluent recirculation. The UV source, as described above, was a low-pressure mercury lamp with a predominant wavelength of 292.3 nm. An anatase structure of TiO_2 was utilized as photocatalyst without any previous treatment. A stock solution at the concentration of 100 mg/L of the dye was prepared with distilled water, without any further purification. The photocatalytic activity of TiO_2 on the degradation of the synthetically prepared effluent was evaluated under laboratory conditions with the UV irradiation inside of the photoreactor. To avoid pH variance was used as a buffer mixture with HCl/NaCl to maintain the pH around 2.0.

Photocatalytic activity was evaluated by the absorbance of the methyl orange with a TiO_2 solution. Measurements were performed using UV-Vis spectrophotometer in the range of 200–600 nm. The degradation of the dye solution was monitored at a wavelength of 510 nm (absorption spectrum maximum). Many factors interfere with the photocatalytic activity. To study the best dye degradation conditions, in this work, the amount of TiO_2 used, and the temperature of the solution in the reactor had been varied to optimize the degradation process as much as possible.

3 Results and Discussion

Following experimental planning, the pH of the solution containing the dye was maintained constant at 2.0 with an HCl/NaCl mixture. The adsorption of the dye on the catalyst was studied by experimenting with recirculation without the UV source. No adsorption of the dye on the catalyst has been observed.

With the use of a hydrostatic bath, studies were carried out to adjust a temperature value that improved the dye degradation process. The best attempt was at 40 °C. As the next adjustment, the amount of photocatalyst in suspension without reserve was varied. As a result (refer to Fig. 1), the amount that has the best result was 2 mL of TiO_2 in a concentration of 0.10 mg/mL. The results were standardized for a percentage of degradation with all the experiments starting at 100%.

After lots of tests with the standardized temperature and varying the quantity of the catalyst in solution, it was possible to reach a very good result compared to other methods and authors. The 0.20 mg of catalyst (average of mass in suspension) was the best result so far. Other tests with smaller quantities were made, but the results

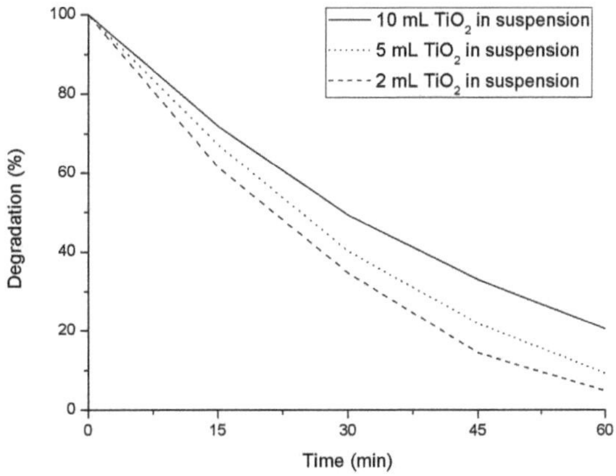

Fig. 1 Degradation of methyl orange dye using different amounts of TiO_2 photocatalyst in suspension (0.10 mg/mL) at a constant temperature

were worst or equivalent of other experiments. Since TiO_2 is cheap and non-toxic, the result reached is significant for the optimization of the process.

4 Conclusion

A process called photocatalytic oxidation helps to guarantee the degradation of organic and inorganic contaminants, and it employs semiconductors such as TiO_2. Several factors may affect the photocatalytic activity.

The present study adjusts some variables in the process of degradation of the methyl orange dye. Since it is an azo compound, it is one of the most harmful environmental dyes. 95% of degradation was reached in 60 min followed by a decrease in the absorption peak of the dye spectrum. Compared to other studies [11, 15] the time of photolysis obtained in this work for methyl orange degradation was much lower. Considering that, it is possible to see potential in AOP in large scale to deal with the waste of food and textile industry.

References

1. Brown, T.L., Lemay, H.E., Bursten, B.E.: Química: a ciência central, 9th edn. Pearson Prentice Hall, São Paulo (2005)
2. Guaratini, C.C.I., Zanoni, M.V.B.: Corantes têxteis. Quím. Nova **23**(1), 71–78 (2000)
3. Carneiro, P.A., Osugi, M.E., Sene, J.J., Anderson, M.A., Zanoni, M.V.B.: Evaluation of color removal and degradation of a reactive textile azo dye on nanoporous TiO_2 thin-film electrodes. Electrochim. Acta **49**, 3807–3820 (2004)
4. Teixeira, C.P.A.B., Jardim, W.F.: Processos Oxidativos Avançados: conceitos teóricos. Caderno temático, vol. 3. Universidade Estadual de campinas (UNICAMP), Instituto de Química—IQ, Laboratório de Química Ambiental—LQA. Campinas (2004)

5. Rizzo, L., Meric, S., Guida, M., Kassinos, D., Belgiorno, V.: Heterogenous photocatalytic degradation kinetics and detoxification of an urban wastewater treatment plant effluent contaminated with pharmaceuticals. Water Res. **43**, 4070–4078 (2009)
6. Huang, C.P., Dong, C., Tang, Z.: Advanced chemical oxidation: its present role and potential future in hazardous waste treatment. Water Manage. **13**, 361–377 (1993)
7. Chiou, C.H., Wu, C.Y., Juang, R.S.: Influence of operating parameters on photocatalytic degradation of phenol in UV/TiO_2 process. Chem. Eng. J. **139**, 322–329 (2008)
8. Duminica, F.D., Maury, F., Hausbrand, R.: Growth of TiO_2 thin films by AP-MOCVD on stainless steel substrates for photocatalytic applications. Surf. Coat. Technol. **201**, 9304–9308 (2007)
9. Fujishima, A., Honda, K.: Electrochemical photolysis of water at a semiconductor electrode. Nature **238**, 37–38 (1972)
10. Hu, Y., Li, D., Zheng, Y., Chen, W., He, Y., Shao, Y., Fu, X., Xiao, G.: BiVO4/TiO_2 nanocrystalline heterostructure: A wide spectrum responsive photocatalytic towards the highly efficient decomposition of gaseous benzene. Appl. Catal. B **104**, 30–36 (2011)
11. Guettai, N., Amar, H.A.: Photocatalytic oxidation of methyl orange in presence of titanium dioxide in aqueous suspension. Part I: Parametric study. Desalination **185**, 427–437 (2005)
12. Nogueira, R.F.P., Jardim, W.F.: A fotocatálise heterogênea e sua aplicação ambiental. Quim. Nova **21**, 69–72 (1998)
13. Zhang, X., Zhou, M., Lei, L.: Preparation of an Ag–TiO_2 photocatalyst coated on activated carbon by MOCVD. Mater. Chem. Phys. **91**, 73–79 (2005)
14. Dai, K., Chen, H., Peng, T., Ke, D., Yi, H.: Photocatalytic degradation of methyl orange in aqueous suspension of mesoporous titania nanoparticles. Chemosphere **69**, 1361–1367 (2007)
15. Qaradawi, S.A., Salman, S.R.: Photocatalytic degradation of methyl orange as a model compound. J. Photochem. Photobiol., A **148**, 161–168 (2002)

Cognitive Ergonomics and the Industry 4.0

Alessandra Cristina Santos Akkari, Mateus Faraj Marques da Rocha and Rosani Franco de Farias Novaes

Abstract Much has been debated about training, skills, and abilities of the Engineer 4.0. However, the cognitive aspects that are intrinsic to this new paradigm are not considered in these analyses. Therefore, this work aimed to study the Industry 4.0 under a macro, based on standardization of information, and a micro vision focused on cognitive ergonomics. Industry 4.0 attributes pointed to an increasing industry complexity in the midst of a technological virtualization process, requiring faster and assertive decisions in the face of a wide range of information. It was found that the Engineer 4.0 will be in charge of the solution of increasingly interdisciplinary problems in a digital environment, requiring denser cognitive efforts and more accurate social-emotional skills, such as communication, motivation, autonomy, perseverance, resilience, collaboration and creativity.

Keywords Cognitive ergonomics · Industry 4.0 · Production engineering
Internet of services · Internet of things · Cyber-Physical systems

1 Introduction

The Internet of Things (IoT), in which people, processes and products are part of the same system, led to the emergence of Industry 4.0. The term Industry 4.0 emerged from the Industry-Science Research Alliance group report presented to the German Chancellor and released at the Hannover Messe in 2013. According to the literature, although new solutions are arising from the field of Industry 4.0, a clear definition of multiple aspects and perspectives from advanced manufacturing and Cyber-Physical Systems (CPSs) is not yet observed, especially in the academic scope [1, 2]. Therefore, systematic literature reviews, as well as exploratory and descriptive studies, are important tools to provide evidence for the decision-making process in Industry 4.0 and to allow an in-depth investigation by the scientific community.

A. C. S. Akkari (✉) · M. F. M. da Rocha · R. F. de Farias Novaes
Science and Technology Center, Mackenzie Presbyterian University, Campinas, Brazil
e-mail: alessandra.akkari@mackenzie.br

Y. Iano et al. (eds.), *Proceedings of the 3rd Brazilian Technology Symposium*,
https://doi.org/10.1007/978-3-319-93112-8_28

There is a worldwide trend to deploy the Industry 4.0, especially in industrialized countries. This tendency is defining a new production era marked by the encompassing of CPSs, which, in turn, are the bases for intelligent machines and industrial facilities to be able to exchange information and make decisions with no need of controllers [3].

The conventional computing will have difficulty staggering with the large data flow and with the complexity of the analysis, demanding more sophisticated technologies in the scope of Industry 4.0. Thus, the automation of manufacturing plants, for example, comes from keywords such as *information* and *integration*; different technologies such as IoT, Big Data; Intelligent Robotics; Industrial Automation and Analytics allow the interconnection between technologies themselves and between these technologies and labor, inputs, energy and production data [4].

Therefore, mental and cognitive abilities, in addition to the mastery and operationalization of the various technologies, as well as the analysis of a variety of workflow, process and environment information knowledge will be required from the Engineer 4.0 to boost quality and accelerate decision-making process in an assertive manner. Thus, the processual study of Industry 4.0 under the focus of cognitive ergonomics, a field of ergonomics specialization, becomes very important.

Section 2 establishes the necessary background to enable the discussion of the cognitive process in an environment dominated by new and daring technological perspectives. Thus, are explored aspects of industry 4.0 as well as ergonomics fundamentals, especially cognitive ergonomics.

2 Background

2.1 Industry 4.0

The Fourth Industrial Revolution began with the integration of the Internet of Things (IoT) and the Internet of Services (IoS) in the manufacturing process. The IoT allows things and objects, such as sensors, actuators, and mobiles to interact and cooperate with its neighbors through smart components to achieve common goals. The IoT is a network where CPSs cooperate through unique addressing schemes. The IoS, in turn, allows companies to offer their services through the Internet, consisting of individuals, service infrastructure, business models and the services themselves [5].

The concept of the Industry 4.0 emerges in the context of the Fourth Industrial Revolution. In October 2012, a 4.0 Industry working group chaired by Siegfried Dais, vice chairman of the Board of Directors of Robert Bosch Healthcare, and Henning Kagermann, a member of German Academy of Science and Engineering, submitted the concept for the integration of advanced control systems with the Internet, allowing communication between people, products and complex systems [6].

The main approach, present in this new concept, is to make direct cooperation and communication between people, machines, equipment, products and logistics systems, using production systems with embedded technology, composed of sensors already integrated to smart actuators, still relying on direct communication with the operating control, called CPS [4, 7].

Indeed, an important component of the Industry 4.0 is the fusion of physical and digital world [5] provided by the CPS. The CPS refers to the integration of computers and physical systems into a single process, in which both feedback and interact, playing a key role in advanced manufacturing. The development of CPS is characterized by three phases: (1) the first generation of CPS includes identification technologies such as Radio-Frequency Identification (RFID) tags, which allowed unique identification, and storage and analysis should be provided as a centralized service; (2) the second generation of CPSs is equipped with sensors and actuators with a limited range of functions; and (3) the third generation CPS can store and analyze data, being equipped with several sensors and actuators and are compatible with the network [7].

Therefore, Industry 4.0 can be understood as a concept that encompasses the intelligent factory which includes automation technology and real-time data exchange, using the Internet, cloud computing and CPSs to execute, predict and to correct process errors faster and more efficiently [3].

2.2 *Ergonomics and the Cognitive Specialization*

According to International Ergonomics Association (IEA), ergonomics is the scientific discipline which deals with the understanding of interactions between humans and other elements of a system, and also is the profession that applies theories, principles, data and methods to projects, in order to optimize human well-being and the overall performance of systems. At the same time, Ergonomics Society (England) defines ergonomics as the study of the relationship between humans and the work, equipment, environment and, particularly, the application of knowledge of anatomy, physiology and psychology in solving the problems that arise from this relationship [8].

So, ergonomics is a science-oriented to a systemic approach and which currently extends to all aspects of human activity and whose objective is to study the factors that can influence the productive system, seeking to reduce its consequences for the worker. As a field of ergonomics specialization, there is physical ergonomics, organizational ergonomics and cognitive ergonomics [9, 10].

Physical ergonomics seeks to study the relationship between physical aspects and work, encompassing the analysis of human anatomical characteristics, anthropometry, physiology, biomechanics, among others. In the scope of organizational ergonomics, there is the optimization of sociotechnical systems, involving humans and management techniques, organizational structures, policies, and processes [9, 10].

Finally, cognitive ergonomics, an object of study of the present work under the focus of 4.0 Industry, studies how human beings interact with several elements of a system, analyzing aspects of mental processes, such as perception, memory, motor response, and reasoning, among others. Thus, the mental workload; decision-making process; human–computer interaction and stress are topics of cognitive ergonomics [8, 9]. It is noticed that many of the items mentioned are directly related to the new demands of Industry 4.0, due to the flow and speed of information transmission, the need for accelerated decision-making and approaches based on the intelligent system, triggering in the worker high demands on mental processes.

3 Research Approach

The present work can be considered an exploratory study, starting with a literature review to identify the guiding principles and concepts of the Industry 4.0 in order to develop a technological mapping and analyze the potential implementation challenges.

The following academic databases were used: Web of Science, Scopus and SAGE Journals; the main research fields included Engineering, Advanced Manufacturing, Industrial Internet, Production Technology, Cyber-Physics, Information Security and Cloud Computing. Criteria for inclusion and exclusion of articles, the definition of information to be extracted from articles, analysis, discussion and presentation of results were considered as stages of bibliographic review.

The analysis from the cognitive ergonomics point of view was based on the identification of the mental requirements to work in the middle of intelligent systems, understanding the necessary qualifications for the Engineer 4.0. In addition, the attributes of Industry 4.0 and its potential relationship with cognitive ergonomics were identified through a process analyses.

4 Results and Discussion

To characterize this emerging industry concept and to understand the nature of its processes, it was studied the main attributes of Industry 4.0. Interoperability is an essential element of advanced manufacturing because CPS and humans are linked to IoT and IoS so that standards will be a key success factor for communication between CPS from different manufacturers. Thus, interoperability means that all CPS within the plant can communicate with each other over open networks. Virtualization, as another attribute, expresses that a CPS can monitor physical processes, so these sensors and data are linked to digital plant models and splicing models.

Decentralization, on the other hand, starts from the point that the growing demand for individual products makes it increasingly difficult to control processes centrally. Embedded computers allow CPS to make their own decisions, and only in cases of failure will the tasks be delegated to a higher level. However, to ensure quality and traceability, it is necessary to monitor the entire system at any time. The real-time capability is also a critical request for Industry 4.0 because, in organizational tasks, data need to be collected and analyzed in real time, so traceability of the process can be verified and action taken immediately.

Another attribute, the service orientation, states that the CPS and other resources are available on IoS and can be used by other participants, being offered both internally and across company boundaries. Finally, there is the modularity attribute, in which modular systems can flexibly adapt to changes and expansions by individual modules. Therefore, modular systems may be suitable in the event of seasonal fluctuations or changes in product characteristics. Table 1 presents these key attributes in summary form.

The attributes of Industry 4.0 prove an increasing industrial complexity in the middle of a process of technological virtualization, requiring new skills and competencies of this professional. Moreover, there is a need for a faster and assertive decision-making process in front of a set of information, increasing the cognitive demand of the worker.

Therefore, due to its unique and innovative characteristics, the Industry 4.0 may require a new type of worker. As a result, Industry 4.0 requires optimized training of professionals, qualifying them to work in this type of organization, to guarantee the health, safety, and satisfaction in the work environment, as recommended by ergonomics.

In this context, the correct application of engineering knowledge leads to the critical analysis of the required qualifications of the employee inserted in the context of advanced manufacturing and the industrial Internet. Specifically, the so-called Engineer 4.0 entitles four essential requirements: (1) interdisciplinary training; (2) adaptability; (3) sense of urgency; and (4) good interpersonal relationships. In fact, it was observed that the Engineer would have to break and surpass the search only for technical solutions to a problem, requiring the interaction with professionals of diverse areas, of creativity and the adaptive capacity.

Table 1 Key attributes of Industry 4.0

Attribute	Characteristics
Interoperability	Communication and connectivity of cyber—physical systems, intelligent plants and humans
Virtualization	Virtual copy of intelligent factories (data, simulations, models)
Decentralization	Decisions without human intervention
Capacity in real time	Collect, analyze and respond to data in a short time
Orientation service	Providing services through Cloud Computing

Adapted from [1]

5 Conclusion

The Fourth Industrial Revolution involves a new paradigm, based on advanced manufacturing and industrial internet. Therefore, it was found that the demand for mental and cognitive processes in Industry 4.0 scope will certainly be different from that required in traditional industry, due to the flow and speed of information transmission, need for accelerated decision-making and approaches based on intelligent systems, demanding, from the professional, many mental activities, in order to enter the area of study of cognitive ergonomics. In addition, the lack of standardization of information on the subject matter, especially in the academic field, is a consensus, hindering further investigations by the scientific community.

Particularly in the area of Engineering, much has been debated about the Engineer 4.0 training. This study revealed that the engineer would be connected to the solution of increasingly complex problems in a digital environment, requiring denser cognitive efforts and more accurate socio-emotional skills, such as communication, motivation, autonomy, perseverance, self-control, resilience, collaboration, and creativity.

References

1. Shafiq, I., Sanin, C., Szczerbicki, E., Toro, C.: Virtual engineering object/virtual engineering process: a specialized form of cyber physical system for Industrie 4.0. Procedia Comput. Sci. **60**, 1146–1155 (2015)
2. Herman, M., Pentek, T., Otto, B.: Design principles for Industrie 4.0 scenarios. Technical University of Dortmund, Faculty of Engineering, Audi Endowment Chair Supply Net Order Management (2015). http://www.snom.mb.tu-dortmund.de/cms/de/forschung/Arbeitsberichte/Design-Principles-for-Industrie-4_0-C. 15 Sept 2017
3. Toro, C., Barandiaran, I., Posada, J.: A perspective on Knowledge Based and Intelligent systems implementation in Industrie 4.0. Procedia Comput. Sci. **60**, 362–370 (2015)
4. Brettel, M., Friederichsen, N., Keller, M., Rosenberg, M.: How virtualization, decentralization and network building change the manufacturing landscape: an Industry 4.0 perspective. Int. J. Mech. Ind. Sci. Eng. **8**, 37–44 (2014)
5. Kagermann, H.: Chancen von Industrie 4.0 nutzen. In: Bauernhansl, T., Hompel, M.T., Vogel-Heuser, B. (eds.) Industrie 4.0 in Produktion, Automatisierung und Logistik. Anwendung. Technologien und Migration, pp. 603–614 (2014)
6. Astarloa, A., Bidarte, U., Jiménez, J., Zuloaga, A., Lázaro, J.: Intelligent gateway for Industry 4.0-compliant production. In: 42nd Annual Conference of the IEEE Industrial Electronics Society, Florence, pp. 4902–4907 (2016)
7. Bauernhansl, T.: Die vierte industrielle Revolution. Der Weg in ein wertschaffendes Produktionsparadigma, In: Bauernhansl, T., Hompel, M.T., Vogel-Heuser, B. (eds.) Industrie 4.0 in Produktion, Au-tomatisierung und Logistik. Anwendung. Technologien und Migration, pp. 3–35 (2014)
8. IEA. International Ergonomics Association: The discipline of ergonomics (2000). http://www.iea.cc/ergonomics. 15 Sept 2017
9. IIDA, I: Ergonomia: projeto e produção. 9th edn. Sao Paulo, Edgard Blücher (2003)
10. Falzon, P. (ed.): Ergonomia. Sao Paulo, Edgard Blücher (2007)

The Contribution of the University Extension for Solid Waste Management

Ana Claudia Mendes de Seixas and Giovanna Ramos Maccari

Abstract With the massive increase of the urbanization process, the pursuit of sustainability all around the world, situates the cities in evidence regarding various kinds of problems, mainly related to the proper disposal of solid waste. The integrated management of this problem, refers to the decent destiny to it, also a generation's decrease through its source and recycling of this waste. The first major Brazilian mark, was the creation of the Law 12,305 of 2010, establishing the National Policy on Solid Waste. Assuming these facts, the present work, aims to promote the environmental education of the solid waste's reuse with the target audience of seventh-year primary students from Pio XII Application College, located in Campinas—SP, seeking an improvement of the cities habitants lifestyle. Initially, the individual ecological footprint was analyzed for consumption investigations and reflection of concepts related to their daily life, relating to Selective Collection and composting of organic waste. Including the help from the information materials, confectioned by the participants of the project with the target audience. It was possible to understand these themes, arousing great interest on their part and generating good communication and friendly dynamics in the workshops. After that, it was possible to identify the behavior turnover of the target students, acting as multipliers by extending the practices from their homes and communities to the environment, making themselves more aware, showing extreme importance for their socialization process, critical thinking about the environment and the importance of preserving it for the well being of the community. This essay has the purpose to make children more aware and responsible for their actions, capable of reflecting on reality and facts to improve people's lifestyle, personal and community, and can be transmitted to the present and future generations.

Keywords Management · Solid waste · University extension

A. C. M. de Seixas (✉) · G. R. Maccari
Pontificia Universidade Católica de Campinas, Campinas, SP 08544, Brazil
e-mail: acmseixas@gmail.com

G. R. Maccari
e-mail: grmaccari@gmail.com

Y. Iano et al. (eds.), *Proceedings of the 3rd Brazilian Technology Symposium*,
https://doi.org/10.1007/978-3-319-93112-8_29

1 Introduction

With the Industrial Revolution in the eighteenth century, mankind's ability to interfere in the nature gives a great and advanced leap. This evolution resulted in a huge environmental damage, mainly in finite resources use, such as fossil fuels, and since there was no immediate threat of resource depletion, the mankind has delayed the adoption of sustainable techniques and procedures.

According to Ref. [1] precisely in the period of the eighteenth century there was an increase in the production of various goods, intensifying the consumption and bringing with it the great generation of wasting, which occurs currently, that its majority is difficult to decompose.

In the mid 1970s, with the movement called "New Environmentalism", there was a growing concern about the depletion of the natural resources and overproduction, fueled by uncontrolled consumption, generating much of the planet's pollution [2]. This unbridled consumption results in one of the biggest problems facing the world today, the generation of solid waste of all classes.

In Brazil, the first major step on solid waste is recent. In 2010, the National Solid Waste Policy (Law No. 12,305 of 2010) was established as: "an economic and social development instrument characterized by a set of actions, procedures and measures to enable solid waste to the business sector, for reuse in its cycle or in other production cycles, or other environmentally appropriate final disposal" [3]. According to PNRS, "solid waste" is defined as: "Material, substance, object or well-disposed resultant of human activities in society, which final destination is proceeded, whether it is proposed to proceed or is obliged to keep up, as solid or semi-solid, as well as gases contained in containers and liquids that particularities make it unfeasible for them to be introduced into the public sewage system or bodies of water, or which require technically or economically unviable solutions in view of the best available technology" [3].

In addition to the concept of waste, it is also important to highlight the definition of "reject", also according to the PNRS (2010): "solid waste that, after exhausting all possibilities of treatment and recovery by available and economically viable technological processes, do not present other possibility than the final disposition environmentally appropriate". As a result, these should be the only types of waste destined for landfills [3].

Currently, the management of this solid urban waste (MSW) is a major and important challenge for all of the cities that seek sustainability, making it a concern lately. According to the United Nations, the current generation of waste in the world is around 12 billion tons/year and by 2020 the estimated volume is 18 billion tons/year.

The management and inadequate disposal of solid wastes are the major cause of socio-environmental impacts, including soil degradation, exposure of water bodies and springs, floods, contribution to air pollution, proliferation of vectors of sanitary importance in urban centers and in unhealthy conditions on the streets and areas of final disposal [4].

That's why, the debate on the theme "Environmental Education" becomes an important tool to make humans more aware that maintaining biodiversity is fundamental for our survival, to act responsibly, conserving the environment for present and future, thus, bringing the balance of human relationship with nature [5]. In order to Environmental Education transform, harmful values, which contribute to the degradation of the environment through the lack of management of solid waste, a permanent and continuous education is necessary. Therefore, the preparation for the necessary changes consists in the collective understanding of the functioning of the environment, how we depend on it, how we affect it, how we can promote sustainability, and clarify that nature is not an inexhaustible source of resources, in crises that threaten the future of the planet [6].

Consequently, the present Extension Project involves the mapping of the solid waste generated in Pio XII Application College, followed by an adequate environmental management through the development of new mechanisms, emphasizing the reuse, which is able to aid in adaptations, environmental laws and policies, share different realities with autonomy with a mutual learning in a creative way, leading to a change and transformative action, fostering autonomy in the community and awakening in our stakeholders a greater awareness of issues social policies.

The extension project counts on the participation of an extension teacher and an alumni extensionist from the College of Environmental Engineering, in which Solid Waste Management is interconnected and related to the pedagogical project from the Environmental and Sanitary Engineering course at Campinas. In addition, there is a direct participation of the target audience, approximately 10 seventh-year students from the Pio XII College of Application Campus-SP, with a concern to improve their knowledge and the environment they live, providing a greater awareness of the Solid Waste Management, bringing contributions of their experiences and experiences and, in general, act as multipliers, by extending the practices to their residences or surrounding communities. This follows a thought from Ref. [7] that says: "Teaching is not transferring knowledge but creating possibilities for its own production or its construction."

2 Objectives

The objective is to develop socio-educational activities from Environmental Education with the target audience in a creative and innovative way, increasing the themes addressed, in order to raise awareness and perception regarding the integrated solid waste management generated by unbridled consumption.

3 Methodologic

The workshops were held at the PIO XII Application College, located in the city of Campinas, São Paulo, each week, with the duration of 1 h. There was the concern of the community development, mobilization and participation. The activities aimed at raising the awareness of the target public and the community regarding the issue of solid waste in the College, mainly by enabling the implementation of an integrated management system for all waste generated, organic (Composting organic waste) or recyclable (Selective Collection).

The work has begun with the realization of the individual ecological footprint (Fig. 1), an environmental accounting methodology that evaluates the pressure of the human population consumption on natural resources, for consumption analysis and reflection on concepts related to their daily routine, for example, the act of recycling at home. In addition, the school's dumps where the work is being developed were analyzed by those involved in the process with the orientation of the extension student and supervised by the responsible teacher (Fig. 2).

Firstly, the theme Composting was approached from the realization of a Composite, elaborated by each student participating in the Project made with three PET

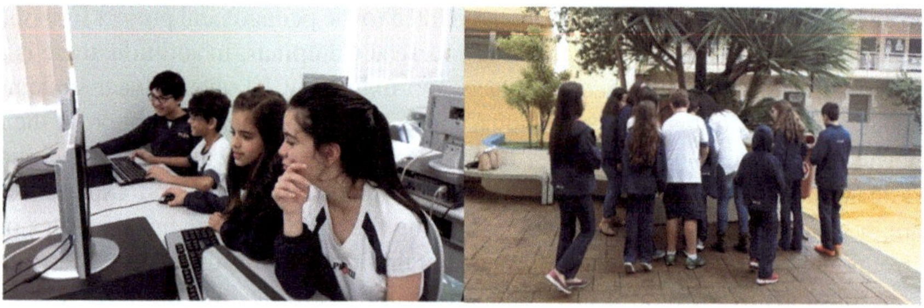

Fig. 1 Students with computers to realization of the individual ecological footprint and analyzing solid wastes in the college (*Photos* Seixas 2017)

Fig. 2 Students doing composting organic waste (*Photos* Seixas 2017)

bottles that would be discarded and the compound made with vegetal soil, organic residues, sawdust, and water (Fig. 2).

Each week the compound was analyzed for the identification of its biological process (decomposition of residues), temperature verification, appearance, odor of the compound and formation of the liquid formed by the process. After four weeks, the compound was finished and suitable for use as an organic fertilizer from the liquid originated in the biological decomposition action.

In addition to Composting, the paper recycling workshop was held, in which the students reduced these residues from the grinding with water, staining and molding of this material, be it in sieves for the production of sheets of paper or in silicone forms, making it a decorative material (Fig. 3).

Together with Recycling Paper, the theme 'Selective Collection' was also developed from the elaboration of an object organizer and a bag created from the reuse from the materials, cardboard boxes and tetra pack packaging, respectively. The preparation of both was executed by cutting and gluing the materials that would be discarded with fabrics, making them useful for the target audience (Fig. 4).

In addition, Tetra Pack packaging has also been reused for another purpose. From the cut of the same, the packaging became a container to conserve food hygienically (Fig. 5).

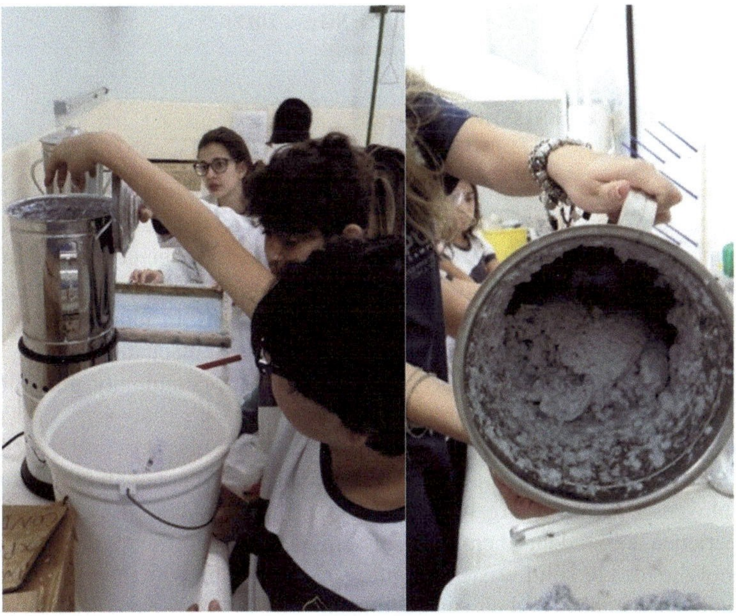

Fig. 3 Students doing the paper recycling workshop (*Photos* Seixas 2017)

Fig. 4 Students assembling the bag and organizer of objects from Tetra Pack (*Photos* Seixas 2017)

Fig. 5 Students packing to preserve food **(Photos: Seixas, 2017)**

4 Results and Discussion

In the socio-educational workshops, with the help of didactic material and expository content, the communication and knowledge were promoted, through dialogue, on the Composition and Selective Collection of solid waste in the community and its surroundings, supporting and encouraging the public to take the knowledge acquired in the offices to their families and to their circle of relations; In addition, it has made it possible for the participants to improve their human and social condition, being critical citizens through this process of intervention in the community, thus allowing the realization of the great transforming action in each one's life.

From the analysis of the dumps located all over the places, it was possible to verify that the waste was not properly deposited, proving that not all of the students and employees separate the organic residues from the recyclables, for example. With this experience, the students themselves disseminated to their friends and throughout the school the idea of solid waste separation, collaborating with theirselves and the community's awareness of their surroundings.

With the Composting workshop, at the end of the process, it was possible to plant Ipê (a typical Brazilian plant whose scientific name is *Tabebuia roseo-alba*) seedlings from the liquid produced by the biological decomposition action (Fig. 6), which were taken to the students' homes, involving their relatives about of the theme.

Fig. 6 Students with the Ipê seedlings planted in their final compound (*Photos* Seixas 2017)

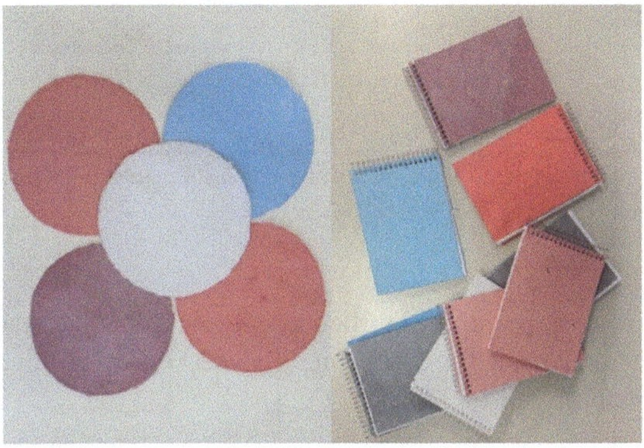

Fig. 7 Sheets of paper and small notebooks from the recycling of the material (*Photos* Seixas 2017)

In addition, in the recycling paper workshop, it was possible to manufacture new colored sheets, decorative objects and small notebooks for students (Fig. 7).

At the end of the workshops, the students made an informative material about the Paper Recycling workshop (Fig. 8). This material was used to absorb knowledge by the students and was it spread this idea to the community to the environment, family and friends.

The University Extension has the extreme importance for the professional and human development of the individual who participates, whether in the form of an extension student or as a target audience, because from this parallel strand of graduation it is possible to absorb concepts that cannot be learned in classroom and it causes a great effect in teaching and raising awareness in the target audience. In addition, participation in events such as Congresses, Meetings and Symposiums makes possible a greater autonomy in the subject, through articles, expanded abstracts and presentations besides being very important for the academic formation.

This allows a more conscious attitude with the actions of raising awareness, increasing the critical capacity and professional maturity to face real problems, cre-

Fig. 8 Information materials regarding Composting organic waste and the paper recycling workshop (*Photos* Seixas 2017)

ating opportunities to experience the theoretical and conconceptual content of its course, as well as prepare it for companies focused on economic, environmental, social, and sustainability.

5 Final Considerations

Throughout the Extension Project, it was possible to observe the behavioral change of the students involved through the awareness of the Integrated Management of Solid Waste and Environmental Education, bringing contributions of their experiences, making them act as multipliers of the idea, by extending the practices to their residences or surrounding communities. This project encouraged actions that promote citizenship, as well as democratic values of equal rights and participation, allied to respect for the person and the sustainability of interventions in the environment.

For this reason, Environmental Education is fundamental for the formation of the individual. From the encouragement in childhood, the small acts of sustainability changes the way of thinking from the individual, causing anyone a reflection of the acts before consuming something rampantly, for instance.

An example of this is the reduction of the organic residues generated in each residence, with the reduction of the same destined to the landfills, it is already possible an improvement in the life of everyone. With the use of this material as an organic fertilizer, in addition to reducing waste, the family can grow an organic vegetable garden in their home to improve their nutrition.

The informative materials generated, with practical information, besides being a challenging factor for the target audience, allowed a multiplier effect when they were divulged to the participants' acquaintances.

Acknowledgements The Pro-Rectory of Extension and Community Affairs of the Pontifical Catholic University of Campinas for all support, which from the first moment made feasible and contributed to make Extension Work a reality. The Direction of the College of Application Pio XII of Campinas (SP) that make possible and support the Extension Project.

References

1. Leite, D., e Araujo, J.: Aspectos da política pública municipal de resíduos sólidos em Fortaleza. s/d Disponível em (2016). http://www.anppas.org.br/encontro_anual/encontro3/arquivos/TA674-08032006-152340.DOC. Acesso em: 31 de Março de 2016
2. Guimarães, A.P.F.V., Camargo, S.A.F.: Consumo e sustentabilidade: um desafio para a administração pública. Disponível em (2011). http://www.ambito-juridico.com.br/site/?n_link=revista_artigos_leitura&artigo_id=11359. Acesso em: 16 de novembro de 2016
3. Brasil: Política Nacional dos Resíduos Sólidos- PNRS. Decreto-lei nº 12.305, de 2 de agosto de 2010. Disponível em: http://www.planalto.gov.br/ccivil_03/_ato2007-2010/2010/lei/l12305.htm. Acessado em 20 outubro de 2015
4. Jacobi, P.R., Besen, G.R.: Gestão de resíduos sólidos em São Paulo: desafios da sustentabilidade. Estud. av. **25**(71), 135–158 (2011)
5. Pereira, F.A.: Educação ambiental e interdisciplinaridade: avanços e retrocessos. Brazilian Geograph. J.: Geosci. Human. Res. Medium Ituiutaba **5**(2), 575–594 (2014)
6. Dias, G.F.: Educação ambiental: princípios e práticas. São Paulo, Gaia (1992)
7. Freire, P.:. Pedagogia da Autonomia: saberes necessários à prática educativa. 34ª edição. Paz e Terra, São Paulo (2006)

Using i-Tree Canopy to Estimate and Value Ecosystem Services of Air Pollutant Removal

Tatiane Ferreira Olivatto

Abstract Environmental monitoring processes have become a demand in the most diverse branches of society. In cities, monitoring techniques support urban planning, oversight, and decision-making process. At the same time, ecosystem services valuation has been used more frequently to include the environmental variable in the environmental management process. Atmospheric pollution is an aspect that is often monitored and also the subject of valuation studies. This paper estimates the potential for atmospheric pollutants removal and their equivalent monetary value in two different scenarios: preserved and degraded. To achieve this objective, the i-Tree Canopy Suite, from the i-Tree toolkit, was used. The procedure verified that the benefits in the preserved scenario are equivalent to more than UD$320,000, removing an average of 8000 t of atmospheric pollutants. In the degraded scenario, the loss was about UD$10,000, adding an average of 250 t of carbon dioxide that was previously stored in trees. Although easy handling, the use of this software presents some limitations, such as results depending on the user's ability to classify points and satellite images quality. Also, some degree of generalization was verified by the software's methodology; resulting in the need for a specific database when the desired result requires a more technical and less statistical precision. Therefore, despite these limitations, the use of i-Tree software has shown promising to estimate and value the benefits generated by ecosystem services.

Keywords Ecosystem services · Ecosystem services valuation · i-Tree Canopy
Ecological economics · Green design · Air pollutants removal · Environmental economy · Environmental management

T. F. Olivatto (✉)
Centre of Exact Sciences, Environmental and Technologies, Pontifical Catholic University of Campinas, Campinas, SP, Brazil
e-mail: tatianeolivatto@yahoo.com.br

Y. Iano et al. (eds.), *Proceedings of the 3rd Brazilian Technology Symposium*,
https://doi.org/10.1007/978-3-319-93112-8_30

1 Introduction

Deforestation is widely identified as the driver of the environmental problems we currently face, especially global warming. Livestock farming, logging, agriculture, burning, mining and urban expansion are recognized as leading causes of deforestation in the world. In addition, various environmental experts associate the exclusion of environmental costs from market prices and unsustainable resource use along with population growth and poverty as essential causes of these environmental problems [5]. In this context, forest cover areas monitoring and ecosystem services valuation emerge as promising techniques to face environmental issues.

The 225th article from the Constitution of the Federative Republic of Brazil, 1988, ensured that everyone, including future generations, has the right to an ecologically balanced environment and that it is a responsibility of society and public authorities [1]. An efficient and relatively low-cost method that leads to an ecologically balanced environment is the preservation and recovery of green areas.

Therefore, instruments that allow the inclusion of the environmental variable in the activities of the public, private and third-sector spheres are needed [6]. This task can be complicated when environmental services are not analyzed with a distinct value besides being often interpreted as costs rather than investments. For this reason, the valuation of ecosystem services can be seen as an instrument for supporting decision-making in the public management process [7]. Estimating and valuing ecosystem services are valuable tools for evaluating the function of green areas inserted in the urban context. In the case of green areas inserted in the urban context, the data resulting from this quantification can base awareness initiatives on the importance of these areas for human well-being and helps the processes involved in environmental management [7]. Valuation, although less widespread, supports payment programs for environmental services and bases decision-making on public policies [6].

Tree canopies have an important role in urban environments. They bring an array of benefits including psychological and aesthetic effects on humans and climate, mitigation of air pollution, support biodiversity and other environmental services. These benefits increase health, well-being, and quality of life to the urban residents [2]. The public domain software i-Tree can examine the potential for these benefits in urban areas. Moreover, the data provided by the i-Tree can be used to improve understanding of the cities' surface areas and their environmental aspects [8]. Another interesting use of the i-Tree software is environmental monitoring, enabling control of areas of ecological interest and following land use changes.

In this work, the potential for atmospheric pollutants removal and their equivalent value were estimated using the i-Tree Canopy Report tool, from the i-Tree toolkit. In order to provide a better understand of application possibilities, two different areas were studied: an urban park and a private property. The first one is a preserved area and the second one is a degraded region.

1.1 Aims and Objectives

- Undertake a land cover classification of the studies areas using the i-Tree Canopy program;
- Quantify the potential for atmospheric pollutants removal, carbon sequestration and storage provided by the vegetation in both cases: preserved (in Portugal Park) and before being removed (in the private property);
- Estimate the monetary value of tree benefits regarding air pollutants removal.

This paper is structured as follows: Sect. 2 describes the use of the i-Tree Suite in canopy assessment. Section 3 investigates two Brazilian case studies with the help of the i-Tree canopy report feature. Finally, some discussion and future investigation are introduced in Sect. 4.

2 i-Tree Canopy Assessment

The i-Tree Suite is an open source software from the USDA Forest Service that provides urban and rural forestry analysis as well as benefits appraisal tools. The tools available in i-Tree help community, private companies and government organizations by quantifying the structure of trees and forests, and the ecosystem services those healthy environments provide [11].

The i-Tree software (Fig. 1) includes nine main tools for various purposes, including i-Tree Eco, i-Tree Landscape, i-Tree Hydro, i-Tree Design, i-Tree Canopy, i-Tree Species, i-Tree Streets i-Tree Vue and i-Tree IPED [9]. These tools and along with the complete information and manuals about them can be found at i-Tree official website [10].

Fig. 1 i-Tree logo [11]

Table 1 Environmental pollutants covered by the i-Tree canopy report [10]

Abbr.	Description
CO	Carbon monoxide removed annually
NO_2	Nitrogen dioxide removed annually
O_3	Ozone removed annually
PM 2.5	Particulate matter less than 2.5 μ removed annually
SO_2	Sulfur dioxide removed annually
PM 10	Particulate Matter greater than 2.5 μ and less than 10 μ removed annually
CO_2 (SEQ)	Carbon dioxide removed annually in trees
CO_2 (stor)	Carbon dioxide stored in trees (note: this benefit is not an annual rate)

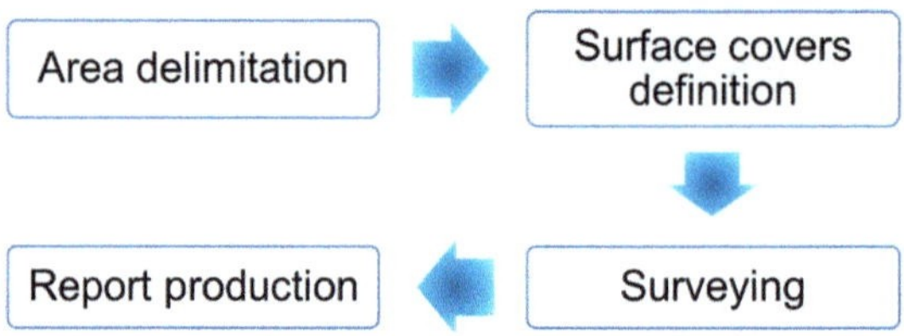

Fig. 2 i-Tree canopy methodology flow-chart

All tools are connected to a complete database with information on species, climatology, hydrology and soil structure. Currently, this database covers only the United States, Australia, the United Kingdom and Canada. Heretofore, only the i-Tree Canopy tool has been fully developed for use in Brazil. The program developer suggests the use of the i-Tree Canopy to quantify and value the environmental services provided by the trees and more specifically, the ability to remove atmospheric pollutants, store and sequester carbon dioxide [11].

The i-Tree Canopy uses a simple methodology to quickly produce a statistically valid estimate of the land cover use through aerial images available in Google Maps. This suite provides two outputs: the quantity of pollutants removal and their equivalent monetary value [9]. Table 1 shows the pollutants covered by the software.

The methodology involved in the use of the i-Tree Canopy is described in 4 basic steps as seen in Fig. 2.

Fig. 3 Study areas location [4]

Fig. 4 Case study 1: Area 1 in 1940 (left) and 2017 (right) [3, 4]

3 Case Studies

The software was used to examine land cover types and access ecosystem valuation with respect to two study areas located in Campinas, Sao Paulo. Figure 3 shows Area 1 (red color, upper right), a green area, known as Portugal Park (or Taquaral Lake). It also depicts Area 2 (yellow color, left bottom corner), which is a recently deforested private property.

For case study 1, the characteristic that differs most from the Area 1 and Area 2 is the evolution of the vegetation cover. Area 1 is an urban park used for recreation, sports, and cultural activities. Figure 4 describes its evolution with an increase in green surface all over the years.

Area 2 is a private property, belonging to a logistics company. Last year, due to deforestation for construction purposes, the vegetation cover has been completely removed (see Fig. 5).

Fig. 5 Case study 2: Area 2 in 2015 (left) and 2017 (right) [3]

Table 2 Suggested surface cover classification [6]

Cover class	Description	Area 1	Area 2
Hard surface (H)	Non-plantable	X	
Aquatic surface (W)	Water bodies	X	
Bare soil (B)	Potentially plantable	X	
Grass and shrub (G)	Developing canopy cover, understory	X	X
Tree (T)	Canopy cover	X	X

4 Results

The software was used to examine land cover types with respect to five cover classes for Area 1 and two cover classes for Area 2. Table 2 specifies which classes were used in each case. This cover classes were suggested by Ref. [6].

The i-Tree Canopy can export a report containing cover class percentage estimation and the tree benefits (amount and monetary value). The results refer to annual values, except by carbon dioxide stored in trees [9].

The land area of Portugal Park as assessed by the i-Tree is approximately 800,000 m^2. The results indicate that in 2017, tree canopy was the predominant land cover, followed by the aquatic surface, grass/shrubs, hard surface and bare soil respectively (refer to Fig. 6).

The land area of the private property assessed in i-Tree is approximately 23,000 square meters. The results indicate that in 2015, before deforestation, tree canopy covered approximately 38.9% of the area, being the remaining grass/shrubs (see Fig. 7).

Figure 8 (Area 1) and Fig. 9 (Area 2) show the i-Tree assessment results (ecosystem services regarding pollutants removal amount and monetary value).

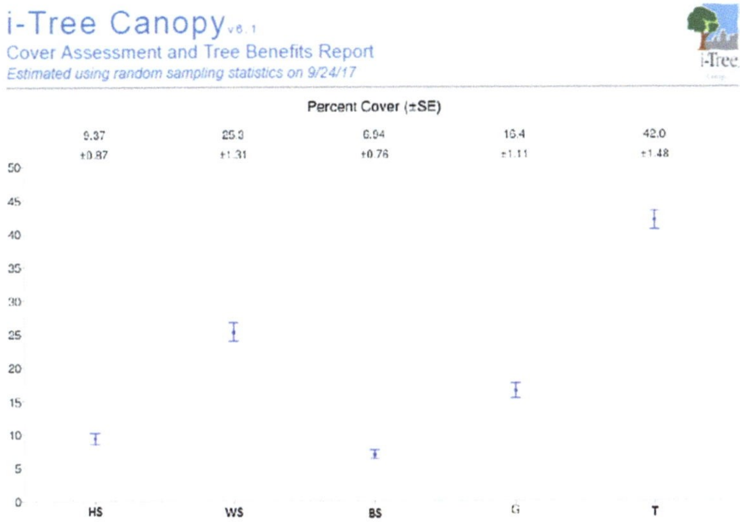

Cover Class	Description	Abbr.	Points	% Cover
Hard Surface	Non-plantable	HS	104	9.37 ±0.87
Water Surface	Water bodies	WS	281	25.3 ±1.31
Bare soil	Potentially plantable	BS	77	6.94 ±0.76
Grass e shrub	Developing canopy cover, understory	G	182	16.4 ±1.11
Tree	Canopy cover	T	466	42.0 ±1.48

Fig. 6 Case study 1: i-Tree canopy report corresponding to land use

Quantitatively speaking, the carbon dioxide stock is the most significant ecosystem service considered by the i-Tree Canopy. In general, the ecosystem services in atmospheric pollutants removal and their monetary values are 8225.94 t and US$321.440.01 for Area 1; and 265.72 t and US$10,439.83 for Area 2.

5 Final Considerations

Overall, the i-Tree Canopy was a very useful tool to quantify and value ecosystem services related to air quality. However, the tool has some limitations that, depending on the purpose, may be relevant.

The main limitation is a degree of generalization resulting from the software methodology [8]. When the purpose requires more statistical analysis, and also, completely remote, the i-Tree Canopy proves appropriate. However, when the purpose requires a more technical and accurate analysis and also allows some field work, the i-Tree Eco tool of the i-Tree toolkit is the most appropriate.

Regarding the quality of the results, the accuracy of the analysis performed was dependent on three main factors. First, the human ability to correctly classify a point into its correct class. Secondly, the quality of Google Earth imagery (e.g., resolution,

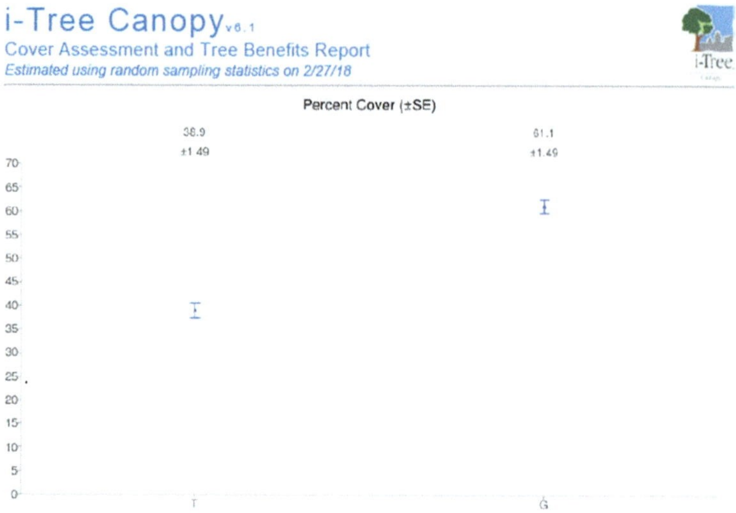

Cover Class	Description	Abbr.	Points	% Cover
Tree	Tree	T	414	38.9 ±1.49
Grass	Grass	G	650	61.1 ±1.49

Fig. 7 Case study 2: i-Tree canopy report corresponding to land use

Tree Benefit Estimates

Abbr.	Benefit Description	Value (USD)	±SE	Amount	±SE
CO	Carbon Monoxide removed annually	$2.66	±0.09	28.35 kg	±1.00
NO2	Nitrogen Dioxide removed annually	$4.58	±0.16	154.61 kg	±5.46
O3	Ozone removed annually	$238.42	±8.41	1.54 t	±0.05
PM2.5	Particulate Matter less than 2.5 microns removed annually	$492.86	±17.39	74.82 kg	±2.64
SO2	Sulfur Dioxide removed annually	$0.80	±0.03	97.43 kg	±3.44
PM10*	Particulate Matter greater than 2.5 microns and less than 10 microns removed annually	$173.09	±6.11	515.78 kg	±18.20
CO2seq	Carbon Dioxide sequestered annually in trees	$12,227.76	±431.45	313.53 t	±11.06
CO2stor	Carbon Dioxide stored in trees (Note: this benefit is not an annual rate)	$308,299.84	±10,878.32	7.91 kt	±0.28

Fig. 8 Area 1: i-Tree assessment results (i-Tree canopy report)

shadows). Finally, the number of points classified. As more points are categorized, the precision increases and, consequently, the standard error decreases. In addition, if the number of classes is considerably high, more points will be required to achieve a suitable standard error.

The methodology used in this work concluded that the ecosystem service for pollutants removal in Portugal Park reaches the monetary value of more than US321,440.01. On the private property, the results indicate that the previous vegetation used to remove 265.72 t of air pollutants (equivalent to US$10,439.83). Although very significant, ecosystem services cover more than air quality maintenance only. Unfortunately, i-Tree Canopy tool, the only one in an i-Tree suite that can already be

Tree Benefit Estimates

Abbr.	Benefit Description	Value (USD)	±SE	Amount	±SE
CO	Carbon Monoxide removed annually	$0.30	±0.01	1.69 kg	±0.07
NO2	Nitrogen Dioxide removed annually	$0.28	±0.01	6.32 kg	±0.24
O3	Ozone removed annually	$22.96	±0.88	61.91 kg	±2.38
PM2.5	Particulate Matter less than 2.5 microns removed annually	$42.06	±1.62	3.15 kg	±0.12
SO2	Sulfur Dioxide removed annually	$0.03	±0.00	1.92 kg	±0.07
PM10*	Particulate Matter greater than 2.5 microns and less than 10 microns removed annually	$14.97	±0.57	20.67 kg	±0.79
CO2seq	Carbon Dioxide sequestered annually in trees	$602.65	±23.15	15.45 t	±0.59
CO2stor	Carbon Dioxide stored in trees (Note: this benefit is not an annual rate)	$9,756.58	±374.79	250.17 t	±9.61

Fig. 9 Area 2: i-Tree assessment results (i-Tree canopy report)

used in Brazil, calculates only air quality benefits. This limitation lies in a lacking Brazilian database. Therefore, the development of this database is crucial for the use of other tools from the i-Tree suite and, consequently, for others benefits calculation and valuation.

References

1. Brazil. Constitution: Constitution of the Federative Republic of Brazil 1988. Brazilian Federal Register, Brasília (1988). http://www.planalto.gov.br/ccivil_03/constituicao/constituicaocompilado.htm. Last accessed 14 Sept 2017
2. Brazil. Law n° 12.651, de 25 may 2012. Brazilian Forestry Code. Brazilian Federal Register, Brasília (2012). http://www.planalto.gov.br/ccivil_03/_ato2011-2014/2012/lei/l12651.htm. Last accessed 2017/09/26
3. Campinas City Council. Portugal Park Aerial Imagery: 200, Anchieta Avenue, Campinas, Brazil: Cartographic Information Department (1940)
4. Google Inc.: Google Earth Pro. Version 7.3.0.3832 (64-bit) (2017). https://www.google.com/earth/desktop/. Last accessed 27 Feb 2018
5. Miller, G.T., Spoolman, S., Society, N.G.: Environmental Science, 13th edn, p. 552 p. Cengage Learning, Pacific Grove, CA (2015)
6. Rogers, K., Evans, G.: Valuing the natural capital: i-Tree technical report, p. 54p. Highways Agency, Exeter, UK (2015)
7. Romeiro, A.D.: Sustainable Development: An Ecological Economics Perspective, vol. 26, No. 74, pp. 65–92. Estudos Avançados, São Paulo (2012). http://www.scielo.br/scielo.php?script=sci_arttext&pid=S0103-40142012000100006. Last accessed 23 June 2017
8. Tolffo FA.: Payment for Ecosystem Services as an environmental management tool for urban space. 289f. Master's thesis, School of Public Health, University of São Paulo, São Paulo (2015). http://www.teses.usp.br/teses/disponiveis/6/_6139/_tde-09122015-112922/pt-br.php. Last accessed 17 Aug 2017
9. USDA Forest Service: i-Tree User's Manual (2016). https://www.itreetools.org/resources/manuals/Ecov6_ManualsGuides/Ecov6_UsersManual.pdf. Last accessed 13 Jan 2017
10. USDA Forest Service: i-Tree Canopy Technical Notes (2011). https://canopy.itreetools.org/resources/iTreeCanopyMethodology.pdf. Last accessed 31 Jan 2017
11. USDA Forest Service: What is i-Tree? (2006). https://www.itreetools.org. Last accessed 13 Jan 2017

Detecting and Counting of Blood Cells Using Watershed Transform: An Improved Methodology

Ana Carolina Borges Monteiro, Yuzo Iano and Reinaldo Padilha França

Abstract The necessity for practicality, reliability, and agility has stimulated the creation of new tools for the health area. The blood cell counting is an important exam that helps to diagnose various diseases besides to be an important field of study in biomedical engineering. Currently, the red blood cells count (RBC) and white blood cells (WBC) have been performed by the automated method. However, still does not dispense the use of the manual method, through the preparation of blood smears, in specific cases or only for confirmation of diagnostics. In the last years, the Matlab® software has been used to develop algorithms to facilitate cell counting. Thus, the present work implements an algorithm based on Watershed transform, where its results show a better performance of 34% in execution time in comparison with other works developed, as well as a computational performance of 1.98 s, as also showing high reliability in your digital count.

Keywords RBC · WBC · Watershed transform · Biomedical engineering · Blood smear

1 Introduction

The human blood consists of approximately 45% of plasma and 55% of cells: red blood cells, leukocytes and platelets [1]. The erythrocytes are anucleated cells with biconcave disc-shaped, formed by two α chains and two β chains, responsible for the transport of oxygen through four iron molecules. The erythrocytes are also called

A. C. B. Monteiro (✉) · Y. Iano · R. P. França
School of Electrical and Computer Engineering (FEEC), University of Campinas – UNICAMP, Av. Albert Einstein - 400, Barão Geraldo, Campinas, SP, Brazil
e-mail: monteiro@decom.fee.unicamp.com

Y. Iano
e-mail: yuzo@decom.fee.unicamp.com

R. P. França
e-mail: padilha@decom.fee.unicamp.com

Y. Iano et al. (eds.), *Proceedings of the 3rd Brazilian Technology Symposium*,
https://doi.org/10.1007/978-3-319-93112-8_31

red blood cells and measure under normal conditions average 7 μm of diameter [2, 3]. The red blood cell count (RBC) show differences in concentration according to patient characteristics, such as: age (children 4.5 to 4.7 × $10^6/mm^3$ blood) and gender (men: 4.5 to 6.0 × $10^6/mm^3$ and women: 4.0 to 4.6 × $10^6/mm^3$ blood) [4, 5].

When there is a decrease in the number of red blood cells (erythropenia), blood loss or anemia is suspected, which may be due to food deficits or genetic factors. In turn, the increase in the number of red blood cell (erythrocytosis) may suggest polycythemia, which causes various damage to the body, such as stroke and infarction [6].

The leukocytes are also called white blood cells, being the body's defense cells, and are classified into five cell types: neutrophils, eosinophils, basophils, monocytes and lymphocytes, which are differentiated by the morphological characteristics as number of lobes, type of cytoplasmic granules, shape, coloration (blue/orange) and size (variable between 7 and 15 μm) [6, 7].

The white blood cell count (WBC) vary from 3500 to 10,500 leukocytes/mm^3 of blood, being an increase in this parameter (leukocytosis) suggestive infections or acute or chronic inflammations [7].

In this way, blood cell counting help health professionals to suspect or directly conclude a diagnosis of pathologies. These diseases may be of blood origin (genetic anemias, leukemias, or polycythemias), tissue injury (infections or inflammation) or even due to invasion by external etiological agents, such as malaria, mononucleosis, helminths, allergenic agents, bacterial infections, viral infections, among others [3, 8].

Thus, many methodologies have been developed in the last years to facilitate and solve problems as described, with the use of MATLAB® software, can be observed in [8–15, 16]. In these studies, the extraction and counting are based on the morphology of the blood cells through the distance found between each pixel, or through the Hough transform realizing the red cell counting through the detection of geometric characteristics, to find the circular center of the image, or through segmentation of erythrocytes, in order to determine the cell boundary, among others. However, some techniques result in poor performance and low accuracy when applied to smears with large agglomerates of cells.

The algorithm presented in this paper reaches all these points, solving in less processing time and computational cost associated the high precision in the detection in both RBC and WBC, as well as a better execution performance in comparison to other works as [5, 8–11, 14, 16, 17].

The present work is organized in the following structure: in Sect. 2 discusses the problematic of the subject and the Sect. 3 presents the proposal of this paper. In Sect. 4 the results are shown, and finally, in Sect. 5, the conclusions are presented associated with the potential of the research.

2 The Problematic

Blood cell counts can be done by automated or manual methods. The automated methods are faster, however more expensive [18]. Such method uses a 4 ml sample of blood with anticoagulant EDTA or heparin, and may be based on the principles of impedance pulses, flow cytometry, absorbance, or fluorescence marking, among others [18, 19].

However, the existing automatic methods do not annul the use of manual methods, when the equipment's are unable to analyze blood samples, in cases of children under 5 years old and patients upper 75 years old; cancer patients; patients in grave condition; patients with leukocytosis or leukopenia, cases of color variations, size and shape of red blood cells, among others [18].

The manual methods are cheaper, however more time consuming, because require the sliding of a few microliters of blood on a glass lamina and subsequent use of dyes, which allow the visualization of cellular structures [19]. Only the final portion of the lamina is used to perform the count, because the anterior portions have clustered and/or overlapping cells, preventing a reliable count. Besides of counts and observations of blood cell morphologies, health professionals perform calculations to conclude the diagnoses, which generally consider the total counts of RBC and WBC [3, 6, 19].

The complexity of medical images requires systematic segmentation, which be accurate and presents a reliable method [20]. The implementation of digital image processing of blood cells helps in clinical decisions, through of obtaining results in a way faster, easier and more agile, besides of the reduction of costs with the acquisition of equipment.

Thus, the Watershed transform is a technique for digital image segmentation and processing and that can be applied in several areas of knowledge. This technique interprets the images where each pixel corresponds to a position and the levels gray relative to each of these pixels determine their altitudes, being a powerful tool found to solve the problems approached in this work [4, 12, 15, 20–22].

3 The Proposal

The algorithm performs the counts of red and white blood cells through a digital image of a peripheral blood smear. The processes of counting and differentiating blood cells are performed based on the recognition of cell nuclei, size differentiation in pixels and color intensity present in each cell type analyzed. Following the segmentation and image processing logic, the algorithm is executed, with objectives of precision and computational costs.

The experiments were conducted by blood cells smear, stained and analyzed under an optical microscope under the objective lenses of 40 and 100, resulting in an increase of the image of 400 and 1000 times, respectively. Posteriorly, there was the acquisition of digital images of the fields of the lamina containing red blood cells without agglomerations and the presence of leukocytes.

The image processing involves some steps such as filtering and treatment of the image, in terms of quality for the next steps of the algorithm, the conversion of RGB images to gray scales and binary scales, have with the goal to solve the illumination problems normally found in images of optical microscopy.

In this step, morphological operations were made followed of the use of the Watershed transform, which involves steps such as binary erosion, dilation, opening, closing, and reconstruction of the image [4].

The Watershed transform acts on the processing, calculation, and recognition of distances between nuclei, avoiding possible erroneous counts of overlapping cells. The logic of the algorithm is represented in the diagram of Fig. 1 and its execution in Figs. 2 and 3, in images with different cell quantities and digital quality.

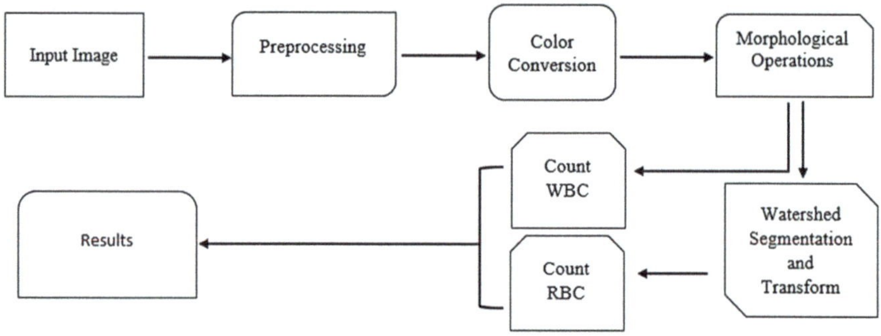

Fig. 1 Diagram representing the algorithm

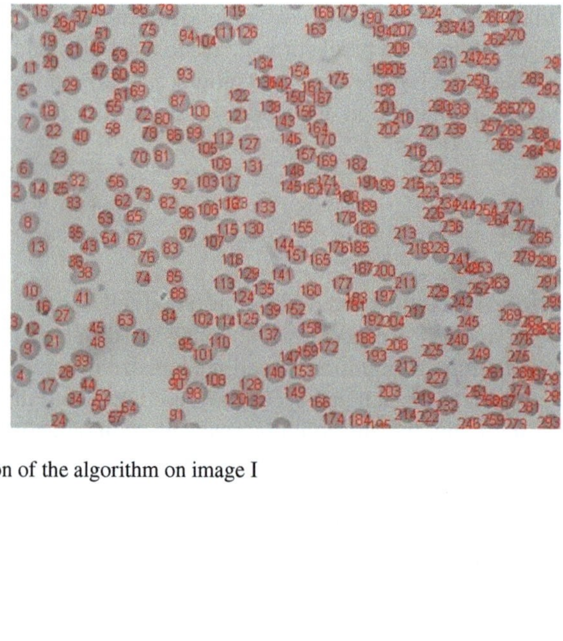

Fig. 2 Execution of the algorithm on image I

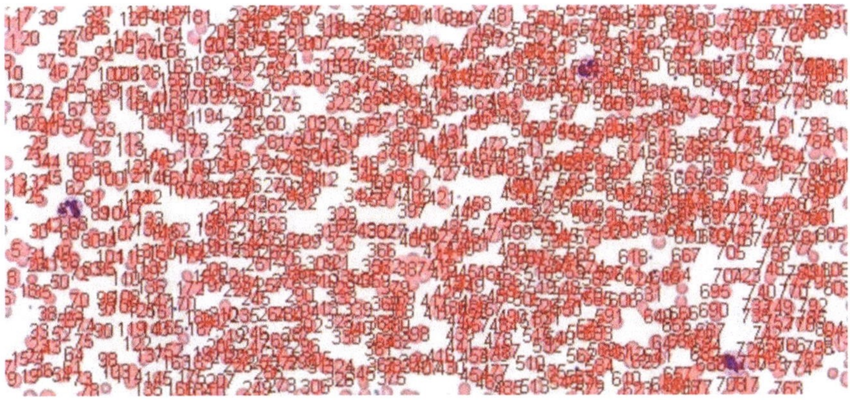

Fig. 3 Execution of the algorithm on image II

4 Results

In this section, the results will be presented on the evaluations of the algorithm in processing time and the time spent for execution. To obtain the same were performed executions of each command used, on physical machines with different hardware configuration, consisting of an Intel Core i5 processor and 8 GB RAM, and another with an Intel Core i3 processor and 4 GB RAM.

The tic, toc, and sim commands were used via the command line at the Matlab prompt, which measures how long time the algorithm spends on its execution (in seconds).

Was also used the cputime command, which returns the total cputime (in seconds) used by the algorithm from the moment was started.

The tic and toc function measures the elapsed time the algorithm spend, that is, its execution time. The cputime function calculates its computation time. So together, show the performance and efficiency of the algorithm.

Thus, the first execution of the algorithm was analyzed on the described commands, because it is in first execution that the variables are allocated and the memory reserved for the execution of the same, referring to Figs. 6 and 7.

For this study, 18 digital images were used, being 10 images for RBC and eight images for WBC. These samples were submitted to manual and digital counts by the algorithm, having better accuracy than [5, 8–11, 14, 16, 17], as shown in Figs. 4 and 5.

Aiming at the applicability of this work, the executions of the proposal were performed on machines with different hardware, because both possibilities are possible in the daily life of a laboratory. In this way, according to [17], an execution time of 3 s per sample was obtained, whereas in the present study a better performance of 15.3% was obtained in its first execution and 34% in the following ones, which correspond to 2.54 and 1.98 s, respectively, being the average time of the first execution of all the

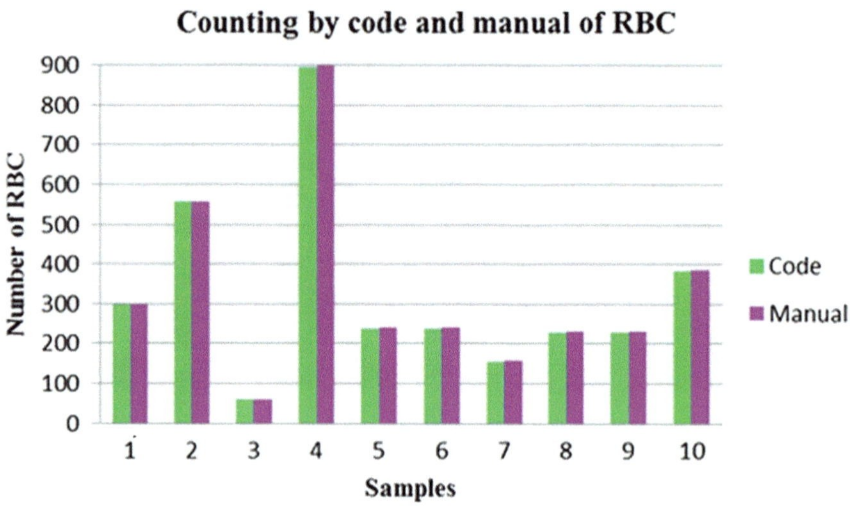

Fig. 4 RBC manual and by algorithm

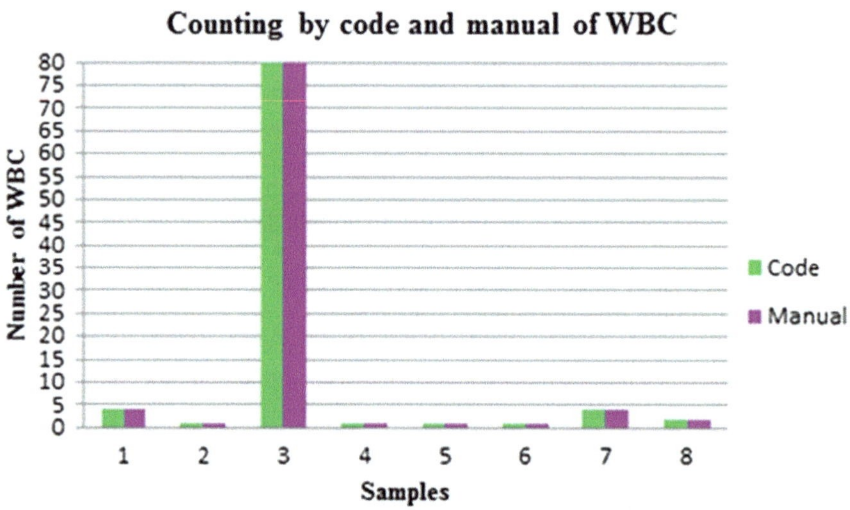

Fig. 5 WBC manual and by algorithm

samples. These results refer to executions performed on an Intel Core i5 machine, according to the data presented in Fig. 6.

The executions performed on the Intel Core i3 machine also demonstrated satisfactory results than those found in the study [17], wherein its first execution, there was a 34% improvement in performance, and 42% in the following ones, corresponding to 1.98 and 1.74 s, respectively, also consisting of the average time over all samples. These results are presented in Fig. 7.

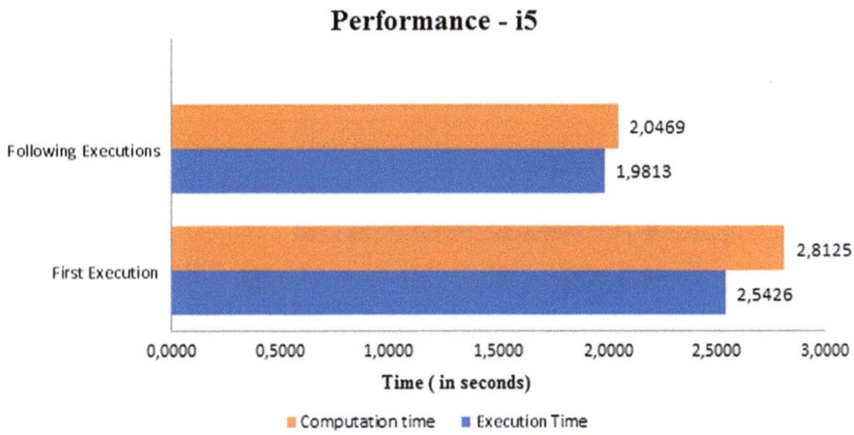

Fig. 6 Execution time spent by the algorithm (i5)

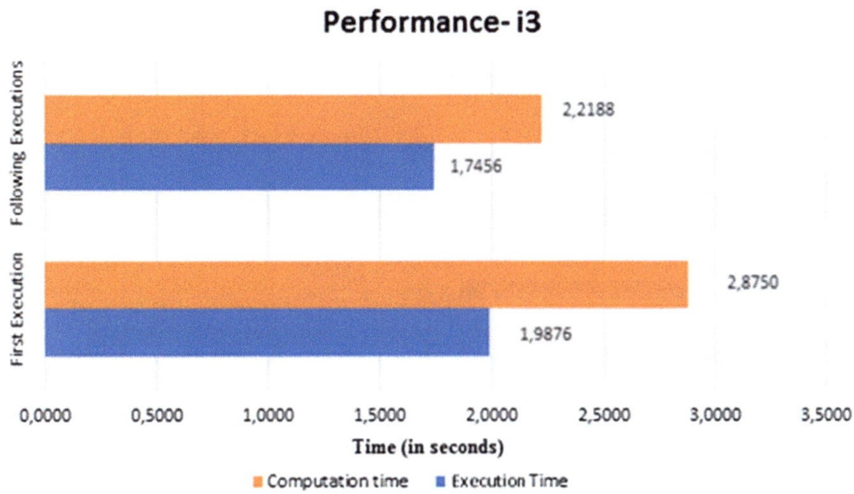

Fig. 7 Execution time spent by the algorithm (i3)

Were selected five images of the total used in this study, with the objective of analyzing the performance of the algorithm on the different characteristics that each of them has, for example, the size, the quantity of red and white blood cells, not being the first execution for this case, and so, detailing further the performance of the algorithm on the samples. The results on this optics are presented in Figs. 8 and 9.

Was noted in Figs. 8 and 9 that in machine i3, the computation time was higher, such factor is derived that the i3 processor has two cores less than the i5. Besides the fact, the i5 machine has twice as much RAM. However, even with these differences,

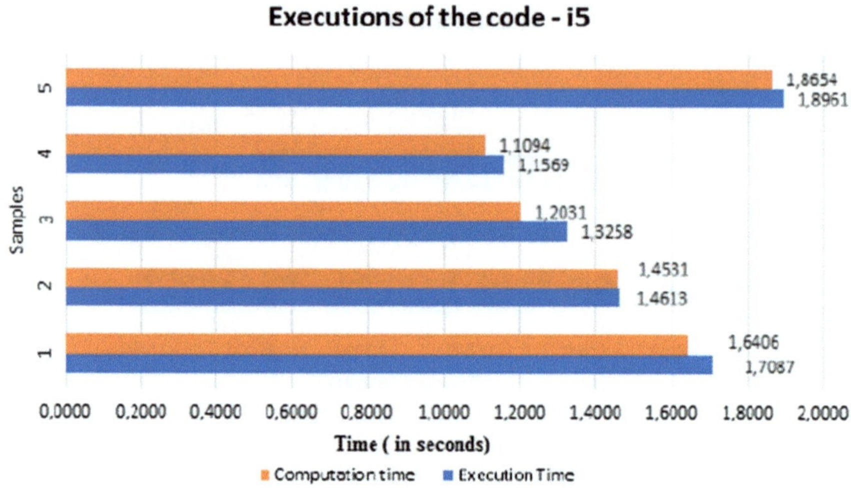

Fig. 8 Performance of the algorithm (i5)

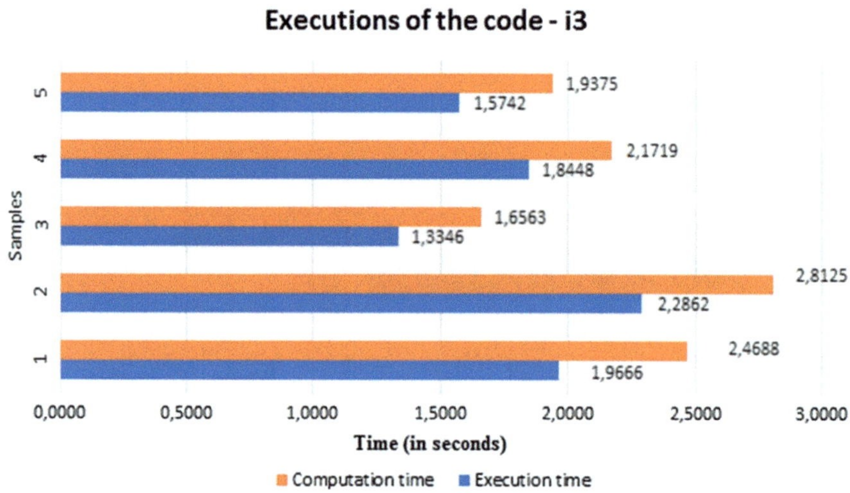

Fig. 9 Performance of the algorithm (i3)

the results are still satisfactory regarding the performance of the proposal of this study.

5 Conclusions

In all the scenarios analyzed and on all samples, the algorithm developed in this study obtained better execution results when compared to other papers performed, also solving the problems encountered during the counts of red blood cells and leucocytes, succeeding in a satisfactory performance, both in efficiency and reliability, which are paramount during the conclusion of a laboratory diagnosis.

The good performance of the proposal on the machines with different hardware configurations concludes that the algorithm is feasible for the different realities of the laboratories.

The area of health is a broad field directly linked to medical diagnoses through images, so the proposal of this study also provides that the results suggestive of more serious pathologies can be stored in digital files for future consultations, dispensing the creation of a physical space, in this case for hospitals.

The development of this efficient algorithm for counting red and white blood cells, using current techniques of image processing and segmentation, contributes to medical studies.

References

1. Nasreen, N., Kumar, C., Nabeel, A.P.: Counting of RBC using circular hough transform with median filtering. Bonfring (2015). ISBN 978-93-85477-33-1/2015
2. Monteiro, A.C.B., et al.: Sickle cell anemia, a genetic disorder characterized by the change in shape of red blood cells. Saúde em Foco, Edição nº: 07 (2015)
3. Turgeon, M.L.: Clinical hematology theory and procedures, 4th edn. Lippincott Williams and Wilkins, Philadelphia (2004)
4. Mohammed, E.A., et al.: Chronic lymphocytic leukemia cell segmentation from microscopic blood images using watershed algorithm and optimal thresholding. In: 26th IEEE Canadian Conference of Electrical and Computer Engineering (CCECE) (2013)
5. Sahastrabuddhe, A.P., Ajij S.D.: Blood group detection and RBC, WBC counting: an image processing approach. Int. J. Eng. Comput. Sci. **5**, 18635–18639 (2010). ISSN 2319-7242
6. Bernadette, F., Rodak, G.A., Fristma, K.D.: Hematology—Clinical Principles and Applications. Elsevier, New York (2015)
7. Ciesla, B.: Hematology in Practice, 2nd edn. Davis Company, Philadelphia (2012)
8. Mazalan, S.M., Mahmood, N.H., Razak, M.A.A.: Automated red blood cells counting in peripheral blood smear image using circular hough transform. In: First International Conference on Artificial Intelligence, Modelling and Simulation (2013)
9. Arivu S.K., Sathiya, M.: Analyzing blood cell images to differentiate WBC and counting of linear and non-linear overlapping RBC based on morphological features. Elixir Comp. Sci. Eng. **48**, 9410–9413 (2012)
10. Guitao, C., Cai, Z., Ling, L., Jun D.: Detection of red blood cell in urine micrograph. In: The 3rd International Conference on Bioinformatics and Biomedical Engineering (ICBBE) (2009)
11. Heidi, B., D. T., Sriram K., Tania, S.D.: Improved red blood cell counting in thin blood smears. In: Proceedings of ISBI, pp. 204–207 (2011)
12. Hemant, T., Sexena, S. Vadak, N.: Segmentation using morfological watershed transformation counting blood cells. Int. J. Comput. Appl. Inf. Technol. **2**(III) (2013)

13. Kaur, R., Kaur, H.: Comparative analysis of white blood cell by different segmentation methods using knowledge based learning. Int. J. Adv. Res. Elect. Electr. Instrum. Eng. (An ISO 3297: 2007 Certified Organization **3**(9)) (2014)
14. Mogra, M., Srivastava, V.: A comprehensive review of analysis of counting blood cells using different image processing algorithms. Int. J. Eng. Sci. Invention **3**(6), 29–31 (2014). ISSN (Online): 2319–6734, ISSN (Print): 2319–6726. www.ijesi.org
15. Sobhy, N.M., Salem, N.M., Dosoki, M.: A comparative study of white blood cells segmentation using Otsu threshold and watershed transformation. J. Biomed. Eng. Med. Imaging Soc. Sc. Educ. UK **3**(3) (2010)
16. Soltanzade, R.: Classification of three types of red blood cells in peripheral blood smear based on morphology. in: Proceedings of ICSP (2010)
17. Mahmood, N., Humaimi, M.M.A.: Red blood cells estimation using hough transform technique. Sig. Image Process: Int. J. (SIPIJ) **3**(2) (2012)
18. Failace, R., et al.: Hemograma, manual de interpretação. 5ª. Porto Alegre: Artemed, Guitao Cao, Cai Zhong, Ling Li and Jun Dong. Detection of red blood cell in urine micrograph. In: The 3rd International Conference on Bioinformatics and Biomedical Engineering (ICBBE) (2009)
19. Hoofman, R., et al.: Hematology: Basic Principles and Practice, 6ª es. Elsevier, Canadá (2013)
20. Cuevas, E., et al.: A Comparative Study of White Blood Cells Segmentation Using Otsu Threshold and Watershed Transformation, vol. 2013, Article ID 137392. Hindawi Publishing Corporation. Computational and Mathematical Methods in Medicine (2013)
21. Vincent, L., Soille, P.: Watersheds in digital spaces: an efficient algorithm based on immersion simulations. IEEE Trans. Pattern Anal. Mach. Intell. **13**(6), 583–598 (1991)
22. Rafael, C.G., Richard, E.W.: Digital Image Processing, 2nd edn. Prentice Hall (2002)

General Aspects of Pathophysiology, Diagnosis, and Treatment of Sickle Cell Disease

Ana Carolina Borges Monteiro, Yuzo Iano and Reinaldo Padilha França

Abstract Since its first report, sickle cell disease (SCD) has attracted the attention of many health professionals and researchers, because its affects about 24.6% of the world population and can cause the death of up to 90% of patients in Africa Sub-Saharan Africa with SCD before 5 years of age. Hemolysis, vaso-occlusive crises, pain crises, chronic inflammation, risk of infarction, thrombosis and stroke are realities constantly faced by the patients affected by this hematologic disorder. The stages of the pathophysiological process, the pathways of activation and molecular aspects are still not fully understood, hence, the patients affected dependent on medical follow-up, periodic examinations, use of medications and even periodic blood transfusions. The objective of the present study was to carry out a bibliographic review aiming at a direct and clear approach to the pathophysiological aspects of the disease discovered and to link them with laboratory tests, drugs used and the recent discoveries made through the research with MicroRNAs.

Keywords Sickle cell disease · Diagnosis · Hydroxyurea · MicroRNAs

1 Introduction

In 1910, sickle cell disease (SCD) was described by Dr. James Herrick in a case report of a 20-year-old patient with severe anemia accompanied by lower limb ulcers, pulmonary complications, and jaundice [12]. Due to the elongated erythrocyte morphology, similar to the sickle shape, in 1945, the disease was named sickle cell disease

A. C. B. Monteiro (✉) · Y. Iano · R. P. França
School of Electrical and Computer Engineering (FEEC), University of Campinas – UNICAMP, Av. Albert Einstein - 400, Barão Geraldo, Campinas, SP, Brazil
e-mail: monteiro@decom.fee.unicamp.com

Y. Iano
e-mail: yuzo@decom.fee.unicamp.com

R. P. França
e-mail: padilha@decom.fee.unicamp.com

Y. Iano et al. (eds.), *Proceedings of the 3rd Brazilian Technology Symposium*,
https://doi.org/10.1007/978-3-319-93112-8_32

[21]. Years later, SCD was described as result of a mutation in the sixth codon, in the polypeptide chain of β—globin on chromosome 11, thus being classified as a genetic disease of autosomal recessive inheritance. This mutation is responsible for the substitution of a single nitrogenous base: adenine (A) by a thymine (T), leading to the erroneous production of glutamic acid instead of valine. Hence, there is the formation of an anomalous hemoglobin, called HbS [7, 35].

In healthy individuals, there are three distinct types of hemoglobin: hemoglobin A (HbA), hemoglobin A2 (HbA2) and fetal hemoglobin (HbF) [21, 38]. Thus, the HbS gene can be combined with HbA, resulting in the manifestation of the heterozygous genotype (HbAS) [11]. There are also cases in where there is the combination of HbS with other hemoglobinopathies, such as HbD, beta-thalassemias, HbC, among others [21, 28]. This association is in a way beneficial process, because it causes a dilution effect, which reduces the contact between the SCD erythrocytes, difficulting the formation of a polymer [21].

Some authors consider that SCD be a protection against malaria. Research has shown that in heterozygous phenotypes, the morphological alteration of erythrocytes results from a defense process against Plasmodium sp. SCD erythrocytes act facilitating the phagocytosis process of the parasite and through the low availability of oxygen, reducing the development process of the etiologic agent of malaria. However, certain stages of this mechanism remain obscure, being so-called "Malaria Hypothesis" [8].

The phenotype of sickle cell anemia presents its first manifestations between 6 months and 2 years of age, due to the decrease in the availability of HbF, which slows the polymerization processes of HbS, difficulting the early visualization of the signs and symptoms of hematological disorder [7, 34]. During first infancy, when HbF decreases, the first characteristic symptom is pain crisis. In addition, due to functional asplenia, children with SCD when compared to healthy children have a higher risk of developing infections, such as those caused by Streptococcus pneumoniae and parvovirus B19 [7].

The pathophysiology of SCD is initiated in situations with low oxygen tension, where HbS undergoes polymerization. This especially occurs in organs where blood flow is slower or in areas where the terminal blood supply is limited. However, this fact does not avoid that lungs and brain can be affected [16]. The polymerization and deoxygenation of HbS are responsible for the morphological alteration, rigidity and decrease in the erythrocyte deformation capacity [15, 35]. This process includes the morphological changes of the red blood cells, hemolysis, chronic inflammation, cells with high adhesion, increased oxidative stress and dysfunction/activation of the endothelial cells, which cause an acute vaso-occlusion process [31, 32].

The vessel occlusion is a multicellular and multi-stage process, being initiated through the interaction between erythrocytes, activated leukocytes, endothelial cells, platelets, and plasma proteins [15, 39]. The adhesion between leukocytes and erythrocytes activates the endothelial pathway leading to events of hypoxia, oxidative stress, reduced availability of nitric oxide (NO) and inflammation [15, 31]. The occlusive vessel process associated with ischemia and reperfusion with consequent injuries and endothelial activations results in a chronic inflammatory process, which is maintained through the reduction of NO availability, elevated levels of inflammatory cytokines and the presence of oxidative stress [30].

In addition, the morphological alteration of erythrocytes alters the functionality of the sodium and potassium pump, leading to loss of water and potassium, generating erythrocytes denser and more susceptible to polymer formation. Membrane alterations are triggered by the rearrangement of the spectrinactin protein, phosphatidylserine externalization, with subsequent generation of free radicals and acceleration of apoptosis process. Due to hypoxia, infarction of several types of tissue can occur, especially those where there is acid pH [20]. This morphological alteration of the erythrocytes leads to obstruction of the blood flow in the capillary vessels and hemolysis, which can lead to bone marrow necrosis, splenic alterations, vasoconstriction, acute thoracic syndrome, among others [3, 15].

The activation of endothelial cells is responsible for increased expression of surface adhesion proteins, such as selectin, vascular cell adhesion molecule-1 (VCAM-1), intercellular adhesion molecule-1 (ICAM-1), fibrinogen, fibronectin and CD36 [13, 36]. In this way, endothelial activation favors the capture of erythrocytes and leukocytes, which in turn also have increased adhesive properties [21].

The SCD individuals have activated leukocytes in the blood vessel, facilitating adhesion to the vascular endothelium, especially when there is inflammation. SCD leukocytes have twice as many adhesive properties as leukocytes from healthy subjects [21, 40]. Once adhered the leukocytes have the ability to perform secondary adhesions with erythrocytes leading to physical obstruction of the microcirculation [40]. Research has shown that SCD patients also present a numerical increase in the amount of neutrophils and eosinophils, but the participation of eosinophils in the process has not yet been clearly established [39].

Thus, the inflammatory process of the airways in SCD patients, is characterized by the intense participation of eosinophils, being the lungs more vulnerable to vessel occlusion due to its anatomy. Research suggests that asthma is associated with acute chest syndrome (ACS) and pain crises in SCD patients, especially children. SCD children diagnosed with asthma have the first episode of acute chest syndrome earlier. They require more transfusions and are affected by episodes of more frequent pain, consequently they need a longer hospitalization time [4]. The pathophysiological relationship, as well as the triggering factors of asthma attacks in SCD patients are still not fully understood. At the moment only treatments during asthma crisis have shown results [33].

Increased mechanical fragility and loss of erythrocyte elasticity are directly proportional to HbS concentration. Thus, another remarkable mechanism is hemolysis. Extravascular hemolysis is caused by SCD erythrocyte phagocytosis, whereas intravascular hemolysis is a consequence of lysis of SCD erythrocytes [1]. Thus, patients with this pathology usually present splenomegaly, due to the repetitions of vaso-occlusive episodes, hence may have fibrosis and spleen atrophy [15, 21].

This paper has the objective of carrying out a review of the literature on the pathophysiology of SCD, as well as the presentation of the new findings and treatments performed until the moment. This work was carried out through researches in books and articles about the theme, aiming to present to students, researchers and health professionals a concise and clear view of the main aspects of the disease.

2 Results and Discuss

It is estimated that 222785 of SCD births around the world occur every year. About 24.6% of the world's population is affected by SCD, being populations in sub-Saharan Africa, parts of the Mediterranean and midwestern India most affected [19, 21]. However, due to population migrations, this pathology also reached Europe and the Americas, being classified as the most common hereditary hematological disorder of the United States [18, 25]. In developed countries, in 1967, the mean mortality from SCD was around 20 years of age, being resulted from infections [29]. In the 1990s, the life expectancy of SCD men was 42 years and 48 years for women [26]. Between the 1980s and 1990s, there was a decrease of 68% in the mortality of SCD children between 0 and 3 years old [41].

Currently, cohort studies conducted between 2000 and 2007 demonstrate the absence of SCD deaths before 18 years of age in developed countries [27]. Developing countries have limitations on mortality data for these patients. Retrospective studies show that due to the presence of infections and differences in health access in urban and rural areas, 50–90% of SCD children are afflicted by premature death in sub-Saharan Africa [28]. Thus, neonatal screening tests are extremely important because its help in the early detection of SCD. The child death affects about 25–30% of SCD children worldwide. One of these factors is the great possibility of congestion in the red pulp by sequestration of SCD erythrocytes, resulting in auto-splenectomy [25].

The diagnosis of SCD is performed by blood count total, solubility test, fetal hemoglobin and hemoglobin A2 measurement, and neonatal screening. However, the solubility test is not indicated for children less than 6 months of age, because during this period there are low HbS levels and high concentration of HbF [7].

Confirmation of SCD is only obtained with the detection of HbS by electrophoresis [21]. Hemolytic anemia present in SCD is associated with moderate reductions in hematocrit, hemoglobin, and count erythrocytes. The hemogram of these patients demonstrates normocytic and normochromic anemia. Biochemical tests demonstrate unconjugated hyperbilirubinemia; high lactate dehydrogenase concentrations; and low concentrations of haptoglobin [2].

During the last decade, interest in microRNAs (miRNAs) has increased exponentially after the discovery of its importance in biological processes such as development, cell proliferation, and apoptosis [24]. The miRNAs were first mentioned in the year 1993, when Rosalind C. Lee noted that the Lin 4 gene was responsible to regulated of *Caenorhabditis elegans* developmental events, through the down-regulation of the Lin-14 protein level, which is responsible for the first stage of larval development of *Caenorhabditis elegans* [17].

Currently, 2578 human mature miRNAs have been identified, but the majority biological function remains it is estimated that about one-third of the human genome is post-transcriptionally controlled through miRNAs and changes in its expression levels may contribute to the pathophysiology of various diseases, including SCD and associated diseases such as asthma, acute chest syndrome, thrombosis, stroke, among others [10]. The miRNAs are classified as small non-coding RNAs, located mainly in the introns, responsible for silencing through mRNA cleavage or protein translation blockade [10, 24].

Research done in recent years has shown that comparing healthy individuals with SCD patients, it is possible to note changes in the expression profile of miRNAs between them. Among these alterations 40 miRNAs are differentially express in SCD platelets, suggesting the existence of a direct connection of the expression of the miRNAs with the platelet functions, being able to influence other blood cells and vascular endothelial cells [14]. In this context, studies with SCD have related the 146a, 155 and 21 miRNAs with the immune system. The miRNA-146a has an anti-inflammatory function by suppressing NF-kB activation, while miRNA-21 is responsible for the expression of NF-kb and STAT3. However, miRNA-155 has a pro-inflammatory action [37].

During a hemolytic crisis, SCD patients may need of red blood cell transfusion, when the hematocrit has a decrease of greater than or equal to 20%. Blood transfusion helps in the recovery process; however, it exposes the patient to the risk of contracting blood-borne diseases. In addition, with the blood transfusion, there creation of antibodies against less immunogenic blood groups occurs, hence there is the difficulty of blood compatibility for future transfusions [39].

Another recurrent problem blood transfusion is iron overload, which can be detected by laboratory tests such as iron levels, ferritin level, iron binding capacity and transferrin. [21]. Iron toxicity is directly linked to iron not bound to transferrin, leading to its deposition in organs, triggering the onset of hepatic cirrhosis and hepatic insufficiency and hyperpigmentation of the skin [5, 21]. In such cases, the indicated treatment is the use of iron chelating drugs such as deferoxamine; deferiprone, and deferasirox [5].

In order to avoid vaso-occlusive crises and consequent hemolytic crisis and prolonged hospitalizations, many drugs have been researched and developed. Currently the most commonly used drug is hydroxyurea, a hydrocarbon that is responsible for impeding the exit of G1/S phase cells from the cell cycle, being widely used in the treatment of neoplasias of the hematopoietic system [34]. In SCD, hydroxyurea acts to increase HbF synthesis and increase NO bioavailability. As a consequence, there is inhibition of platelet aggregation and decreased expression of vascular endothelial adhesion molecules, reducing patient mortality by 40%. However, some patients present unsatisfactory responses to hydroxyurea treatment, especially when the use of this drug is continuous, leading to a decrease in efficacy [8].

Another therapeutic approach used is increased synthesis and availability of HbF. The 5-azycytidine, has a preventive effect on the first complications of SCD by increasing HbF, leading to a reduction of up to 20% of symptoms [6]. In turn, 5-Aza-2′-deoxycytidine is a drug with lower toxicity and lower tumor induction potential when compared to 5-azacytidine [8]. The sodium phenylbutyrate showed satisfactory results in the short term, however, in the long term, it generates tolerance to patients [9]. Thalidomide withdrawn from the market in the 1960s, due to its poor fetal formation, was once again the target of research due to its inhibitory properties of TNF-α. Derivatives of thalidomide such as lenalidomide and pomalidomide also have the ability to induce HbF synthesis [38].

During the occlusive vessel process, the bioavailability of arginine is low. Thus, intervention through supplementation with arginine helps in the rapid reversal of vaso-occlusive processes [15, 22, 23]. Arginine, when metabolized by arginase, is converted to nitric oxide, which acts directly on the vascular endothelium. Nitric oxide has vasodilatory action, decreases platelet aggregation and endothelial activation and neutrophil adhesion, causing increased blood flow and consequent reoxygenation, favoring the dispersion of the polymers [8].

Inhalation of nitric oxide gas has been shown to be effective during vaso-occlusive crises and is of great utility in the reduction of acute chest syndrome. However, the high cost and difficulty of manipulation hinders the access of this therapy by a large part of the population of developing or underdeveloped countries [8, 22, 23]. In addition, phytomedicines over the years have proven quite effective. The ethanolic extract of *Cajamuscajam* sp. plant has been shown to reduce the degree of erythrocyte deformation during hypoxia [8, 21].

3 Conclusions

Even with more than a century of its discovery, certain pathophysiological factors of SCD remain obscure, as a new guide for pathophysiological understanding, the MiRNAs present great potential for new discoveries. The development of future treatments the miRNAs has proved very promising, since a treatment at the molecular level improved the quality of life since childhood, dispensing the use of drugs and the needs of blood transfusions. However, this reality of miRNA manipulation is still a distant future, because the studies are still recent and there is still much to be clarified.

More current alternatives are expected to launch more effective drugs in the market, as well as better attention to these patients through periodic exams, appropriate treatment with hydroxyurea, genetic counseling, dissemination of the disease in developing and underdeveloped countries, among others. These measures are fundamental, since the earlier the detection of SCD, the better the living conditions of these patients.

References

1. Alison, A.C.: Notes on sickell—Cell polymorphism. Ann. Hum. Genet. **19**(1), 39–14 (1954)
2. Arend, W., et al.: Tratado de medicina interna, 22ª. Elsevier, Rio de Janeiro (2005)
3. Ballas, S.K., et al.: Beyond the definitions of the phenotypic complications of sickle cell disease: an update on management. Sci. World J. **2012**(Article ID 949535) 55 p (2012)
4. Boyd, J.H., et al.: Asthma is associated with increased mortality in individuals with sickle cell anemia. Haematologica **92**, 1115–1118 (2007)
5. Cançado, R.D.: Iron overload and iron chelation in sickle cell disease. Rev. Bras. Hematol. Hemoter. **29**(3), 316–326 (2007)
6. Charache, S., et al.: Treatment of sickle cell anemia with 5-azacytidine results in increased fetal hemoglobin production and is associated with nonrandom hypomethylation of DNA around the 6-fi-globin gene complex. Aug Med Sci **80**, 4842–4846 (1983)
7. Di Nuzzo, D.V.P., Fonseca, S.F.: Sickle cell disease and infection. J Pediatr (Rio J). **80**(5), 347–54 (2004)
8. Dos Santos, J.L., Chin, C.M.: Sickle cell disease: challenges and advances in drug discovery. Quim. Nova **35**(4), 783–790 (2012)
9. Dover, G.J., Dover, S.B., Charache, S.: Induction of fetal hemoglobin production in subjects with sickle cell anemia by oral sodium phenylbutyrate. Blood **84**(1), 339–343 (1994)
10. Há, T.Y.: MicroRNAs in human diseases: from lung, liver and kidney diseases to infectious disease, sickle cell disease and endometrium disease. Immune Netw. **11**(6), 309–323 (2011)
11. Haldane, J.B.S.: The rate of mutation of human genes. Hereditas **35**(S1), 267–273 (1949)
12. Herrick, J.B.: Peculiar elongated and sckell shaped red blood cell corpuscles in a case of severe anemia. Arch Int Med **6**(5), 517–521 (1910)
13. Hordijk M.L., et al.: A randomized comparison of electrocautery incision with Savary bougienage for relief of anastomotic gastroesophageal strictures. Gastrointest Endosc. **70**(5), 849–55 (2009)
14. Jain, S., et al.: Expression of regulatory platelet microRNAs in patients with sickle cell disease. PLoS One **8**(4), e60932 (2013)
15. Kato, G.J., Gladwin, M.T., Steinberg, M.H.: Deconstructing sickle cell disease: reappraisal of the role of hemolysis in the development of clinical subphenotypes. Blood Rev. **21**, 37–47 (2007)
16. Labie, D., Elion, J.: Molecular and cellular pathophysiology of sickle cell anemia. Pathol. Biol. (Paris) **47**(1), 7–12 (1999)
17. Lee, R.C., Feinbaum, R.L., Ambrost, V.: The *C. elegans* heterochronic gene lin-4 encodes small RNAs with antisense complementarity to and II-14. Cell **75**, 843–854 (1993)
18. Lervolino, L.G., et al.: Prevalence of sickel cell trait in national neo—Natal screening studies. Rev. Bras. Hematol. **33**(1), 49–54 (2011)
19. Makani, J., et al.: Health policy for sickel cell disease in Africa. Experience from Tanzania on interventions to reduce under five mortality. Trop. Mes. Int. Health **20**(2), 184–187 (2015)
20. Manfredini, V., et al.: A fisiopatologia da anemia falciforme. Rev. Informa **19**(1/2) (2007)
21. Monteiro, A.C.B., et al.: Sickle cell anemia, a genetic disorder characterized by the change in shape of red blood cells. Saúde em Foco, Edição **07** (2015)
22. Morris, C.R., et al.: Arginine therapy: a new treatment for pulmonary hypertension in sickle cell disease? Am. J. Respir. Crit. Care Med. **18**, 168 (2003)
23. Morris, C.R., et al.: Arginine therapy: a novel strategy to induce nitric oxide production in sickell cell disease. Br. J. Haematol. **111**, 498–500 (2000)
24. Pereira, T.: Introdução ao mundo dos microRNAs. GeneSeq (2015)
25. Platt, O.S., et al.: Mortality in sickle cell disease: Life expectancy and risk factors for early death. New Eng. J. Med. 330(23) (1994)
26. Quinn, C.T., et al.: Improved survival of children with sickle cell disease. Blood **115**(17), 3447–3452 (2010)

27. Rajaratnam, J.K., et al.: Neonatal, postnatal, childhood, and under—5 mortality for 187 countries, 1970–2010: a systematic analysis of Progress Toward Millennium Development Goal. 4. Lancet **375**(9730), 1988–2008 (2010)
28. Rodger, G.P., et al.: Hematologic responses of patientes with sickle cell disesase to treatment with hydroxyurea. New Engl. J. Med. **332**(15) (1990)
29. Scoot, R.B.: Health care priority and sickel cell anemia. JAMA **214**(4), 731–734 (1970)
30. Setty, B.N.Y., Stuart, M.J.: Vascular cell adhesion molecule-l is involved in mediating hypoxia-induced sickle red blood cell adherence to endothelium: potential role in sickle cell disease. Blood **88**(6), 2311–2320 (1996)
31. Shet, A.S., et al.: Sickle blood contains tissue factor–positive microparticles derived from endothelial cells and monocytes. Blood **102**(7) (1996)
32. Solovey, A., et al.: Tissue factor expression by endothelial cells in sickle cell anemia. J. Clin. Invest. Am. Soc. Clin. Invest. **101**(9), 1899–1904 (1998). Inc. 0021-9738/98/05/1899/06
33. Steinberg, M.H.: Management of sickle cell disease. N. Engl. J. Med. **340**, 1021–1030 (1999)
34. Steinberg, M.H.: Predicting clinical severety in sickell cell anemia. Br. J. Haematol. **129**(4), 465–481 (2005)
35. Steinberg, M.H., et al.: Disorders of Hemoglobin. Genetic, Pathophysiology and Clininal Manegement. Cambridge University Press, pp. 231–356 (2001)
36. Stuart, M.J., Nagel, R.L.: Sickle-cell disease. Lancet **364**(9442), 1343–1460 (2004)
37. Tang, G.N., et al.: MicroRNAs involved in asthma after mesenchymal stem cells treatment. Stem Cells Dev. **25**(12) (2016)
38. Trompeter, S., Roberts, I.: Haemoglobin F modulation in childhood sickle cell disease. **144**(3), 308–316 (2001)
39. Turgeon, M.L.: Clinical Hematology Theory and Procedures, 4th edn. Lippincott Williams and Wilkins, Philadelphia (2004)
40. Turhan, A., et al.: Primary role for adherent leucocytes in sickle cell vascular occlusion: a new paradigm. PNAS **99**(5), 3047–3051 (2002)
41. Yanni, E., et al.: Trends in pediatric sickell cell disease—Related matality in United States, 1983–2002. J. Pediatr. **154**(4), 541–545 (2009)

Emergency Response Cyber-Physical System for Flood Prevention with Sustainable Electronics

Vania V. Estrela, Jude Hemanth, Osamu Saotome, Edwiges G. H. Grata and Daniel R. F. Izario

Abstract Inundations lead to massive fatalities and loss of assets. Notifying communities of the arriving hazards gives the general public enough time to remediate, protect possessions, and abandon a place if needed. Emergency Response (ER) structures can be modeled as a Cyber-Physical System (CPS) with control units, sensors, and actuators for environmental observing. This work aims at the instrumentation level of Visual Sensor and Actuator Networks (VSANs) using sustainable electronics and reengineering to ease the complications engendered by the present economic crisis and scarce financial resources. This approach examines ways that salvaged electronic components can assist far away and economically deficient locations. The intricacy of these systems and the indispensable self-sufficiency within the perspective of a developing country while keeping the framework maintainable and accessible to workers without technical skills presents challenges.

Keywords Flood detection · Motion estimation · Emergency response · Salvaged electronics · Cyber-Physical system · Sensors · Humanitarian engineering · Green design · Computer vision · Surveillance

V. V. Estrela (✉) · E. G. H. Grata
Department of Telecommunications, Universidade Federal Fluminense, Niterói, RJ, Brazil
e-mail: vania.estrela.phd@ieee.org

E. G. H. Grata
e-mail: edwigesghgrata@gmail.com

J. Hemanth
Department of ECE, Karunya University, Coimbatore, India
e-mail: jude_hemanth@rediffmail.com

O. Saotome
CTA-ITA-IEEA, Sao Jose dos Campos, SP, Brazil
e-mail: osaotome@gmail.com

D. R. F. Izario
Faculty of Electrical and Computer Engineering, UNICAMP, Campinas, SP, Brazil
e-mail: daniel_izario@hotmail.com

Y. Iano et al. (eds.), *Proceedings of the 3rd Brazilian Technology Symposium*,
https://doi.org/10.1007/978-3-319-93112-8_33

(a) (b)

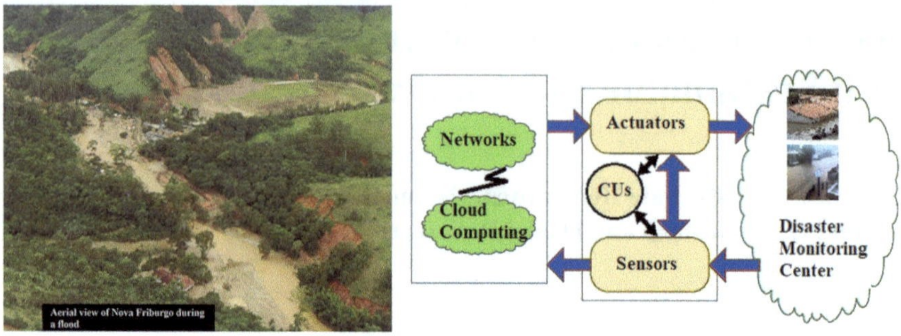

Fig. 1 FDM: **a** flooded area in Nova Friburgo [4], and **b** relationship between its parts

1 Introduction

Robustness and resiliency to natural and human-made catastrophes depends significantly on (i) the ability to recognize risk, (ii) adapt to circumstances, and (iii) to mitigate hazard, which all entails an associated cost. Disasters can be triggered by repetitive or foreseeable activities and circumstances. Emergency Response (ER) handles the threats to public safety, individual/collective health, and welfare, thus protecting the nature, possessions, and valuable infrastructures.

This work studies floods when there is water overflowing from hydric bodies (e.g., oceans, rivers, lagoons, man-made hydric structures, and so on) due to a rainwater buildup on saturated ground, or a level rise caused by some loss of control. Loss of life and huge economic impact can happen as a consequence of the inundation of neighboring lands belonging to rural and residential areas. Flood Detection and Monitoring (FDM) or modeling algorithms can use In-Situ Sensors (ISSs), including still cameras [1–3], to monitor flood-prone regions. ISSs acquires data about an object or phenomenon from close range (in place or proximity sensors).

FDM ISS measurements allow reusing more intensively electronic parts, which is a sustainable practice. At present, regions subject to floods should consider social approaches to resilience (like cooperation between community members via smartphones) as well as the opportunities created with technology. A Cyber-Physical System (CPS) can simplify modeling and mitigate disasters because its architecture has Visual Sensors and Actuators Networks (VSANs) to decrease and/or remediate the flood impacts like the one that took place in 2011 in Nova Friburgo, Rio de Janeiro (RJ) state, Brazil as displayed in Fig. 1a [1, 2, 4]. The physical world comprises the environmental occurrences to be supervised or controlled. CPSs work on information distributedly and communicate with their target environment via sensors and actuators that interface and convert other forms of energy to electric signals and vice versa (Fig. 1b).

The humanitarian and economic impacts of floods show an astonishing increase. As massive urbanization and global mobility withstand, more at-risk areas appear.

Thus, prevention, Disaster Management (DM), adaptation, and remediation should include ongoing risks and be permanently inserted in urban as well as rural planning and policies [2]. Nowadays, regions should consider methods to assure resilience using science and technology. A CPS can improve considerably natural disaster resilience via community sensing with VSANs, thus lessening the impact of a catastrophe.

Previous works [3, 5–7] addressed the problem of detecting changes using photodetectors and Passive Infrared (PIR) sensors from salvaged equipment and scattered throughout a flood-prone area without very complicated interfacing.

Section 2 explains an ER framework relying on CPSs and ISSs. Section 3 explains Optical Flow (OF), which detects motion rather than processing the complete images. Section 4 discusses OF Sensors (OFSs), which amounts to hardware platforms that can implement flood detection via OF. Section 5 presents the architecture of VSANs. Lastly, Sect. 6 closes this manuscript with the conclusion and future work.

2 Emergency Response CPS to Prevent Flood

CPS can deliver fast ER via a large number of sensor and actuators nodes controlled by Control Units (CUs) in the areas subject to natural and human-made disasters. Nonetheless, this rapid response entails the nodes to collectively evaluate the situation and rapidly inform the central authority or Disaster Management Center (DMC) even in the frequently changing environments. As a result, an ER-CPS requires robustness, efficient resource utilization, adaptivity, and timeliness [1, 3, 7].

ER and DM have always drawn attention because of their public implications. Emergency management should adopt emerging ICTs and social media. Strong cooperation amongst emergency management professionals, local/national authorities, and the community is essential. The efficient use of forewarning, hazard response, and recovery mechanisms are required [3, 7] to provide efficient search and rescue efforts. New technologies having physical awareness can be integrated into the infrastructures to manage ER and disaster recovery.

ISS Networks (ISSNs) appear in several contexts, e.g., meteorological stations and continue to advance as CPSs become increasingly better. New uses and designs appear at a fast pace making feasible to provide network-wide data to users with high quality in near real time, by-products from data fusion, and data assimilation into models in a short time.

Resource scheduling in VSANs is challenging and it has a paramount role in CPS operation, which entails management of the interaction between actuators and sensors. It is essential to schedule or to manage control actions properly with a CU to observe as well as manage the entire CPS (refer to Fig. 2).

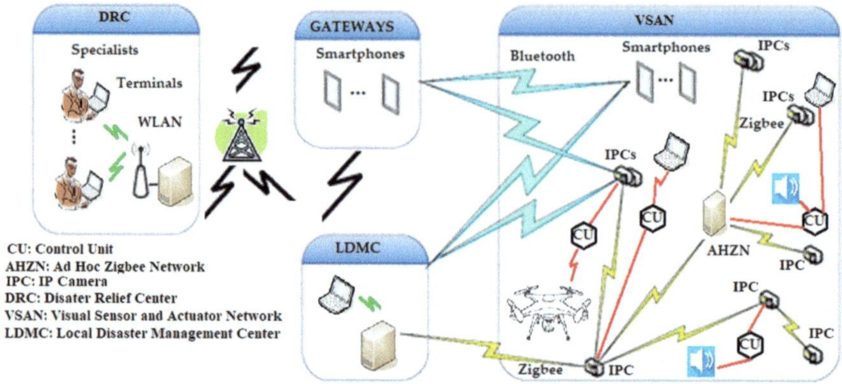

Fig. 2 Framework of the proposed Flood Detection and Monitoring (FDM) ER-CPS

3 Optical Flow

Several different OF algorithms have been realized differently relying on intensity changes among consecutive frames from a scene, assuming constant brightness alterations or correction factors corresponding to the image acquisition system, and scene light [8–12]. The intensity $\boldsymbol{I}_k(x, y)$ at an image frame k at a pixel location $\boldsymbol{r} = (x, y)$ is considered perfectly registered when

$$I_k(x, y) = I_{k-1}(x - \mathrm{d}x, y - \mathrm{d}y), \tag{1}$$

where $\boldsymbol{d} = (\mathrm{d}x, \mathrm{d}y)$ is the Disparity Vector (DV) with the corresponding disparity field (DF) or Motion Vector (MV) in the image space describing the transformation mapping adjacent frames $\boldsymbol{I}_k$ onto $\boldsymbol{I}_{k-1}$ whose intensity difference image is given by

$$I(x, y) = I_k(x, y) - I_{k-1}(x, y), \tag{2}$$

For small offsets $\boldsymbol{d}$, this relationship is obtained from a Taylor series expansion which holds

$$\Delta\mathbf{I}(x, y) \approx -d_x(x, y)\frac{\partial \mathbf{I_{k-1}}(x, y)}{\partial x} - d_y(x, y)\frac{\partial \mathbf{I_{k-1}}(x, y)}{\partial y}, \tag{3}$$

and it shows that the deformation between two frames is encrypted in the brightness differences. This expression yields an ill-posed problem since only the parallel component to the image brightness gradient $(\partial \mathbf{I_{k-1}}(x, y)/\partial x,\ \partial \mathbf{I_{k-1}}(x, y)/\partial y)$ of the MV can be found. With the help of a local window, the problem can be regularized, assuming that the disparity field is constant over a certain area (smoothness constraint) or using a global regularization methodology. Ideally, the performance should only be limited by the radiometric noise. The OF technique is effective to measure strain from photogrammetry, but it fails when the images have different view angles, the surface roughness at the pixel scale is large (e.g., with high-resolution imagery of important areas) or when the displacement field is locally discontinuous. OF is very sensitive to intensity variations due to object deformations, which can be modeled by OF methods relying on models taking into consideration higher order deformations with minor contrast variations, and adequate to locally affine distortions.

When there is motion, the objects closest to the observer seem to move the fastest. A camera chip perceives that as pixels changing position faster, thus implying OF and the sensors in the next Section explore this principle.

4 Optical Flow Sensors

This text considers the adequate architecture for an ER-CPS with ISSs for FDM using salvaged electronic parts from old equipment with reengineering. The CPS paradigm is very opportune to model, delineate, and simulate the prevention, remediation, and adaptation to risks in underprivileged regions. On-site VSANs can lessen the shortcomings brought in by the current economic crisis and scarcity of funds at the instrumentation level.

4.1 Optical Flow Sensor (OFS) from Computer Optical Mice

Real-time surveillance can rely upon as many video cameras as needed to examine a region. The channels on the screen may have to be relocated, so that each channel is allotted to a window or employ the automatic channel framing alternative on the website.

The principal part of a computer mouse is presently an OF Sensor (OFS), which is a motion sensor that measures the linear relative displacements under the optical mouse. OFSs are sensitive to changes in height and the type of terrain/trajectory, besides depending heavily on the surface where measurements take place. Notwithstanding these restrictions, OFSs allow numerous uses because of their versatility and low cost.

For an OFS like, for example, the Avago Technologies' ADNS-3080 [13], which contains a low-resolution video camera. This IC determines the relative dislocation of the images' micro shadows by means of a Digital Signal Processor. The sensor calculates the OF following an analysis of the sequence of image frames of the surface facing the sensor to evaluate the motion measured at different resolutions, for example, 800 Counts per Inch (CPI). Usually, an optical mouse relying on the ADNS-3080 combines (i) an OFS, (ii) an external LED that illumines the surface opposite the sensor, and (iii) two small convex lenses to focus the light to improve the picture obtained by the OFS. Combining all these parts requires optimization to obtain focused surface images from a diminutive distance (in the interval going from 2.3 to 2.5 mm), even though replacing the original set of lens magnify this range.

The ADNS-3080 contains a 30 × 30 imaging array with high frame rates and a fast shutter where each cell is a gray-level pixel. Its Serial Peripheral Interface (SPI) bus helps to access the content of the internal registers besides to control other sensor actions. In customary operation, the displacement of the flat surface beneath the sensor comes from contents of two registers that stores and yields the relative movement measured.

4.2 Video Sensing

Surveillance can involve multiple cameras to observe a disaster-prone region. The images can come from crowdsourcing and IP cameras stated in tall places.

Crowdsourced footage acquired at near ground level with smartphones and social media hints afford knowledge about real-time flooding recognition. Crowdsourcing imagery from the same inundated areas and other ones got under dry conditions can be combined to give analytical results for flooding comparison and detection [3]. This strategy requires some steps to normalize the images with and without inundation. Then, the imagery is registered taking into consideration the images without flood as the ground truth. Afterwards, registration, algorithms can extract the flooded area in the crowdsourced imageries from context features such as time, geo-location, environmental/weather conditions, and the image types, which impact photos and videos. The resulting metadata assists context identification. Social media cues are used for further refinement and evaluation. Algorithms can detect water reflections from nearby landmarks and the clouds/sky above. Smartphones can function both as data gatherers and as alarms as they communicate with several CUs, sensors, and alarms. This work does not cover building tagger and tagging communities in a crowdsourcing setting, the digital resource infrastructure and control classifications, and the typifications, and definitions related to the visual data that arises from people interactions via smartphones [3, 7].

IP cameras fixed on lighting poles or other tall structures can survey the flood-prone area in real time with as many cameras as needed. It is necessary to organize the channels on the display by allocating to each channel a window or setting some type of website automatic channel displaying option. People in charge of security can monitor the video and give technical suggestions to the public entities to take some kind of action, subject to internal procedures used by each agency but computer vision techniques can also replace the current security systems, which reduce the operational cost and possible human errors. These procedures involve computational resources for image processing but offer intelligent tools capable of recognizing patterns and recognizing patterns. Legacy equipment such as camcorders can be transformed into IP cameras to avoid disassembling, unnecessary contamination, to load landfills, and so on (Fig. 3).

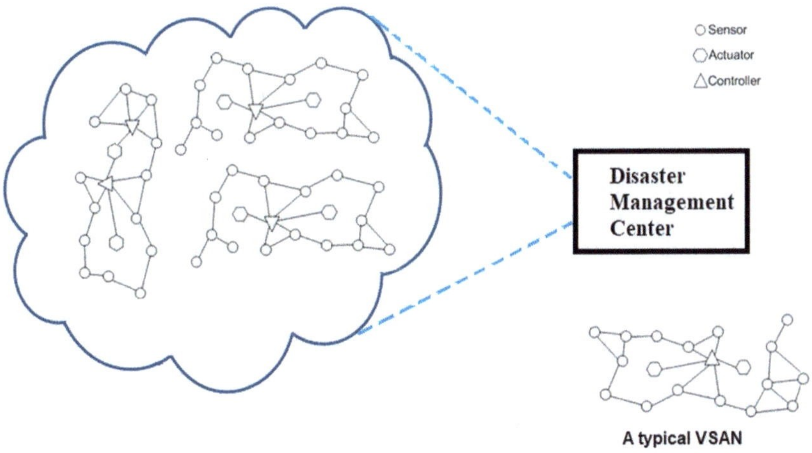

Fig. 3 Relationship between CUs, sensors, and actuators in VSANs

5 Visual Sensors and Actuator Networks (VSANs)

OFSs tend to be habitually sensitive to weather conditions. Their temporal and spatial resolution are higher than what space/airborne systems can deliver, and consequently integrate better with other physical parameters measured in situ. A single camera can monitor displacement patterns qualitatively [10, 14, 15].

Certain environmental conditions, for example, floods can be inexpensively monitored by means of salvaged electronic sensors harvested from computer mice. The same rationale holds for actuators and CUs. The Real-Time Communication (RTC) functionality allows for date and time stamping of each data point or data set to seek correlations or relations within the data. The growing accessibility of sensors, actuators, and computation platforms facilitate to interface and partake work to effortlessly and quickly generate functional solutions.

With proper and secure selection, electronic parts can be reused in low-cost applications, where tolerances are not too austere. Regardless of the low price of electronics parts, in emerging countries, they are not straightforwardly found outside big cities. Hence, such recycling can diminish the effect on landfills and cut the burden of remanufacturing the materials from otherwise discarded electronic equipment. Accessibility with low-cost boosts environmental awareness and care. Furthermore, there is a better management of surroundings and resources.

VSANs can be built using salvaged components, where each node contains several types of ISSs [1, 3, 7, 16, 17]. A simple VSAN can rely on optical hardware from optical mice to recognize motion via OF and landmarks. With the help of some reengineering, reused microprocessors, and microcontrollers can be turned in CUs to interface the communication modules such as ZigBee and Bluetooth [17]. VSANs with low-power low-cost energy harvesting technologies, assorted types of sensing, data processing, and adequate communication capabilities can get images, analyze/treat them, and exchange mined data with each other and to some disaster management centers or base stations for proper mitigation. Regrettably, the considerable amount of information gathered from several VSANs and processed by the CUs put constraints on the supervision strategies. There are several challenges involved in the design and implementation of VSANs [16].

The number of sensors required to investigate different phenomena may reach the order of hundreds or thousands. Depending on the type of necessary actions, the number of actuators can also be enormous.

A typical VSAN controller has a processor or microcontroller, memory, and communication units that processes and exchanges collected data. A VSAN spends long idle periods listening to the channel while trying to keep its energy consumption as small as possible, with the intention of working for a long time. VSANs should be designed with low-power characteristics despite the need to have circuits for data capture, processing, and communication.

Examples of the VSANs actuators are servo-mechanisms such as sound/visual alarms, pumps, water drainage systems, robotic arms, Unmanned Aerial Vehicles (UAVs), etc., that interacts with the flood-prone environment.

6 Conclusion

This article concentrates for the most part on the development of an ER-CPS with VSANs for flood monitoring and mitigation. Hence, underprivileged regions can avoid, remediate, and adapt themselves to natural and man-made catastrophes. It tackles the instrumentation level of VSANs built in with sustainable electronics and reengineering to lessen the complications instigated by the present-day economic crisis and restricted funds.

This work examines the way salvaged electronic components can be recycled and assists distant and economically deficient localities. It is worth mentioning that dumping electronic equipment in landfills is very unsafe. Reprocessing electronic parts to obtain and reuse materials can be difficult and energy-consuming, principally in developing countries.

Ideally, ISSs, crowdsourced information, and RS methods should handle different flood types and site features. It is challenging to evaluate the benefits and limitations of fusing knowledge from different technologies.

With a camera experiencing translation, the resultant OF magnitude will hinge on (i) the distance between the camera (observer) and the target as well as (ii) their speed difference. The range is not a straight OF by-product since the magnitude of the OF vectors is proportional to the camera speed and inversely related to the distance to the objective. Therefore, what can be observed is the ratio between speed and distance. OF is also susceptible to camera rotation but this flow is not related to the objects features distance in the scene and must be discarded. The best camera position is perpendicular to the motion direction for maximum Signal-to-Noise Ratio (SNR) because the OF maximum occurs perpendicularly to the motion path. If the observer speed is constant, the observed OF change in magnitude will reflect variations in the distance to the probed target. When the observer's speed is known, it can help to adjust the OF vector lengths for distance [18].

Although this text focuses on computer vision techniques, other types of sensors should be added in a VSAN.

References

1. Harrison, C.G., Williams, P.R.: A systems approach to natural disaster resilience. Simul. Model. Pract. Theory **65**, 11–31 (2016). https://doi.org/10.1016/j.simpat.2016.02.008
2. de Oliveira, N.S., Rotunno Filho, O.C., Marton, E., Silva, C.: Correlation between rainfall and landslides in Nova Friburgo, Rio de Janeiro-Brazil: a case study. Environ. Earth Sci. **75**, 1358 (2016). https://doi.org/10.1007/s12665-016-6171-7
3. Estrela, V.V., Saotome, O., Loschi, H.J., Hemanth, J., Farfan, W.S., Aroma, J., Saravanan, C., Grata, E.G.H.: Emergency response cyber-physical framework for landslide avoidance with sustainable electronics. Technologies **6**, 42 (2018). https://doi.org/10.3390/technologies6020042
4. Guerra, A.J.T., Graeff, O., Jorge, M.C.O.: Floods and landslides in Brazil—a case study of the 2011 event. Geogr. Rev. **26**, 38–41 (2011). https://doi.org/10.13140/rg.2.2.29550.41280
5. Tsai, C.F., Young, M.S.: Pyroelectric infrared sensor-based thermometer for monitoring indoor objects. Rev. Sci. Instrum. **74**, 5267–5273 (2003). https://doi.org/10.1063/1.1626005
6. Priyanka, S., Baskar, K.: Control of solar LED street lighting system based on climatic conditions and object movements. J. Inf. Knowl. Res. Electr. Eng. **3**, 480–486, ISSN: 0975–6736 (2014)
7. Estrela, V.V., Saotome, O., Hemanth, J., Cabral, R.J.R.: Emergency response cyber-physical system for disaster prevention with sustainable electronics. In: Proceedings of the 10th ACM International Conference on PErvasive Technologies Related to Assistive Environment (PETRA '17). ACM, New York, Greece, pp. 238–239 (2017). https://doi.org/10.1145/3056540.3064966
8. Marins, H.R., Estrela, V.V.: On the use of motion vectors for 2D and 3D error concealment in H. 264 AVC video. In: Dey, N., Suvojit, A., Patra, P.K., Ashour, A. (eds.) Handbook of Research on Applied Video Processing and Mining. IGI Global, Hershey, PA, USA (2017). https://doi.org/10.4018/978-1-5225-1025-3.ch008
9. Coelho, A.M., Estrela, V.V.: EM-based mixture models applied to video event detection. In: Principal Component Analysis-Engineering Applications, Infotech, USA (2012). https://doi.org/10.5772/38129
10. Fernandes, S.R., de Assis, J.T., Pacheco, M.P., Estrela, V.V., Medina, I.: Desenvolvimento de uma Ferramenta Computacional para o Processamento de Imagens Estereoscopicas. In: Proc. Congresso Nacional de Matemática Aplicada e Computacional. SBMAC (2007)
11. De Jesus, M.A., Estrela, V.V.: Optical flow estimation using total least squares variants. Orient. J. Comput. Sci. Technol. **10**, 563–579 (2017). https://doi.org/10.13005/ojcst/10.03.03
12. Coelho, A.M., Estrela, V.V.: A study on the effect of regularization matrices in motion estimation. Int. J. Comput. Appl. **51**, 17–24 (2012). https://doi.org/10.5120/8151-1886
13. Avago. ADNS-3080 and ADNS-3088 High Performance Optical Sensor. Available online: http://www.alldatasheet.com. Accessed on 15 Feb 2018
14. Avouac, J.-P., Leprince, S.: Geodetic imaging using optical systems. In: Reference Module in Earth Systems and Environmental Sciences. Treatise on Geophysicsm 2nd edn, vol. 3, pp. 387–424. Elsevier, Amsterdam (2015). https://doi.org/10.1016/b978-0-444-53802-4.00067-1

15. Stumpf, A., Malet, J.-P., Allemand, P., Pierrot-Deseilligny, M., Skupinski, G.: Ground-based multi-view photogrammetry for the monitoring of landslide deformation and erosion. Geomorphology **231**, 130–145 (2015). https://doi.org/10.1016/j.geomorph.2014.10.039
16. Al Najjar, M., Ghantous, M., Bayoumi, M.: Visual sensor nodes. Video Surveillance for Sensor Platforms, pp. 17–35 (2013)
17. Eady, F.: Hands-on ZigBee: implementing 802.15.4 with microcontrollers. Newnes, Burlington, MA, USA (2007)
18. Nourani-Vatani, N., Borges, P., Roberts, J., Srinivasan, M.: Topological localization using optical flow descriptors. In: Proceedings of the IEEE International Conference on Computer Vision Workshops (ICCV Workshops). IEEE Press, New York, pp. 1030–1037 (2011)

MIX
Papier aus verantwortungsvollen Quellen
Paper from responsible sources
FSC® C105338

If you have any concerns about our products,
you can contact us on
ProductSafety@springernature.com

In case Publisher is established outside the EU,
the EU authorized representative is:
Springer Nature Customer Service Center GmbH
Europaplatz 3, 69115 Heidelberg, Germany

Printed by Libri Plureos GmbH
in Hamburg, Germany